SIGHTSEEKING

REVISITING NEW ENGLAND
The New Regionalism

SERIES EDITORS:

Lisa MacFarlane
University of New Hampshire

Dona Brown
University of Vermont

Stephen Nissenbaum
University of Massachusetts at Amherst

David H. Watters
University of New Hampshire

This series presents fresh discussions of the distinctiveness of New England culture. The editors seek manuscripts examining the history of New England regionalism; the way its culture came to represent American national culture as a whole; the interaction between that "official" New England culture and the people who lived in the region; and local, subregional, or even biographical subjects as microcosms that explicitly open up and consider larger issues. The series welcomes new theoretical and historical perspectives and is designed to cross disciplinary boundaries and appeal to a wide audience.

Richard Archer, *Fissures in the Rock: New England in the Seventeenth Century*

Nancy L. Gallagher, *Breeding Better Vermonters: The Eugenics Project in Vermont*

Sidney V. James, *The Colonial Metamorphoses in Rhode Island: A Study of Institutions in Change*

Diana Muir, *Reflections in Bullough's Pond: Economy and Ecosystem in New England*

James C. O'Connell, *Becoming Cape Cod: Creating a Seaside Resort*

Christopher J. Lenney, *Sightseeking: Clues to the Landscape History of New England*

Priscilla Paton, *Abandoned New England: Landscape in the Works of Homer, Frost, Hopper, Wyeth, and Bishop*

Adam Sweeting, *Beneath the Second Sun: A Cultural History of Indian Summer*

SIGHTSEEKING

*Clues to the Landscape History
of New England*

CHRISTOPHER J. LENNEY

University of New Hampshire
PUBLISHED BY UNIVERSITY PRESS OF NEW ENGLAND
HANOVER AND LONDON

University of New Hampshire
Published by University Press of New England,
37 Lafayette St., Lebanon, NH 03766

Designed by Dean Bornstein

LIBRARY OF CONGRESS CATALOGING-IN-PUBLICATION DATA

Lenney, Christopher J.
   Sightseeking : clues to the landscape history of New England /
Christopher J. Lenney.
      p. cm.
Includes bibliographical references and index.
   ISBN 1-58465-205-5 (alk. paper)
   1. New England—History, Local.   2. Landscape—New England—History.
3. New England—Description and travel.   4. Human geography—New
England.   5. Regionalism—New England.   6. New England—Antiquities.
7. Historic sites—New England.   I. Title.
   F4 .L46 2003
   917.404'44—dc21   2002154135

Maps, figures, and illustrations
by Mila Natalie Zelkha

To My Mother and Father

CONTENTS

FOUR: ROADS

FIVE: HOUSES

SIX: GRAVESTONES

SEVEN: EPILOGUE

APPENDIX

FIGURES

PREFACE

I N 1804 the traveler Timothy Dwight noted in his diary "When we had passed the line which divides Massachusetts from New York, the appearance of the country in many respects was changed in an instant." "The houses became ordinary and ill repaired," while the people (generally intoxicated) were "rude in their appearance and clownish in their manners."[1] More than a century later, the photographer-antiquarian Wallace Nutting made much the same observation about the New England–New York border: "the demarcation is almost as distinct as the color on the map."[2] Even today, a through-hiker on the Appalachian Trail passing from Kent, Connecticut, to Webatuck, New York, senses a similar *dépaysement*. Such interregional comparisons are familiar and valuable, but the concern of this book is with subtler, intraregional ones. Comparing apples and oranges may have touristic shock value, but it defines things more by what they are not than by what they are. Comparing apples to apples not only cultivates eye and palate, for the botanically minded, it can also afford insight into the origin of the genus. Family resemblances are best worked out in the family album, and so also with the New Englandness of places.

This book was created *ex nihilo*, not in the sense that it is wholly original (parallels, if not predecessors, are many), but because it was called forth by a conceptual gap in the New England bookshelf. It is the slow ten-year concretization of an elusive book that I could never lay finger on, but that I came nonetheless to see ever more precisely in my mind's eye. I wrote it ultimately because I wanted to read it, because I felt deeply that the subject was one worthy to be written of and read. I wanted to keep some record of my researches, both within doors and without: a record that, in my better moments, could serve to raise a weary round of thought into an upward spiral, and in my worse, to admonish bewildered travelers not to pass that way again. I wanted as well to redeem years spent in the world's largest open-stack library, and add my pebble to the cairn.

This book is not primarily offered as an original work of research; rather, it summarizes, re-indexes, and reinterprets "the known," (however, in truth, little-known). Leaves from every town history could stuff its pigeonholes; it gives home to many stray observations gleaned from scholarly footnotes, neglected histories, guidebooks, news items, old photos, and maps. Above all, it creates a place where creative, amateur landscape study, across the board and from the ground up, is valid. The classification and description of even six aboveground artifacts achieves a critical mass of interconnections, parallels, extrapolations, and conjectures that quickly expands beyond the compass of one volume. While *Sightseeking* deals directly with but one percent of what we see, this figurative fraction is the part that stands poetically for the whole. For the writer, this synecdoche runs deeper: the illustrative examples herein compress much biography and many privileged hours under New England skies. In exchange for days of pleasure afoot and awheel, the crucifixion of footnotes has been suffered as both an honest and necessary part of the bargain.

These pages eclectically distill the fruits of many authors' research, authors to whom I am everlastingly indebted, as much for the pleasure of their insights as for the stuff of this book. Any misconstruals, misinterpretations, or misapplications of their ideas are entirely my own. I wish to thank my first reader, Leslie MacPherson Artinian, who commented and encouraged; Elizabeth Loutrel, who read selected chapters; Mila Zelkha, who drew with flair; Lynn Sayers, who cheerfully shared her computer expertise and enthusiasm for getting things right; the helpful staffs of the Harvard University Library (especially those of Widener, Loeb Design, the Map Collection, and my own Lamont), as well as those of the public libraries of Lexington, Belmont, and Concord. I wish also to thank all who listened, both at the beginning, and at the end, especially the secret sharers who in the final stages made the electronic analogue of the "inexorable sadness of pencils" all the more bearable. Lastly, I would be remiss not to mention my unknown neighbors at the top of the street who neatly discarded the word processor on which the first draft was written.

Cambridge, Massachusetts   C. J. L.

SIGHTSEEKING

{ *One* }

PROLOGUE

SIGHTSEEKING, briefly defined, is systematic sightseeing, and sight-seeing is but each traveler's highly conditioned manner of looking at the world. The intent of this book is to recondition that outlook. No traveler can deny the thrill of borders: the way in which perceptions heighten and sensibilities expectantly shift gears at national boundaries, even state lines. Or deny that the past, to the degree that it can still be glimpsed across its uncrossable frontier, is itself another country,[1] with landscapes that cast an eerie spell of their own. The thrill of borders—whether those of space or time—serves as our instinctive invitation to the fundamental theme developed in this book: spatiotemporal variation, a theme abstract only in name.

It is almost a tautology to say that landscape history is the study of space over time. The recorded moves of a chess game, or last week's set of weather maps, are also in their own way studies of space over time: the first measured in minutes over square inches, the second in days over thousands of square miles. Implicit in enough such recorded spatiotem-poral sequences is the whole dynamic of a system, whether the rules and strategy of chess, the meteorology of highs and lows, cold and warm fronts, or the workings of landscape history. For the study of a single town, a set of historic topographic maps might suffice, but to comprehend the land-scape history of a region is a far more vast and complex undertaking. One classic British approach interprets landscape thematically, in multiple map-layers: solid geology, drift geology, landforms, soils, drainage, vegetation, agriculture, road and rail net, settlement pattern, and industry.[2] Each layer overlays and underpins the layers next below and above, moving from the natural to the cultural landscape, and building up historically from the geologic past to the present day. The layers, of course, are never so neat, but ooze together like a decuple-decker sandwich. While our approach will be similarly chronological and thematic, the subject of *Sightseeking* lies chiefly in the upper layers of this sandwich, in the cultural landscape of New England since English settlement. Even when so defined, the "spa-

I

tiotemporal variation" of our inquiries still operates within a compass of 60,000 square miles over 350 years: a broad, deep swath, indeed.

The scope of this still vast mandate will be schematically reduced to a focused investigation into six landscape artifacts—common, immovable, man-made features—chosen for their proven richness as historic indicators: Placenames, Boundaries, Townplans, Roads, Houses, and Gravestones. For reasons that will become increasingly apparent, our emphasis will be on pre-1850 "vernacular" artifacts, although no firm cutoff is imposed. Recognition of these artifacts, like the recognition of chess pieces or weather fronts in the analogies above, gives us clear points to fix on and interpret in the swirl of change.

Landscape history has sometimes, only slightly facetiously, been described as "above-ground" archeology, and certainly the methodological (and inspirational) debt owed this latter discipline is very great. The recurrent touchstone for our analysis will be the clay pot, the quintessential artifact of classical archeology. Briefly, an artifact is an object fabricated or adapted by humans for their own use or pleasure. Its particular form proceeds from a culturally inculcated image or "template" that exists in the mind of the maker. It possesses stylistic and technical characteristics, or "attributes," that enable it to be classified and chronologically sequenced with similar objects, much as one might sequence twentieth-century automobiles decade by decade on stylistic and mechanical criteria. Over time, its popularity waxes, then wanes, in a predictable "battleship-shaped" curve, while its status can decline from urban and high-style to rural and low-style.[3] When mapped, its pattern of distribution can testify to its origin and diffusion, limn out a former frontier line, a cultural sphere (or axis) of influence, or trace a route of inland migration. Crucial to the sightseeker is the strong intrinsic appeal of the artifact itself—whether house, gravestone, or placename—because while analysis may be dispassionate, the vigilant accumulation of evidence is a labor of love.

Kurathian Hypothesis

While many archeological methods will be creatively applied to landscape artifacts, our analysis is equally beholden to dialect geography, and in particular to what I call the Kurathian Hypothesis. This asserts that the distribution of vernacular artifacts follows subregional lines that reflect original points of settlement (hearths) and subsequent internal migration streams (settlement paths), both of which have strong geographical determinants. The scholarly foundations of this theory were laid in 1939–1943 with the publication of the *Linguistic Atlas of New England* (*LANE*) and its accompanying *Handbook of the Linguistic Geography of New England*, under the editorship of Hans Kurath. Here were first laid down and documented the dialectal subregions of New England based on historico-linguistic evidence. In time, cultural historians began to see broader implications for such linguistic research. John T. Kirk's contribution was succinct but catalytic. In his 1972 study, *American Chairs: Queen Anne and Chippendale*, he asserted "dialect regions correspond to furniture regions" and appended a careful synopsis of Kurath's New England subregions as expounded in the *LANE*. In 1978 William N. Hosley Jr. echoed and widened the scope of Kirk's observation to include gravestones, in a brief footnote to his Dublin Seminar paper on the Rockingham [VT] stone-carvers. Fred Kniffen (1965), on a national scale, correlated folk house-building traditions with dialect regions, while Thomas Visser (1997) similarly invoked dialect boundaries in his work on New England barns.[4] It is a small matter for us to broaden this thesis further still to embrace all vernacular artifacts, and landscape artifacts in particular.

Let us return now to Kurath and the *LANE*, the massive undertaking that underpins this whole edifice of theory. Between 1931 and 1933, a team of nine fieldworkers surveyed 416 informants in 213 towns, gathering and mapping data on more than 800 lexical and phonetic items: enough to fill 734 atlas sheets bound in six folio volumes. Equally important for our purposes, a concise synopsis of New England settlement history was prepared, illustrated by maps that charted both its flow-paths and outward spread over time. As much on the strength of this historical data as on the linguistic evidence, Kurath proposed, although seems never to have precisely mapped, something like eight dialectal subregions. A close reading of the *Handbook* (with chapters written by divers hands) reveals contradictions both in their number and nomenclature. Even in 1939 traces of such subregions were fast fading; subsequent work in the 1960s for the *Dictionary of American Regional English* (*DARE*) reduced the number to five. (Kurath could be bolder still, paring his count to three, when viewed within the larger context of the eastern United States.)[5]

Of the eight New England subregions postulated by the *LANE*, six lay to the east: Boston, Plymouth, Narragansett Bay, Worcester County–Upper Connecticut Valley, New Hampshire Bay–Merrimack Valley, Maine. The remaining two lay to the west: the Lower Connecticut Valley, and the Western Fringe. The east-west cleavage reflects the spheres of influence of the two dominant early settlement cores or hearths, Massachusetts Bay and the Connecticut River Valley, spoken of, for short, as the Bay and the River. The Bay was strongly focused on Boston; the River was more diffuse. The main streams of inland settlement that sprang forth from these hearths carried their respective cultural influences deep into the New England interior. Linguistically, this east-west cleavage is strongly marked by broad A versus flat A in words like glass and calf, by R-dropping versus R-keeping, and by the usage of signal dialect terms such as *buttonwood* versus *buttonball* for sycamore. The map of the *LANE* subregions (fig. 1) is derived from that of Craig Carver, but modified at points where it differs from my own reading of the *Handbook*. (The influence of Boston pervaded eastern New England; the subregional boundary merely suggests the core area.) As fundamental to the sightseeker as these subregions are the Settlement Paths and Settlement Frontiers mapped in figures 2 and 3 (for sources of these maps, see notes)[6]. Traces of all three spatiotemporal patterns are deeply ingrained in the landscape, and in a theoretical world, artifacts would neatly array themselves under their influence like iron filings in a magnetic field.

The Kurathian Hypothesis implies that cultural subregions are validated by the congruent distribution of several artifacts. In linguistics this is accomplished by mapping and encircling the distribution of key dialect words as "isoglosses," and then identifying the overlap of several similarly located isoglosses as a "bundle." The thicker the bundle, the more conclusive the subregion. For example, Kurath identified *quahog, apple slump,* and *(Rhode Island) johnnycake* as a Narragansett Bay–Eastern Connecticut isogloss bundle, and *stoop, belly-gut(ter),* and *pot-cheese* as a Hudson Valley bundle.[7] The obscure, largely rural vocabulary is to be noted. Subregions of the cultural landscape should likewise be identifiable by the overlap of several mapped artifact distributions, or *isoscenes*, to coin a term. Isoglosses and isoscenes both represent distributions of cultural artifacts, and the leap between words and things is not so great as it might seem. For example, the east-west divide between purlin-framed and rafter-framed roofs runs very close to the isogloss between broad A and flat A in the pronunciation of calf.[8]

Because the true distribution of even well-recognized landscape artifacts is imperfectly known, it will be often necessary to render these isoscenes nebulously, with what I think of as bubbles or balloons, since their area may expand or contract as evidence accumulates. The isoscenic bundles, to judge by their linguistic counterparts, cannot be expected to comprise

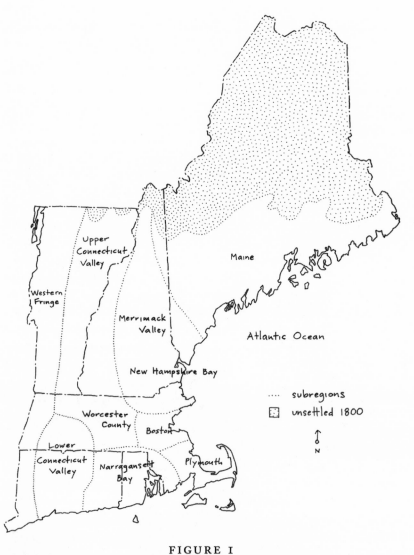

Upper
Connecticut
Valley

Maine

Western
Fringe

Merrimack
Valley

Atlantic Ocean

New Hampshire Bay

subregions

unsettled 1800

N

Worcester
County

Boston

Lower
Connecticut
Valley

Narragansett
Bay

Plymouth

FIGURE I

Kurathian Subregions

Adapted from Carver, Craig 1987, 25, fig. 2.3; incorporating data from
Kurath 1973.

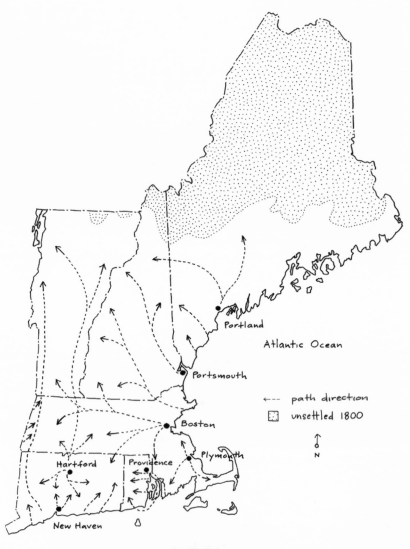

Portland

Atlantic Ocean

Portsmouth

- - - path direction

⬚ unsettled 1800

↑
O
N

Boston

Hartford

Providence

Plymouth

New Haven

FIGURE 2
Settlement Paths
Adapted from Kurath 1973, plate 1.

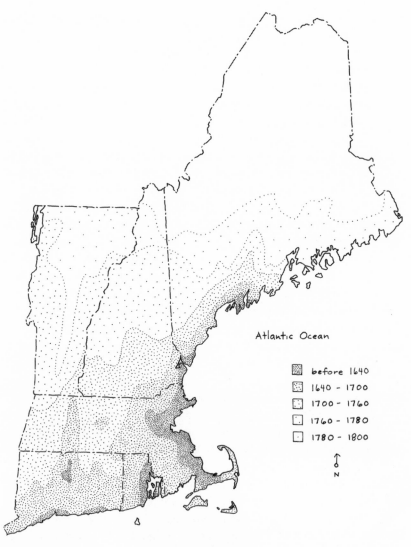

Atlantic Ocean

▨ before 1640
▨ 1640 - 1700
▨ 1700 - 1760
▢ 1760 - 1780
▢ 1780 - 1800

↑
○
N

FIGURE 3
Settlement Frontiers
1620–1800
Adapted from Wood 1977, 35, fig. 1.5.

numerous or even commonplace artifacts. Diagnostic subregional artifacts rarely make up a visible majority of their type. Often they are costly, prestige items (like the Connecticut Valley scroll-topped doorway), because the distribution of such items demarcates the social and business circles of the provincial elite, circles themselves indicative of subregional boundaries. They may be in some way impractical or ill-adapted to wide emulation, or occupy a niche well-filled elsewhere. They are almost axiomatically out of the ordinary, because "ordinariness" is more the mark of regionalisms, not subregionalisms. Indeed, the obscure evidence for subregions is precisely why they are subregions and not regions: by nature, their demarcation can be but comparatively weak. But in the same way that the *Quahog–Apple Slump–Johnnycake* bundle delimits a speech community that shares other, more elusive features, cultural subregions defined by isoscenic bundles would be expected to have been historically amplified by more ephemeral cultural features (as colors of housepaint or perhaps shapes of haystacks). Nor should we prudently expect that landscape evidence in and of itself will prove sufficient to validate the vernacular subregions: it is but a single, slowly perishable layer in a multilayered cultural cake. Even Timothy Dwight (1752–1817), the great itinerant eyewitness to this world we have lost, plainly conceded that—compared to those in Europe—the varieties to be encountered by the descriptive traveler in New England and New York "will not be so marked as to strike his eye with much force or yield to his readers any great gratification."[9]

Indeed, as well-documented distributions of landscape artifacts are so rare, lessons learned from better-mapped cultural phenomena are most welcome. Each such distribution is a spill-stain, suggestive in whole or part of how any other cultural phenomenon, set in motion in the same time and place, might overspread the area. It delimits a zone of relatively free cultural interchange. A subregion in effect is a socioeconomic organism, a more or less closed circulatory system around which all manner of artifacts were pumped. The heart of that subregional economy long ago ceased to beat, to be succeeded over time by regionalization, nationalization, and globalization. But the artifacts it circulated, to the degree to which they survive, still lie exactly where they fell, and with them we can attempt to limn out the limits of the system as it once was. More conventional written records can be enlisted to reconstruct this system, but the landscape record has an immediacy, power, and pleasure all its own. For the classical archeologist the Roman wine trade is traced out more compellingly in strings of unearthed amphorae up and down Europe than in any bills of lading.

Windsor Chair

Since it was a chair historian who early equated artifact and dialect distributions, perhaps a brief look at the history of one well-researched chair will afford insight into the practical applicability of the Kurathian Hypothesis. Nancy Goyne Evans devoted almost half her monumental *American Windsor Chairs* to the New England Windsor, from its arrival in late Colonial Rhode Island (principally from Philadelphia) until its demise around 1850. Over this near century, countless thousands were turned out in New England workshop and factory by some nine hundred known makers. The Windsor is composed of roughly twenty parts and takes on a multiplicity of forms: high and low; sack-backed, bow-backed, and fan-backed; side chair, armchair, and rocker. A single chair might employ four or five woods that, taken together, can botanically attest to its place of origin: basswood, birch, and ash in the Upper Connecticut Valley; white pine, birch, chestnut, or hickory in Rhode Island.[10] A little arithmetic will show that there are thousands of combinations of these characteristics alone.

Attribution in the Windsor is particularly challenging because it is a widely made and fairly standard article, subject to subtle subregional and individual influences. Its very portability makes the provenance of unbranded or unlabeled work tricky. Much like speech, the Windsor chair has its own complex vocabulary and accents; indeed, while Evans seems never to mention Kurath, she owes him a great methodological debt. Her insights are essentially diffusionist, and diffusion was along Kurath's migration and commercial routes. The history of each chair-making town and the biography and genealogy of each artisan are as important to her as such data were to him. The one thing Evans lacks is Kurath's cartographic mindset: aside from aids to pinpoint places of manufacture, maps do not figure in her method. Nor does an explicit enumeration of subregional attributes, although these are constantly invoked: Boston oval-swelled stretchers, Rhode Island–Connecticut hook-eared crests, Worcester County half-spindle work, and so forth. Such elements sometimes combine to form distinctive local subtypes, like the Nantucket fan-back armchair or Rhode Island bow-back,[11] but more often an unknown chair exhibits attributes from a variety of sources and is ascribed to a subregion on the strength of its preponderant affinity with the known work of known makers in a known place. It is as though, Kurath's atlas in hand, one attempted, on the basis of his habits of speech, to pinpoint a New Englander's hometown and those of his parents as well.

Evans, however, never overtly divides the Windsor chair into subregional styles (as Kirk implies could be done) and her own introductory remarks are ambiguous on this score. Regional classification is "expedient," time-honored, but on close study often seems "artificial."[12] Yet elsewhere, "regional studies are the keys to understanding American Windsor furniture. They provide insights on craft structure; pattern evolution; the roles of geography, technology, and territorial expansion in craft diffusion; the widespread appeal of the Windsor; and its perfect suitability as a major commercial product."[13] Part of this ambivalence must result from the way in which she organizes her data strictly along artificial state and county lines. Only the Rhode Island–Connecticut border emerges as a trans-state entity, and even the well-recognized Connecticut Valley is not treated as a coherent whole. Probably Evans was loath to map because she knew early on what Kurath ultimately knew: that the data were so complex. Internal migration and commercial and cultural intercommunication made it so. There is a sober lesson here. Many of the *LANE*'s atlas sheets, while interesting in themselves, seem to confuse, if not refute, and at all events, do not clarify the concept of subregionalism. Aside from a few select features, the patterning is at best elusive.

Sightseeking Defined

While it might seem that artifacts are something serendipitously "found," more often they are specifically sought: indeed, even at times precisely devised to solve a landscape historical problem, like a key to open a lock. (Hence sightseeking.) For example, how sensitively does the landscape artifactually register the advent of new modes of transport: canals, turnpikes, railroads? Wheat prices might drop overnight, but what suddenly introduced landscape artifacts would permanently record this revolution, not vaguely, but precisely? Solutions to such problems involve controlled exercises in historical imagination: in essence, patient "thought experiments." It can be readily imagined that on the Great Plains, whole nineteenth-century towns were uncrated lock, stock, and barrel from the freight cars of the railroads that gave them existence. In old-settled New

England, however, things are rarely so clear-cut. The railroad reached Concord MA in 1844. Dairy herds tripled and upcountry cattle drives ceased; strawberries and asparagus were fruitful and multiplied; even the shire-town drama of September court week lost luster as judges and lawyers commuted. Yet despite this transformation, if there is an enduring stratum of "railroadiana" embedded in the fabric of Concord village, we must train our eyes to see it.[14]

A historical vignette will amplify this point. In 1803 the Middlesex Canal joined Boston with the Merrimack, and the very next year Bulfinch began work on the Charlestown State Prison employing the novel material, granite, barged down from Chelmsford. Before then, only King's Chapel and the Hancock House had been built of boulder-hewn Braintree granite; the chapel was admired as a wonder of the age, never to be outdone due to the dearth of suitable stone. The prison became the bridgehead for mighty works to come, for in its yard convicts hewed and dressed the stone for many later projects. Between 1804 and 1818 Charles Bulfinch designed some eleven granite buildings, mostly civic, all within a stone's throw of the mouth of the Middlesex Canal. The Suffolk County Court House (1811–1812) and the now-named Bulfinch Pavilion of the Massachusetts General Hospital (1818–1823) are among the best known; the most distant, University Hall (1813) at Harvard, required the granite barges to continue up the Charles to College Wharf at the foot of Dunster Street. More than half of these projects were demonstrably of Chelmsford granite, and in all probability they all were. The Quincy and Cape Ann quarries were not opened until ca. 1824; the former served by the innovative Granite Railway of 1826; the latter by stone sloops. Multiple rail lines would open in Boston in 1835, and overnight the picture became vastly more complicated. But in this initial period, the image of these granite wonders floated, as it were, by barge down the Middlesex Canal is a licit, if metaphorical fantasy. They inaugurated what has been called Boston's Granite Age, prefiguring Quincy Market, the Bunker Hill Monument, Castle Island, and Fort Warren, all of which remain great landmarks today.[15] The elucidation of the Transport Revolution Problem is essentially an armchair exercise as few of Bulfinch's buildings survive. Nonetheless, as an example, it well illustrates that landscape study is as much about seeking artifacts as finding them, posing questions as answering them.

While "landscape detective" is a term perhaps even more shopworn than "aboveground archeologist," this book is not a mystery story, and its critical appreciation will be enhanced, not spoiled, by a knowledge of how it ends.

Prologue and Epilogue are meant as companion pieces, prelude and post-lude, in which the themes are introduced, rehearsed, and in some measure resolved. In the five sections that follow, we shall immerse ourselves in landscape artifacts, with a methodological enthusiasm for the things in themselves (artifacts for artifacts' sake), and a patient curiosity for what they may reveal. While proof of the Kurathian Hypothesis might be con-strued as the Holy Grail of sightseeking, truth or falsity is not the sole criterion by which it and the many other hypotheses examined in these pages are to be judged. These hypotheses have among their chief virtues the power to focus and sharpen observation. They provide the visual agenda—the *videnda*—that transforms sightseeing into sightseeking, and landscape into something more than the object of mere passive musing.

Sightseeking is underpinned by a wealth of specific references to both things and places. As an aid to the reader, two typographical conventions have been adopted in the chapters that follow. First, to focus attention on the sightseeker's *videnda*, many of the names of the principal landscape artifacts, particularly those of house-types, have been capitalized (as Salt-box or Three-Decker). And second, to allow for geographical precision without overburdening the text, two-letter postal abbreviations (set in small capitals and without commas) have been used in the state designa-tions of towns, with same-state towns in a series being marked at the end of the series (e.g., Dummerston, Townshend, and Grafton VT). Postal abbreviations have also been used in parenthetic tallies and similar contexts where concise reference to individual states was warranted. Unless context (or common sense) indicate otherwise, all undesignated places should be assumed to be in Massachusetts.

{ *Two* }

PLACENAMES

TOPONYMY is the systematic study of placenames, but the system employed necessarily varies from country to country and depends on the historical and linguistic character of the names themselves. In a country such as Britain, the toponymy is millennial and well documented: it has chronological layers discernible on linguistic evidence, and is highly dense; that is, there are many more names per square mile than in the United States, due to the custom of bestowing (and cartographically recording) names not simply on towns, but houses, farms, fields, and woods. Consequently, the study of British toponymy is highly etymological and requires a thorough knowledge of Old, Middle, and Modern English, as well as the other indigenous and invasive languages of the British Isles, such as Celtic, Norse, and French.

Anglo-Saxon toponyms in particular constitute a remarkably functional nomenclature, compounded of short, significant word-elements that are often descriptive of topography or settlement. These encode valuable data that can be geographically analyzed, linguistically dated, and correlated with the natural, cultural, and archeological features of the site as well as other similarly named sites. In short, they have deep, local, geographic, and historical meaning. Some Old English examples (shorn of diacritics) will serve to illustrate this point.

> Berwick OE *berewic*, "barley-farm"
> Uppingham OE? "village of the upland dwellers"
> Charlcote OE *ceorla-cotu*, "cottages of the churls (peasants)"
> Roffey OE *rah-hege*, "enclosure for roe-deer"
> Sapperton OE *saperetun*, "village of the soapmakers"
> Plaistow OE *pleg-stow*, "sport or play place"[1]

It is helpful to see Anglo-Saxon place-naming in the context of name-giving generally. Anglo-Saxon personal names were characteristically bithematic, as Garmund (spear-protector) or Beorhtsige (bright victory), and children often bore semantically meaningless names that combined elements of each parent's name. Wulfstan (wolfstone) was the son of a mother named Wulfgifu (wolf-gift) and a father named Athelstan (noble stone).[2]

We no longer name children this way, nor do we name places as the Anglo-Saxons did. By and large the study of New England toponymy is not etymological (relic Algonquian placenames aside). Only modest linguistic gifts are called upon to recognize Vermont as Green Mountain and none at all to explicate Marshfield or Rockport. Nor are town names, the best documented aspect of New England toponymy, for the most part locally, topographically, or historically descriptive. Generally they say much about the namer and little about the named. Indeed, nicknames such as Spindle City, Shoe City, Tool Town, and Toy Town are clearly closer in spirit to Anglo-Saxon toponyms and record more about the specific histories of Lowell, Brockton, Athol, and Winchendon MA than do the actual official names, which serve merely to identify and in no way describe these places. The town names do not etymologically encode anything immediately revelatory about the geography, or the economic or cultural history of the localities.

Little wonder that the serious geographer would much prefer the pregnant Anglo-Saxon placename (even the nicknamed place) for his analysis, so much so that the more modest geographical potential of New England town names is not rigorously pursued. Such pursuit is further discouraged by the fact that whatever information the name does encode is more reliably obtained from other sources, and hence without archeological importance. Thus the sightseeker wishing to evaluate town names as spatiotemporal indicators finds himself stymied by a refractory toponymy that is neither etymological, nor intrinsically meaningful, nor solidly grounded in geography. Part of our problem is that British toponymic methods arose in response to the specific nature of British toponymy; such methods, not suited to New England placenames, will consequently yield little. Yet all placenames are nonetheless artifacts and, when studied systematically, can be relied upon to reveal meaningful, if unacknowledged, patterns.

In this connection, the analogy drawn between Anglo-Saxon place-naming and child-naming can help to light our way. Modern motives for child-naming are diverse, but usually the selection is made from a standard name-pool and the decision arrived at by more than one person for more than one reason with the final choice meant to find favor with both family and community at large. Reasons may be to conform to a trend, to confer distinction, to alliterate or harmonize with a last name, to honor a forebear or hero, to soften a rich uncle, to combine names from both sides of the family, to inculcate a virtue, bode well, or prejudice strangers favorably. The standard name-pool constantly changes and reflects tradition (as biblical or saints' names) and current fashion (as names of ancestral, at times mythical, homelands: Erin, Kenya); it may afford imitable templates for new coinages: surnames as forenames (Carson), hyphenated or blended names (Jo-Lynn), even trade names of luxury goods as acceptable girls' names (Tiffany, Chanel). Name selection may wittingly or unwittingly serve as a marker of generational or ethnic identity. Most names have no single, simplistic reason; rather, they rely on a subtle combination of criteria to recommend them. New England town-naming is remarkably analogous to modern child-naming, though the lessons of this analogy are too often ignored. While people well know that a child may be named from a purposefully ambiguous bundle of motives, they prefer the rationale for town names to be cut-and-dried. If two reasons for a town name are adduced, one is right and the other is wrong, and failure to decide between the two is taken as a confession of ignorance.

With this child-naming analogy ever in mind, let us look at the three principal techniques of toponymic analysis available to us. The classic and primary but never self-sufficient method to study town names is to classify them thematically, as for example, towns honoring English nobles (Lenox) or American patriots (Hancock); towns bearing topographically descriptive names (Deerfield) or those of exotic places (Peru), and so forth. This involves compiling lists and cataloging name-histories with scarcely a glance at their true position on the map. While this might seem to be little more than tedious botanizing, chronologically based, thematic analysis with an eye to the origin and diffusion of new town-name types quickly redeems itself from the trivial. Bona fide thematic categories can in fact bring to light important innovative templates that, once established, gave full license to imitation. For example, towns with Indian names, contrary to expectation, prove to be more often of nineteenth- than seventeenth-century origin. Towns named directly and unequivocally for persons are largely of post-Revolutionary vintage: the custom was un-English and only became widespread in the nineteenth century, when antipathy to the practice had broken down. Place-naming is highly imitative and operates within informal and ever evolving rules deducible from the body of placenames at large, much in the way that brand names, whether of cars or candy bars, exert a collective pressure on every new brand name that is coined.

Our second, and oftentimes unorthodox, method of analysis is phonological. This approach examines the sounds of the names, as of their initial or final letters or syllables, in relation to the names of adjacent towns or towns at large. It gives weight to the fact that Granby CT (1786) borders Granville MA (1754) even though the published name-histories of Granby nowhere acknowledge the resemblance. The inhabitants of 1786 surely can be relied upon to have known who their neighbors were. Phonology recognizes that while -on (as in Sharon, Marion) is not etymologically a habitative suffix such as -bury or -field, that nonetheless the remarkable number of town names ending in -on collectively testify that it acted as a quasi-suffix that rang favorably in the town-namer's ear. Phonological analysis is entirely absent from town-name studies, but not surprisingly. It does little for local pride to say that a town name was even in part selected to alliterate with that of a neighbor; even if suspected, only a grand overview of the sheer number of such phonological resemblances in town names would embolden one to credit the possibility. A third technique at the disposal of the student of town names—and most important of place-names generally—is the analysis of distribution patterns.

Because the semantic content of placenames is often of slight geographic value, the study of their distribution can be made to extract from them collectively what they lack individually. Dull and inconspicuous artifacts can take on new luster when so analyzed. The distributions exhibited may be spatial or temporal, and patterns may be hidden depending on from which angle they are viewed. A temporal cluster may translate as a scatter on a map, while a spatial cluster may appear as a scatter on a timeline. For example, New England towns named Auburn, MA (1837), ME (1842), and NH (1845) are spatially scattered but temporally clustered. On the other hand, Massachusetts town names suffixed in -field are spatially clustered in the Connecticut Valley but temporally scattered over almost two centuries (1641–1816). In general, as location in space is the one sure thing about a toponym, and not the year in which it was first applied (evidence for which is usually lacking), it is spatial patterns that will most command our attention.

Difficult but most rewarding to discover are actual cartographic patterns, not random scatters of dots, but arcs, lines, and clusters of place-names that demarcate frontiers, migration routes, cultural hearths, spheres of influence, linguistic boundaries, zones of infiltration, and other facts of settlement history. The existence of such spatiotemporal patterns is all but predictable as settlement expanded westward and northward in discernible stages and as each stage was toponymically marked by the fashions of the time and the cultural heritage of the settlers. Spheres of influence naturally arose around main towns as well as more nebulous attractions between adjacent towns, resulting in localized patterns.

In the pages that follow, we shall examine and classify three broad groups of toponyms: folknames, town names, and street names. Various techniques and trends in the making of placenames will be identified, as well as geographic models for their spread: both those, such as transplantation and diffusion, that leave telltale traces in the name-cover, as well as others that do not, as seen most notably in widely scattered folknames and fad names whose dispersal defies simplistic explanation. The resultant map-patterns (among them what I will call collocations, name-strings, and name-scatters) will be carefully scrutinized, with particular attention paid to the most spectacular of these "name-fields": the Great-Big Line, Reversed Center Arc, and China Syndrome.

Algonquian Toponyms

To begin with the beginning: the strongly etched Algonquian name-cover that is the legacy of the region's first inhabitants. While presented here as a prelude to our first group, folknames, Algonquian toponyms stand as a class apart, exercising a special fascination even as they present virtually insurmountable difficulties of interpretation and analysis. A novel, "meaning-blind" approach will be taken here.

The study of Algonquian placenames has typically been microgeographic in approach: fixated on the semantic content of each individual term in relation to landscape features of the locality it designates. Corrupt transcription by early settlers and our own inadequate knowledge of often extinct tribal dialects bedevil scholarly interpretation; folklore and fabulation have further muddied the waters. Classic etymologies such as those of Massachusetts and Connecticut[3] afford insight into Algonquian toponymy, but such concise and clear-cut derivations are atypical.

> *massa* (great) + *adchu* (hill) + *es* (diminutive suffix) + *et* (locative suffix: at the place of)
> *quinni* (long) + *tuqk* (tidal river) + *ut* (locative suffix)

While Algonquian was a polysynthetic language that could render an English sentence in a single compound word, there is little evidence for Indianesque placenames after the fashion of "You fish on your side, I fish on my side, nobody fish in the middle—no trouble." Lake Chaubunagunga-maug (Webster MA), which has been so interpreted, is more properly and prosaically rendered as boundary fishing-place.[4] Of great interest are the generic toponyms that crop up repeatedly over the landscape, such as the seven variations on Quinnebaug (long pond), the sixteen on (Pi)scatacook (at the river branch), or the five on Shawmut (at the neck or canoe-landing). The Abenaki names of Maine, because of their more recent and presumably more consistent transcription, and because Abenaki is a living language, may be best suited to this sort of semantic and microgeographic analysis.

There remains, however, the possibility of another, macrogeographic approach with other aims, one that avoids semantic issues by being in large measure meaning-blind. The non-semantic study of Algonquian toponyms makes a virtue of necessity: we shall likely never know with certainty what these names meant. Such an approach runs dangers, but it enables even a layman to radically review the old, thin data in a new light, and perhaps raise pointed questions for the scholar. Furthermore, because in our daily experience these names are without semantic content, meaning-blind methods can more closely address the subtle ways in which we actually come to terms with them. You need not know Abenaki, need never have heard the name Tomhegan Pond, yet by unconscious analogy with Skowhegan it may "sound like" a Maine placename to you. (Similarly, one need not know a word of Russian to tumble to the fact that innu-merable Russian city names such as Minsk and Murmansk end in -sk and to wonder whether those strung along Siberian rail lines might signal Russian trans-Ural expansion.)

In even the most general classification, at least nine major Algonquian languages were spoken in reasonably distinct tribal areas of New England. These languages fall geographically and linguistically into two subgroups, a southeastern cluster of dialects that was rimmed on the north and west by Abenaki (found in MA, RI, and CT) and Mahican (found in ME, NH, VT, and NY). Transitional languages, Unquachog on Long Island, and Etchemin on the Maine coast, combined elements of both subgroups.[5] These languages existed as a "chain" of dialects in which mutual intelli-gibility diminished with each intervening link. An anecdotal measure of this continuum is found in two cognate tribal names in Maine and Mas-sachusetts, (W)Abenaki and Wampanoag, both of which literally mean "dawn land," that is, the East, or Easterners. Were the toponymic cover thick enough, might not these tribal-linguistic areas stand out on the map if we examined not so much the sense, but the characteristic forms of the names closely enough?

The density of this toponymic cover is not uniform, but reflects areas of intense Indian occupance and prolonged Indian-European contact, and thus may in its very thickness and thinness testify to something important. Indeed, the climatic suitablity of the various areas of New England to Indian agriculture[6] is roughly registered in the varying intensity of the Algonquian toponymic cover. In theory, prehistoric population movements should be discernible from the wedge-like intrusion of a distant or alien dialect between two others more closely akin. Coastwise trade might be reflected in key words that fanned out across otherwise distinct dialectal boundaries along the shore. Unfortunately, Ives Goddard doubts that the attested corpus of Eastern Algonquian dialects is large enough to permit such analysis in any exact or unambiguous way, even if such methods are viable.[7] Our corpus of toponyms is admittedly even slimmer, yet perhaps some conclusions can be drawn from it.

The classic methodological comparison to be made here is to the Danelaw, the area of northeast England invaded and settled by Danes and Norwegians in the ninth century. The limits of the Danelaw can be demarcated by the distribution of Norse placenames on the map, most notably by village names suffixed by -by and -thorp, but also by an excess of certain un-English phonological characteristics, such as the use of initial sk- or sc-: a sound known in English but less commonly employed.[8] Thus the Danelaw can be effectively mapped out from the form of the names, with little regard to their meaning, an approach that perhaps might be adapted to reveal patterns in the Algonquian name-cover as well. For these inquiries my Bible has been John C. Huden's *Indian Place Names of New England* (1962), a scholarly compendium of some five thousand current and obsolete toponyms with presumed meanings. Locations are approximate, given by county only, but only in the vast counties of Maine is this more than an inconvenience. More regrettably, repetitions may be masked by variant spellings, and the coverage itself is often uneven, due to the accidents of antiquarian name-collecting, but such flaws are unavoidable in a work of this sort. In view of the long name lists involved, my examination has been often by necessity mechanical and only lightly judgmental.

One of the most singular patterns to be noted involves the distribution of placenames beginning in Q (always found as *Qu-*) and K. Both letters represent sounds that are phonologically close and that in terms of initial-letter frequency are nearly equivalent: 164 Q-names and 176 K-names. However, the geography of Q- and K-names as shown below is highly polarized, with most Qs to be found in the south and most Ks in the north, demarcating to a considerable degree non-Abenaki versus Abenaki tribal areas:

MA, RI, CT: 84% Q; 34% K
ME, NH, VT: 16% Q; 66% K

There are two ways to account for this distribution, one phonological and the other orthographical, and both in probability possess some validity. The dichotomy may reflect a true distinction in the speech-sounds found in the Abenaki and non-Abenaki languages, and in certain cases, as in the word for long (*kenne-* versus *quinne-*), K and Q are actually known to be phonological reflexes of the same sound in cognate words. In the alternative view, the overall Q–K difference may result not from what the Algonquian speakers said, but from how their English listeners recorded it. Q and K are exotic letters in English, with antithetical connotations: cultivated versus barbaric, French versus German, Latin versus Greek. Q prevails in southern New England because the orthography is older and intentionally assimilated to English to make it seem less uncouth. K prevails in northern New England because the orthography is largely newer, more systematic, and because a romanticized respect for Indian languages had arisen that made it less necessary to "civilize" them. For example, all eleven recorded *Kw-* names, names in which *Qu-* might have served equally and been more English in appearance, are Abenaki in origin. Furthermore, like breeds like: once K or Q had established itself in an area, it might become the local norm; such might account for the prevalence of K on Martha's Vineyard.

Nowhere in the entire region does a toponymic formula achieve the local dominance that *-et* placenames do in southeastern New England. The locative suffix *-et* is a grammatical case ending typically translated as "at the place of." Huden lists more than five hundred examples of *-et* toponyms that, when mapped, show a staggering 73% concentrated in a core area of southeastern Massachusetts and Rhode Island. This density is so incredible as to prompt both a suspicion that the data may have been overzealously collected there (105 examples in Plymouth County alone) and a conviction notwithstanding that such an ingrained pattern could result only from major tribal-linguistic differences. The major tribes in this core were the Wampanoag, Nauset, and Narragansett. An outer core consisting of eastern Massachusetts and eastern Connecticut contains a further 15%, with the remaining 12% scattered thinly on the periphery. Very few occur more than one hundred miles from the coast; only two, for example, in Vermont. If it is permissible to interpret linguistic distributions as one would those of any other artifact, then this pattern suggests that the usage of -*et* toponyms arose as an innovation in the extreme southeast and diffused coastwise east and west with lesser penetration inland, where this usage of -*et* may never have been fully assimilated.

While no single suffix like -*et* similarly dominates northern New England, a set of three, -*cook* (75), -*keag* (32), and -*he(a)gan* (18), has been tallied that might be regarded as a complex with fairly coextensive distributions and an affinity for Abenaki tribal areas. Taken together, 81% of recorded examples occur in northern New England. -*Cook*, the most numerous, is without doubt itself a phonetic catchall that artificially gathers together several distinct Algonquian elements and arbitrarily excludes these same elements whenever variantly recorded as -*kook, gook*, etc. Whatever its true meanings may have been, it clearly entered the English settlers' toponymic aesthetic: words ending in -*cook* were deemed fitting placenames, and words ending in something like -*cook* might be corrupted conformably. Indeed, if we regard the historic variants on established names, we see that a constant game of musical chairs is being played with suffixes such as -*took, -scot, -keag, -quit*, and -*ket*. Despite, or perhaps because of, all the noise in the data, -*cook* functions remarkably well as a toponymic tracer of the Abenaki tribal-linguistic area. Closer study would undoubtedly reveal other dialectal patterns in the name-cover, but these few examples must serve as our brief but pointed introduction to Algonquian toponymy.

Folknames

Folknames are unofficial, unlegislated names established by popular consensus and then recorded only belatedly—and, one hopes, uncensored—on maps. Unlike town names, which were the plaything of colonial governors, and street names, long claimed as the developer's "perk," locally named hills, swamps, ponds, and brooks more sensitively register the historical, cultural, and demographic data we seek. The essentially unofficial character of these toponyms has meant, however, that many have gone unrecorded, and that still others, while surviving as recorded names, can no longer be located as places. Many were supplanted. Time and the U.S. Geological Survey (USGS) have not dealt kindly with the folkname cover, but other factors are also involved.

Toponymic Density

In comparison with older-settled countries, the New England name-cover often strikes the toponymist as rather thin. By contrast, British land features seem to have been named with the same thoroughness and urgency as the marine ones, suggesting that the medieval landsman related to the terrain much more strongly than the colonial yeoman, and in fact, not unlike the seaman to the physiography of the coast. Economic and human necessity compelled the naming of the seacoast and the medieval landscape; presumably New England was undernamed because fewer such incentives both to bestow and record names prevailed. A rich name-cover would be hampered by a sparser and more footloose population with less transgenerational stability; fee-simple tenure in which consolidated acres were held outright and worked by one man, and in which field names did not constitute a quasi-legal description of the parcel; little or no history of communal ownership and hence no need for the community at large to work and discuss the land by name, tract by tract, acre by acre; rapid turnover in land titles, rural abandonment, overreliance on ephemeral owners' names in placenames with nothing site-specific to recommend them to posterity. Recording of names would be frustrated by poor local maps until the twentieth century and indifference by the exploitation-oriented USGS to an antiquarian role as curator of the toponymic heritage. Furthermore, rapidly changing landscapes due to deforestation, reforestation, urbanization, and suburbanization played havoc with the rationale for names. Contours and the shapes of relief features are often only remarkable when cleared. A ledge prominent when in pasture (as High Knob, Athol MA) may entirely lose its identity when returned to forest, and the collapse of agriculture can in time cancel out the toponymy of an entire farming district.

Hydrographic Names

This dearth in the name-cover is belied by the richness of nautical charts, thickly strewn with islets, shoals, reefs, ledges, half-tide rocks, and so forth.

Such coastal hazards were diligently (and wittily) named and mapped, with the least offshore "thimble" or "nubble" lavished more toponymic and cartographic attention than a hundred more eminent no-name knobs inland. The marine name-cover as charted presents a tight array of obscure origin and uncertain and uneven age. Unlike the land name-cover, it has no implicit chronological grain induced by expanding frontier-lines: the entire coast was cruised and mapped from the time of earliest exploration. Nor is it clear to what degree the maritime toponymy consists of folknames ascribed by mariners or chart-names coined by hydrographers (a certain poetry is no guarantee of folkishness: the canyonlands of the West are proof that the "dullest" nineteenth-century government geologist might rise to astounding romantic flights).

Identification of some salient name trends is possible. Islands might be named for owners whose tenure is datable, as was Thompson's Island in Boston Harbor, where dwelt David Thompson in 1623. Alternatively, their names might perpetuate their earliest uses: Clapboard Island, Cowpen Island, or the numerous "Stage" Islands, where fish were dried. Rocks and islets may bear the recognizable names of local individuals or old families (Bowditch Ledge, Salem; Abbot's Rock, Marblehead; Burnham Rocks, Manchester; Soule's Island, Duxbury MA). Thacher's Island (Rockport MA) recalls its castaways and Blake's Rock (Stonington CT) the hydrographer who first charted it.[9] Reefs and shoals often commemorate ships grounded or wrecked upon them. In 1939, the freighter *City of Salisbury*, laden with exotic animals, broke up in the fog on an uncharted spire of rock off Graves Light in Boston Harbor (now known as Salisbury Pinnacle). Not far distant lies Man-of-war or Seventy-four Bar, which marks where pilot error wrecked the seventy-four-gun French vessel *Magnifique* in 1782.[10] The grounding of the *Queen Elizabeth 2* in Vineyard Sound in 1992 prompted the suggestion that the reef be named Queen's Bottom, but this was deemed to be too racy for the Hydrographic Office. (Rude's Ledge, in honor of the survey vessel that pinpointed the reef, was proffered as the more likely candidate.)[11] Alleghany Rock off Martha's Vineyard, while of uncertain provenance, has the ring of a wreck name.

Dangers might be given names suggestive of treachery or feminine deceit: Satan Rocks (Salem MA), Cutthroat Ledge (Marblehead MA), the Hypocrites, the Cuckolds (Boothbay ME), Haradan's Rock (Annisquam MA), The Hussey (Casco Bay ME). Others are ironic memento mori: Thread-of-Life Ledge (South Bristol ME) or Aquae Vitae rocks (Salem MA). Some metaphors recall that many mariners plied dual trades as

fisherman-farmers: the Cowyard (Plymouth MA), and the Hedge Fence
and the Squash Meadow (Nantucket Sound). Others imply close-knit
communities: Uncle Zeke's Island (Harpswell Sound ME) or Uncle Robert's Cove (Hyannis MA). Adjacent hazards might be gathered together in
laboriously extended metaphors: Goose and Goslings, Cow and Calf, Pig
Rock and the Piglets (Weymouth MA), the Hen and Chickens and the
Old Cock, too (Buzzards Bay MA). Beaver Neck near Newport RI has
Beaverhead at the top and Beavertail Point at the bottom. The fitness of
many names is best brought home on shipboard. The sight of a schooner
gliding ethereally through the narrow interval between Gap Head and
Straitsmouth Island in Rockport MA quickly reveals what this "Gap" business is about. Features were named for their most salient aspect or silhouette when viewed from shipboard: thus Threetree Ledge (near Biddeford
Pool ME) or Two Bush Island (Deer Isle ME). (The Old Sloop Church in
Rockport MA and Mount Ararat in the Provincetown dunes were likely
christened not on land but at sea.) Size and shape are suggested by the
Sugarloaves (Vinalhaven ME) and Twopenny Loaf (Gloucester MA);
soundings by Five-fathom Rock (Boston) and Shovelful Shoal (off Chatham);[12] cost by Ten Pound Island (Gloucester MA).

Metaphorical conceits clearly had mnemonic value in navigation without charts, as when sailing into the New Meadows River in Maine, one
kept the Brown Cow to starboard and the White Bull to larboard. The
Pot was hard by the Ladle, with no safe passage between the two. In 1806,
when one sailed up Moosabec Reach to Machias ME, one kept the Virgin's
Breasts to the larboard and the Ragged-arse (a bare rock) to starboard.
(Sometime between 1826 and 1854 the Ragged-arse was officially bowdler-
ized to Pulpit Rock.) The Virgin's Breasts consist of two islands otherwise
known as Virgin Island and the Nipple. Near Buck's Harbor ME one sailed
between The Ship and Barge.[13] The imagery is vivid, and with reason. In
the workaday world of coastwise mariners, sailing directions were undoubtedly an oral tradition to be memorized. (Compare Mark Twain's
1850s apprenticeship as a river pilot described in *Life on the Mississippi*.)
The fixation on what now seems trivial detail is not simply because there
was no Loran or marker buoys; there was also precious little in the way of
man-made features (such as steeples) to rely upon on many deserted coast-
lines. Even boulders served as important landmarks, as one at Gouldsboro
Bay ME described in the United States *Coast-Pilot* (1874). Big Ben "looks
like a weather-beaten, unpainted barn. It is however a huge, square, gray
rock with perpendicular sides, and is by far the most noticeable feature of

the Bay."[14] Mariners who rely too much on such landmarks have come to be disparaged as "appletreers" or "barking dog navigators."[15]

There is a curious similarity between some of the more blatantly picturesque hydrographic toponyms and the traditional Anglo-American names for taverns and coffeehouses. The inspiration for such names might have come directly from the signboards of such places of seafarers' resort, or coincidentally from some common source of medieval folk imagery that also found expression in coats of arms. In all these contexts, it was the iconic and mnemonic power of the image that recommended its usage. The Rose and Crown is a boot-shaped shoal off Nantucket charted as early as 1675. While the name of no known British danger, the Rose and Crown since at least 1606 has been a perennial and prolific London signboard; it derives from a badge of the Tudors. (It was also not unknown as a ship name.) Hen and Chickens, a marine toponym common to old and New England, is a popular London tavern name (known also from Burlington MA), and incidentally is also tavernkeepers' slang for pewter pots of mixed sizes.[16] Labo(u)or-in-vain is a creek in Ipswich MA; the Sugar Loaf designates each of twin islands in Vinalhaven ME, as well as many mountains; Tumbledown Dick (a supposed reference to the short rule of Richard Cromwell) is a headland, a rock, and a mountain in Maine; Hedgehog is the name of a Maine island and numerous hills and mountains. All are London signboards. Three Turks' Heads was Captain John Smith's name, derived from his coat of arms,[17] for Thachers and nearby islands off Rockport MA, yet Turk's Head is also a prolific signboard. In Britain, the Woolpack is a shoal near Margate and a rock in the Scilly Isles; it is also a tavern name. While no direct connections are implied here, the fund of imagery is remarkably the same.[18]

Hydrographic Transplants

The most salient transplants occur off the southeast coast of New England and derive from the west of England and South Wales. These are, namely, the Bishop and Clerks southeast of Hyannis MA and the Bishops and Clerks off St. David's Head, Wales; the Hen and Chickens off Cuttyhunk MA and off Barnstaple Bay, Devonshire; Bishop Rock Shoal off Jamestown RI and Bishop Rock (famous as the last English landfall) off the southwest tip of the Scilly Isles; Kettlebottom off Newport RI and Kettle Bottom in the Scilly Isles; and conceivably, Old Bartlemy (a traditional corruption of Bartholomew) off New Bedford and Bartholomew Ledges likewise in the

Scillies. Perhaps the most intriguing of these transplants is The Gurnet, or Gurnet Head (now Point), the obscurely named tip of Duxbury Beach in Plymouth Harbor MA. A gurnard or gurnet is a species of fish that was proverbial in Elizabethan times for its oversized head. There are two English Gurnard('s) Heads: one near Land's End in Cornwall; one on the Isle of Wight on the Solent, the major shipping channel to Portsmouth and Southampton. Local authority cites the latter,[19] but both would be within the home waters of West Country mariners who transplanted this significant cluster of hydrographic names to southeastern New England.

Gurnet reappears as the name of a strait in Brunswick ME lying, not surprisingly, within lands granted to the Plymouth Colony in the Kennebec Patent of 1628. How the name The Gurnet came to be applied here to a two-mile inlet, not a headland, is a puzzle, but the usage is of undoubted antiquity, dating at least to the early seventeenth century.[20] (It also was applied to two other narrow island passages nearby, Prince's and Simon's Gurnets.) These Gurnets are by no means the sole toponymic legacy of this lucrative Plymouth fur-trading outpost. Winslow's Rock, in the Kennebec River opposite the city of Bath, was named for Pilgrim Edward Winslow, and Parker's (now Georgetown) Island was named for John Parker, brother of Thomas, mate of the Mayflower.[21] Some complex interrelationship probably also exists between the Maine town names of Falmouth and Yarmouth found here and those on Cape Cod, likewise within the territory of Plymouth Colony. (Falmouth ME appears actually to be slightly older than Falmouth MA.) Cultural links remained strong enough that eighteenth-century Plymouth County–carved gravestones are found in the vicinity.[22] Yet another New England Gurnet, recorded as early as 1660, once denoted the "extreme north-eastern point, or hook of beach" at Edgartown MA, a topographic situation so like that at Duxbury Beach as to suggest the name was also inspired by it.[23] It might also be a direct transplant.

Not far from the Gurnet in Duxbury Bay MA we find the Nummet, which proves not so tough a nut to crack. While not a hydrographic transplant, the term betrays its west of England origins in another way. Nummet(t) or "noon-meat" is archaic for a field-laborer's luncheon, and perhaps it was on this bit of dry ground that longshore fishermen of old customarily took their nooning. As attested by the *English Dialect Dictionary*, it is a West Country word found in Wiltshire, Gloucestershire, Somersetshire, Devonshire, and South Wales. It would seem more than coincidental that these seven West British hydrographic names occur off a corner of New

England noted for seventeen town names drawn from the same region. Such a strong and consistent correspondence is wholly unique to the waters between Plymouth MA and Newport RI and occurs nowhere else on the New England coast. Dangerous shoals off Marion MA bear a provocatively English name, the Bow Bells, but this maritime usage has no known English precedent. Clearly, the name must ultimately derive from the church of St. Mary-le-Bow, Cheapside, London. (To be born within the sound of Bow Bells was the mark of a true Cockney.)

Transplants within New England also occur. The Graves in Rockport ME is a brazen but respectfully distant plagiary of the famous navigational hazard at the mouth of Boston Harbor. Likewise, Smuttynose Island in Brooklin ME lies 130 miles downeast of its more famous namesake in the Isles of Shoals. Both examples illustrate the principle that a landmark name tends to suppress its reusage for many miles around. Some transplants are virtually generic, so ubiquitous as to make their diffusion untraceable. Among the most notable in the British and American repertoire are the Cow and Calf (Fishguard, Wales; Newcastle and Cork, Ireland; Tremont ME; Branford CT), the Sow and Pigs (Blyth Harbour, Northumberland; Cuttyhunk MA; Freeport ME), and the omnipresent Hen and Chickens (Culross, Scotland; off Barnstaple Bay, Devonshire; Greenwich CT; Saybrook CT; Cuttyhunk MA; as well as in Nova Scotia, Delaware Bay, Florida, and the Bahamas). By contrast, one of the most prolific generics, the Roaring Bull(s), sometimes specified as a rock or ledge, has no known English prototype either as a danger or tavernboard, yet occurs at least ten times from New England to the Maritimes: Newport RI; Boston, Marblehead MA; Mantinicus, Vinalhaven, Isle au Haut, Winter Harbor ME; Cape Canso, Pictou, Yarmouth, Nova Scotia. These data suggest that while the Roaring Bull was a homegrown generic diffused by coastwise shipping (probably from Boston or Marblehead), the Hen and Chickens first radiated from Britain to seminal points before becoming generalized along the North American coast. Two other generics with no known English precedents are the Dumpling(s) for an islet or string of islets (Stonington CT, Point Judith RI, Newport RI, Cuttyhunk MA, North Haven ME, Stonington ME) and Gangway rock or ledge (Newburyport MA, Portsmouth NH, Boothbay ME, Friendship ME, Jonesport ME, and near Lubec ME).

It might seem that a transatlantic subculture fostered by the natural mobility and gregariousness of seamen would result in a general concordance in the style of hydrographic names on both sides of the water. However, despite any contrary evidence cited above, such is not the case. Overall, a perusal of the text and indices of the U.S. Hydrographic Office *British Islands Pilot* fails to reveal English sources for the names of marine dangers, neither in unique names, such as the Thread of Life, High Sheriff Rock (Swans Island ME), the Hypocrites, the Cuckolds, etc., nor in generics for islets such as Thimble, Thrumcap, Nubble, nor in names duplicated on the New England coast such as Spectacle Island and Egg Rock. A summary impression is that the New England names are on the whole more accessibly picturesque than those of the British Isles, which often bear ancient names of obscure, and on many coasts, Celtic derivation.

Generic Folknames

Certain folknames—often clever, humorous, or pejorative—unaccountably repeat themselves far and wide over the landscape. The generic folkname is something of a familiar, ready-made garment reached down to cover local situations; the name itself is often more memorable than the supposed circumstances that prompted its application. A modern example well illustrates the point: as told to me, a section of new, expensive houses in Hanover MA ca. 1985 was known as Hamburger Hill because the owners were reduced to eating hamburger to pay for them. The name, of course, is that of one of the bloodiest battles of the Vietnam War; the local pretext for its application has the same hollow ring as that of many nineteenth-century generic folknames: the immediate context does not seem specific or dramatic enough to warrant the genesis of the name, and one is tempted to scoff at the explanation. Generic folknames may occur in scatters or in strings; true clusters are rarer and rendered unlikely because they would require many same-named places within earshot of the inhabitants of a district. Any one, well-known example of a mountain called the Dome, Haystack, or Sugarloaf would inhibit reusage of the name for miles around, and certainly diminish the likelihood of its being cartographically canonized by the USGS.

As communications improved, this "earshot" or horizon of toponymic awareness would expand. An extreme example is the Maine ski resort that bills itself as Sugarloaf USA, which through powerful publicity has made itself at any rate *the* New England Sugarloaf, despite numerous (at least twenty-four) others. The map-patterns that I here call name-scatters and name-strings may testify to different mechanisms of folkname dissemination. Regrettably, the close study of such folkname dynamics is hampered by the lack of comprehensive indexes, especially of archaic names, which are the ones most likely to have distributions explicable in Kurathian terms.

Slab City, the ephemeral folkname for a sawmill village, furnishes an excellent example of a name-scatter impelled by means other than usual settler migration (see fig. 4). "Slab" is the waste wood trimmed from saw-logs, while "City" has been an ironic designation since early colonial times for a busy mill privilege or seat of industry even when remote from population. As a slang term of the widespread and migrant lumber industry, Slab City had an inherent means of diffusion and its ubiquity was inevitable. There were undoubtedly far more unrecorded, uncollected, or obsolete Slab Cities in New England than the nineteen plotted here, and still others may survive only in road names that are not regionally indexed. Slab City conforms well to the theory that a generic folkname-scatter in effect maps the usage of a quasi-dialect word: slab city = sawmill village. (The three Slab Cities in western New York, and one each in Ohio and Wisconsin, effectively limn out the "New England Extended" of westward emigration.)

Another scatter, Puddledock, is more puzzling and may involve both transplantation and perhaps quasi-dialectal diffusion (see fig. 4). The prototype is the Thames-side quay below Blackfriars Bridge in London, which has been called Puddle Dock (or Puddle Wharf) since at least the late sixteenth century. Four other Puddledocks (variously spelled) are found in southeastern England, in Norfolk, East Sussex, and two in Kent, so the name is well represented in a chief source-region of New England emigration. It reached these shores quite early, and is recorded as a defunct locality in Plymouth MA, together with other transplanted London names. Its use for a waterfront neighborhood in colonial Portsmouth NH, would come naturally to anyone engaged in transatlantic trade, and this may have inspired another coastal example at Alna ME, just below Head of Tide on the Sheepscot River. But what of Puddledock Road, forty miles inland at Charleston ME, as well as other, presumably nineteenth-century Puddle-docks that also appear scattered deep in Vermont, at East Fairfield, Granville (site of a lumber camp), and Reading (Felchville)? The latter three are described as "swampy, low-lying places"; that at East Fairfield found further justification (or ex post facto rationalization) in a railroad station built on swampy ground.[24] As it seems unlikely that such a name could be reinvented, was Puddledock, like Slab City, part of the folkname repertoire of a larger speech community (implying others that escaped doc-

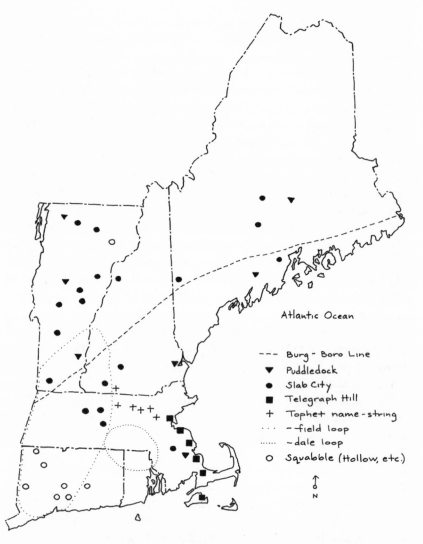

FIGURE 4
Eight Placename Distributions

umentation), or was it transferred from place to place, perhaps from London to Portsmouth to Alna, and then much later, inland? More puzzling still is the recent appearance, or possibly archaic revival of Puddlewharf Lane in Marshfield MA within the last twenty years.

The seven Squabble placenames in Connecticut (mostly Hollows) show an almost classic Kurathian distribution (see fig. 4). Plausibly, they had their start with Squabble Hollow in Derby, the site of the 1680 meeting-house and one presumably chosen with all the usual turmoil that could give birth to a splendid name. Explanations for the other Connecticut Squabbles are lacking, lame (a quarrelsome household), or too late (a fight with the Redcoats) to suggest themselves as the original. Three more (Milford, Bethany, Danbury) lie within 5 to 20 miles, four more if we allow two steps to reach Haddam to the east (the only Squabble truly off the Kurathian beaten path). The next two lie 30 to 50 miles away in northwestern towns (Torrington, North Canaan) settled about 1740 by outmigration. Except Haddam, all are easily interconnected by numbered state highways, which often fossilize colonial travel lines. The only other Squabble in New England is Squabble Hollow below Pudding Hill in Lyndonville VT, 225 miles to the north, at journey's end of the Connecticut exodus into the interior. This last is preposterously said to have been named when, in a time of famine, the gift of a giant pudding was rolled down from hill to hollow, where the people squabbled over it. Certainly this smacks of the ex post facto fabulation commonly used to explain transplanted names.[25]

Name-Strings

Name-strings are coherent, linear distributions that can act as toponymic tracers of settlement movements. One of the best examples is the Tophet name-string (see fig. 4). Tophet is a biblicism and a euphemism for Hell, and its usage in both the Bible and *Pilgrim's Progress* would guarantee that the term was in the vocabulary of virtually everyone in colonial New England; its toponymic usage, however, was surprisingly restricted. Five Tophets, all swamps (the one in Littleton is a swampy Chasm) fall at ten-to-fifteen-mile intervals between Lexington MA and New Ipswich NH (originally MA), along a major colonial axis identified by Kurath, represented by a broad swath of braided through roads between Boston and Keene NH. The only other known Tophets in New England are a brook near North Adams MA, and a hollow in Woodbury CT. The intervals between the five examples increase as one goes westward, perhaps because the settlers' toponymic horizon increased with improved communications, or because population pressures made each successive removal more distant than the last. When mapped the Tophet name-string does plausibly suggest a toponymic tracer of migration movements, and further evidence for such a migration can be found in the placenames of New Ipswich and the adjacent towns of Ashby MA and Greenville NH. Barret Mountain, Emerson Hill, Wheeler Road, Willard Road, Hoar Pond, Davis Hill; Blood Hill, Heywood Road, Davis Road, Barrett Hill and Merriam Hill—all bear names prominent in the history of Concord MA and vicinity, and geo-genealogically connect the settlements at either end of the Tophet name-string. (Perhaps a third of the settlers of New Ipswich were from Concord.)[26]

By way of contrast, Telegraph Hill demonstrates another, albeit unusual way in which a name-string might be systematically laid down on the landscape (see fig. 4). By 1801, a flag "telegraph" or semaphore system with a dozen hilltop signal masts was in place between Edgartown MA and Boston that could carry messages seventy-five miles in ten minutes. Returning East India merchantmen with their valuable cargoes could be redeployed by their Boston or Salem owners to more southerly Atlantic ports directly from Martha's Vineyard without a hazardous voyage around Cape Cod to take orders. By 1819 messages were relayed directly to the cupola-crowned offices of the Semaphore Telegraph Company on Central Wharf in Boston. The memory of this long defunct system is preserved in a string of Telegraph Hills in Edgartown, Woods Hole, Sandwich, Manomet (Plymouth), Duxbury, Hull, and South Boston (alias Dorchester Heights).[27] The only two other Telegraph Hills in New England are near Provincetown MA and East Pittston ME.

As folknames cannot be precisely dated it becomes impossible to reconstruct chronologically the formation of a toponymic artifact such as the Tophet name-string; consequently a simplistic hopscotch of migrations is conjured up to fill the void. However, if we look at a datable sequence of town names such as the Friendship-Harmony name-string in Maine, blind faith in such a scenario is quickly exploded. A straight northerly line of seven towns, more or less contiguous, runs inland sixty-five miles from Friendship on the mid-Maine coast. This sequence of towns, together with the years of their incorporation and probable naming, goes from south to north: [Independence (1848)], Friendship (1807), Union (1786), Hope (1804), Liberty (1827), Freedom (1813), Unity (1804) Harmony (1803). Only two other Maine townships, Amity (1836), and Industry (1803), bear names of this ilk. Independence, in brackets, was the proposed but rejected name for South Thomaston when it was incorporated in 1848; the proposal and rejection signal that the era of this by then well established (and surely locally well recognized) name-string was definitely closed.[28] The chronology of its development (1786–1836) is unstraightforward and anything but predictable. It took forty years to form, yet more than half sprang up in five short years (1803–1807); it did not grow south to north as it might have if a tracer of inland migration; its growth, indeed, was nonlinear: arising in the middle, then at both ends independently, and slowly crystallizing in between. Yet remove the dates, and see how blithely one might analyze it as we have analyzed the Tophet name-string.

Transplanted Folknames

As with hydrographic names, English transplants often also occurred in the folknames of localities. Such names survived not by official sanction, but by winning local approval. Typically, they fit the local facts of geography better, and often pleased by punning. Gravesend, Kent lies at the head of the Thames estuary. Gravesend in Lynn MA was the end of town inhabited by the Graves family (Woodend and Breed's End also came from local family names); in Portsmouth NH, Gravesend Street took its name from Point of Graves, a colonial burying point. Billingsgate, the London fish market, gave its name to what is now Wellfleet for its abundant fishing-grounds. The etymology of Hammersmith, a London borough, was brushed up for the colonial furnace and triphammer at Saugus MA. Cornhill in London was transplanted and in places reliteralized as Cornhill in Boston and Gloucester and Corn Hill in Plymouth and Truro. One expects that Cheapside in Deerfield MA involves an inescapable play on the apparent (though incorrect) etymology of that London market. A copper mine in Granby CT earned the name Newgate (of Dickensian fame) when used as a prison in the Revolution.[29]

Some transplants elude detection because their originals are little known and their local appropriateness is unquestionable. Mount Whoredom, as the west summit of Beacon Hill was called until 1823, likely derives from a hill in London that bore the same name for the same reason. This designation (one fondly recollected by proper Bostonians) may have been military slang: the London Mount Whoredom (so mapped in 1745) was a place of soldiers' resort near the Royal Artillery headquarters in Woolwich. Curiously, the earliest usage of the name in Boston cited by Walter Muir

Whitehill was made by British officers in 1775.[30] Bedford Levels might seem a purely matter-of-fact name for meadow and swamp lands found in Concord MA near the Bedford town-line; the Bedford Level, however, was also the site of a well-known drainage scheme in the English fenland undertaken in the 1640s, a noteworthy engineering achievement in the home district of some of Concord's founders. The unusual word choice, "levels" (itself perhaps humorously suggested by the ditched meadows thereabouts?), strengthens the case for a transplant. Salisbury Plains in Salisbury MA might certainly have arisen without thought of Salisbury Plain of Stonehenge fame, yet some of the first settlers were indeed from Wiltshire.

Under unusual circumstances transplants might run the other way, from New England to Old. The Battle of Bunker Hill likely gave rise to many of the Bunker Hills scattered around New England, but it must be re-membered that Bunker's Hill (as they render it) went down as an English victory at least in the British books. The canon's roar re-echoed across the Atlantic in England, where there are thirteen Bunker's Hills, all of which are believed to derive from the American placename. (English farm fields were often named after battles.)[31] In time, the folk wordplay evident in many of these transplants was turned loose on second-guessing the official origins of town names, as Rutland MA for its red soil that stained the fleece of sheep; Ashburnham MA for its many potash works; New Ashford MA for a fort of ash logs erected there. Indeed, it is generally acknowledged that Bridgewater CT (1803) was named not for any town or person, but for its bridge on the Housatonic, so the line between fact and fancy is not easily drawn.[32] In light of these folk-etymological tendencies, it becomes difficult to dismiss as silly the notion that the "sand-" in Sandwich MA and the "-fleet" in Wellfleet MA were not integral to the popular acceptance of these official, transplanted town names.

-Shire *Names*

Folknames denoting localities that end in *-shire* can be found scattered throughout New England, but for echonymic reasons flourished best on the borders of New Hampshire and in old Hampshire County MA (which until 1811–1812 took in Franklin and Hampden counties as well).

I am aware of no English precedent for this practice. These names are often unflattering of the inhabitants, and occur in clusters, sometimes drawing initial inspiration from an official *-shire* name, and then from each other. Some names (as Pilfershire, Shirkshire) appear several times and raise the old question of whether they should be looked upon as true transplants or instead as a sort of quasi-dialect. Vershire VT (1781) is the only official town name of this ilk, an obvious blend name from Vermont and New Hampshire, which probably reflects the strong Connecticut influence in the Vermont legislature when it was coined. I have come across twenty -*shire* names, some with anecdotal origins. Snumshire NH took its name in the mid-nineteenth century from the trick of speech of a longtime local blacksmith ("Smut" Milliken) who was given to mutter "I snum!" (I do declare!). Trapshire NH derives from an incident in which Simon Sartwell's horse was caught in a trap he set for a bear. Smokeshire VT was befouled by its charcoal kiln.[33] Two Pilfershires have different stories: that in Petersham MA was a corner of town ceded to ("pilfered by") the town of Dana; that in Eastford CT derided its thievish inhabitants. Broomshire in Conway MA derived its name from walnut brooms made there by William Warren. Shirkshire, also in Conway, rhymes neatly with Berkshire, and one imagines the folk there were proverbially lazy. Yet according to local tradition, a Mr. Sherman was refused a full dinner after helping men work on the roads there: again an example of an event seemingly too minor to justify a splendidly clever name.[34] (Shirkshire in Bennington VT was perhaps a transplant.)

These origins are quite plausible. The commemoration of minor incidents in local names (as Spanktown VT for the pluck of a local girl who spanked a drunken farmer)[35] is sociologically consistent with the whimsies and censures of a close-knit community. Snumshire is only a toponymic expression of the same communal humor that nicknamed Moses C. Milliken "Smut" for the soot of his smithy. Like nicknames, the origins of many folknames is anecdotal; the anecdotes may be altered by time and retelling, but explanations should not be rejected simply because they are anecdotal or the event commemorated trivial. Names such as Music Street (West Tisbury MA) from the pianos bought for the daughters of status-seeking sea captains; Tattle Street (Plympton MA) from its gossips, and Pucker Street (Petersham MA) from its snobs; Petticoat Hill (Williamsburg MA) from the sight of the undergarments of seven sisters hung to dry, and Yell Mill (Foster RI) from the deaf millers' colloquies, all testify to this earlier mentality.

By contrast, some anecdotal etymologies strain credulity by betraying an ignorance of the true meaning of the words involved. Three Penny Morris is the odd name of a hill in Petersham MA that tradition holds to have been bought at three cents an acre by a man named Morris.[36] Yet Three Penny Morris is an old English bowling game: are we to believe that the name was a godsent pun, or is this explanation fanciful folk etymology? While skepticism of picturesque name origins is on the whole a healthy thing, an element of the picturesque is integral to folk-naming and the plausibility of a particular explanation is best evaluated by comparison with the body of toponyms at large.

Township Toponymy

Since many towns were named by gubernatorial whim, or for other arbitrary and arcane reasons, the ability of local historical and geographical forces to shape the township toponymy was limited. The New England landscape is a salad of variously motived placenames and to interpret overall map-patterns in ignorance of the true historical value of each and every name is perilous work. Yet as a matter of practicality this is what we must do. If one is ever to take the full measure of what the New England township toponymy encodes, it is necessary to scrutinize its patterns as though they were our sole source of information on settlement history. The assumption that the toponymy of Old England harbors a deeper historical significance than that of New England results in part from its having been archaeologized to death, in quest of simple settlement data for which New England already possesses an adequate written record.

Map-pattern recognition is an informed yet spontaneous act which demands that we interpret first, ask questions later, and be ever ready to discard a cherished theory. The degree to which our reading of town-name map-patterns unravels on deeper research, can shed light on how best to weigh other toponymic evidence—folknames in particular—for which authoritative explanations are lacking. (We have already seen how the chronology of the Friendship-Harmony name-string discourages facile interpretations of name-strings in general.) We shall now turn to the analysis of town names, which while often toponymically bland, are well-documented and manageable in number.[37]

Transplanted Town Names

While innumerable town names were brought over from Old Country to New, the sort of information they encode varies greatly. At one extreme, a transplanted town name can record the place of origin of the settlers, at the other the titular seat of a nobleman whom a royal governor wished to flatter. Occasionally, English placenames were employed connotatively, with an eye to the chief business carried on there. Thus, Cambridge MA (1635) was named for its college; New Castle NH (1693) punningly for its fort; Bristol RI (1680) for its seaport; Portland CT (1841) for its sandstone quarries and Manchester NH (1810) ironically for its barge canal (not, as one might suppose, for its later cotton mills).[38] Systematic transplants on a vast scale are an anomaly. The best illustration involves the pre-Revolutionary name-scheme of the counties of Massachusetts and Maine that in a rubbery way reproduced the relative geography of their English namesakes, an expedient solution to an administrative problem. Boston stood in for London in this scheme; Middlesex adjoined it, Essex lay to the northeast; York, Lincoln, and Cumberland were far north in Maine; Hampshire and Berkshire lay westward. While imperfectly realized, the inspiration is obvious. (The geographic anomaly that placed Norfolk County south of Suffolk is of later date.) Such a measure of calculation is rare in the disposition of transplanted names, is quickly spotted, and could only issue from a central authority.

Much of the earliest town-naming (1620–1660) records the county or town of origin of the first settlers, their port of embarkation, the birthplace of an elder, the parish cure of a minister. Because the first settlers of an area often sprang from the same corner of England, and because later emigrants often cast their lot with their countrymen, the landscape of eastern New England was not simply strewn with far-flung English names. There are discernible clusters of East Anglian, Midland, and West Country names among the early towns, which may reflect the source region of the first settlers (see fig. 5). However, such is the power of suggestion that a Midland-named county like Worcester might have called forth Midland-named towns, and a notable consonance exists between these clusters and their county names. Sometimes remarkably similar collocations of towns can be found on both sides of the water. These might arise by accident, or from willful cleverness on the namers' part, but certainly the existence of two or more towns that bear not simply the names but also replicate the geographical proximity of their English originals gives some grounds to think that the settlers hailed from that district. On the face of it, one Wiltshire borrowing might have as its rationale many things besides the settlers' place of origin, but two adjacent Wiltshire names apparently transported to the New World with propinquity intact strikes one as much harder to explain otherwise. Or so the theory runs. For our purposes,

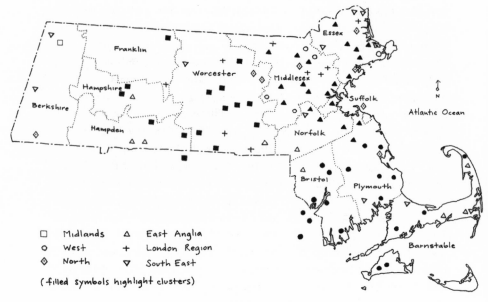

FIGURE 5
English Source Regions
of Massachusetts Town Names
Original research inspired by Wilkie and Tager 1991, 16.

collocations involve towns no more than twenty miles apart. The towns grouped below all preserve a semblance of the proximity of their English originals.

> Massachusetts: Worcester-Leominster; Boston-Lynn; Chelmsford-Billerica; Taunton-Norton-Bridgewater-Somerset-Tiverton (now RI); Dorchester-Weymouth; Amesbury-Salisbury-Newbury-Andover; Dudley-Sturbridge.
> Rhode Island: Warwick-Coventry; Bristol-Newport(?).
> New Hampshire: Portsmouth-Gosport (*obs.*)-Sandown-South Hampton.

While certain collocations are undoubtedly accidental, and others cleverly calculated, some must surely point to a common place of origin for the settlers. However, given the complex nature of emigration, these last must surely be deemed rare curiosities, resulting from specialized circumstances, often far different and more complex than the simplistic scenarios one is apt to imagine in the absence of the facts, as that of a shipload of Dorset men being shepherded by their minister into the wilderness to establish the adjacent settlements of Dorchester and Weymouth MA. While these paired places in both Old and New England lie within ten miles of each other, the Massachusetts towns were separately settled by different enterprises and incorporated five years apart (1630 and 1635). The significance of this collocation is hard to assess. Should it be read as a toponymic tracer of the settlers' place of origin, as a bit of geographically minded wordplay, or be deemed coincidental? We know that common place of origin indeed explains the Amesbury-Salisbury-Newbury-Andover quadruplet, which limns out the west of England home district of the early settlers. The Taunton-Norton collocation has long been noted by local historians, particularly as the etymology of the latter name (i.e., North Town) so admirably suits a town set off from the north end of Taunton in 1711. Norton (i.e., Norton Fitzwarren) lies just northeast of the English Taunton in the county of Somerset, where the two towns are linked in rhyme: "Taunton was a furzy down / When Norton was a market town." (The better-known nursery rhyme "Trot, trot, to Boston / Trot, trot to Lynn . . ." likewise strikes Massachusetts ears as remarkably close to home.)

Yet many collocations are to be explained by means other than settlers' origins. New Roxbury CT (then in MA) was renamed Woodstock in 1690 by Judge Samuel Sewall because of its proximity to Oxford MA. Sewall had visited Oxford while in England the year previous and specifically recorded the rationale for this change in his diary. (Would that more place-namers were compulsive diarists!)[39] Killingly and Pomfret CT were named for adjacent Yorkshire manors held in the family of Connecticut Governor Saltonstall. Wellington and Tolland CT were named for Somerset villages from which the ancestors of Governor Roger Wolcott originated. Gosport NH (the obsolete name of a village on Star Island, Isles of Shoals) was named for its proximity to Portsmouth, as a conscious imitation of old Hampshire in New Hampshire. The contiguous Connecticut River towns of Westminster-Putney-Fullum (i.e., Fullham, now Dummerston) VT were named by Governor Benning Wentworth in 1752–1753 after three Thames-side London villages.[40] These five collocations are calculated wordplay, and not due to any memorialist impulse of homesick settlers.

More common than collocations, however, are scatters or "crazy salads" of Old Country names drawn from far and wide that lack topographic cohesiveness but that statistically may display a demonstrable regionality. Collocations presume that next-town neighbors flocked together to new settlements; crazy salads that they scattered. For example, a random prevalence of East Anglian town names in old-settled lower New England can ultimately be ascribed to the fact that this corner of England was a hotbed of Puritanism (and before that of peasant rebellion). To speak in round numbers, 60% of early emigrants to Massachusetts Bay, as well as 60% of English town names planted there by 1660, derived from an area within a sixty-mile radius of the East Anglian market town of Haverhill, Suffolk, England.[41] Thus the name-scatter in the New Country may furnish puzzle pieces with which to reassemble a map of the settlers' home district in the Old, and in turn lead to speculation on the whys and not simply the whences of their emigration. Lack of regional bias in the names does not of course exclude them from being derived from the settlers' place of origin; it may merely testify to more complex mechanisms of migration.

While our attention has been so far confined to transplanted English geography, a Scotch-Irish frontier also began to be inscribed on the map as early as 1719 with the settlement of Londonderry NH. Presbyterians who were greeted with hostility in Puritan Boston, the Scotch-Irish were encouraged to establish buffer settlements on the northern and western frontier against Indian attack. Yet as with English transplants, Scotch-Irish placenames prove to be variously motivated: one might honor the settlers' hometown and another a Lord Lieutenant. Glasgow (obs.), Colrain MA; Dublin, Limerick (obs.), Antrim, Derry, Londonderry, and Sligo (obs.) NH limn out a prominent arc, yet how to disentangle these from other Scotch or Irish names (as Athol MA) which, while officially ascribed to other causes, may yet have succumbed to a common trend?

In time internal migration led to a similarly transplanted New England geography, although the names involved were often re-recycled English ones. Indeed, by 1635, out-migrants from Dorchester, Newtown (i.e., Cambridge), and Watertown MA had begun briefly to apply these same names to the Connecticut River towns of Windsor, Hartford, and Wethersfield. By the early eighteenth century a periphery of "New" towns, such as New Braintree, New Salem, and New Ipswich (now NH) had begun to spring up in the hinterlands of the Bay Colony. These were typically new grants (often in payment for military service) made to the earlier settled coastal towns, and the historical kinship of the like-named communities may be startlingly corroborated by other local names, such as Appleton Academy and Appleton Manse in New Ipswich NH, Appleton being a well-known ancient name in Ipswich MA (witness the thousand-acre Appleton Farms, an intact seventeenth-century landgrant). Most notable is the wealth of Connecticut names that characterizes the toponymy of Vermont, itself once known as New Connecticut. (The Western Reserve of Ohio, a later "New Connecticut" is blessed with a similar legacy.)

Vermont collocations such as We(a)thersfield-Windsor on the Connecticut River and Cornwall-Salisbury on the other side of the Green Mountains not only preserve the proximity of their downcountry namesakes; they lie due north of them along known migration routes. Woodstock and Pomfret are contiguous in both states. However, of the 60 possible correspondences between Connecticut and Vermont town names noted by Hughes and Morse,[42] only 10 can be definitely documented, and fully 38 are unlikely or provably specious. In 15 cases, contrary to expectation, the Vermont name predates its Connecticut counterpart, as is the case with Bloomfield VT (1830) and Bloomfield CT (1835). (Bloomfield, like Auburn noted earlier, may be a fad name.) Vermont also drew settlers from elsewhere in New England. Londonderry VT was settled from Londonderry NH; when it came to be divided, the eastward daughter town was named Windham VT, after Windham NH, itself the daughter of Londonderry NH.[43] Vermont also exhibits possible collocations of Massachusetts origin, such as Pittsfield-Hancock-Stockbridge, Plymouth-Bridgewater, and Braintree-Randolph. Of these, the first may be genuine, the second is dubious, and the third an uncanny coincidence. For despite the fact that Braintree and Randolph are contiguous in both Massachusetts and Vermont, Randolph VT was named in 1780 and Randolph MA not incorporated until 1793, making the obvious explanation anachronistic. Because the dates are so close, and because the grantees of Braintree VT are known to have come largely from Braintree MA, one is tempted not entirely to dismiss the possibility of reverse influence.

The Temple-Wilton collocation in Maine resulted from settler migration from the adjacent same-named towns in New Hampshire.[44] Maine, as a former "district" of Massachusetts, is rife with Massachusetts collocations: Lexington-Concord, Grafton-Upton, Brighton-Cambridge, Waltham-Dedham, and Burlington-Lincoln. Whether these reflect gubernatorial or legislative whim, absentee grantees' hometowns, or true settlers' origins, only investigation could tell.

All The King's Men

Ever since Captain John Smith sought to interest the young Charles I in his transatlantic ventures by allowing him to name the unsettled domains of New England on his 1614 map, New England place-naming was prone to toadyism. To Smith's fine delineation of the Gulf of Maine, the prince added several dozen names, some to real natural features, but many to nonexistent towns. The River Charles he named for himself, Cape Ann(a) for his mother; Cape Elizabeth and the yet-unsettled Plymouth (the Pilgrims, familiar with the map, adopted the name)[45] also survive more or less as charted. A half-dozen other fanciful, randomly placed towns with remarkably prescient names, among them Boston, Hull, and Ipswich, decorate the wilderness, and one wonders whether the first planters later sought to fulfill this royal cartographic fantasy, or whether these places (mostly ports) were in some other way destined to be honored. (The preindustrial hierarchy of English towns and cities differed from that of today.) However, it was only much later, with the imposition of royal governors in New Hampshire (1679) and Massachusetts (1691) and the subsequent erosion of local control that town-naming became truly sycophantic. (Because Connecticut and Rhode Island retained greater autonomy, town-naming was not so marked by royalist whims.) The titular seats of (politically sympathetic) English peers began to crowd out the native heaths of Puritan dissenters. Settlers' petitions to the General Court in Boston seeking the incorporation of new towns were often enticingly submitted in blank, giving scope to the vanity of higher powers with the bestowal of a name. Others saw their requested names stricken and new ones imposed, as happened at Harvard and Petersham MA.[46] The name itself was not the only sop; in Massachusetts and New Hampshire both, the granting governor customarily was reserved a 500-acre tract or "farm" in each new town as his special perquisite.

The rush to sell off the remaining Province Lands at the close of the French and Indian Wars enabled Governor Francis Bernard to leave his toponymic fingermarks all over the Massachusetts landscape. Bernardston was named for himself; Berkshire County (1761) likely recalls his birthplace; Paxton honors his friend the Boston Commissioner of Customs; Shutesbury his wife's uncle, a former governor. Tyringham was the name of the English family whose ancestral seat Bernard inherited at (Nether) Winchendon: both were remembered in town names. Governor Benning Wentworth of New Hampshire showed himself an even more accomplished township-granter and toady. Bennington VT he named for himself (Wentworth NH, also in his honor, was his nephew's doing); Hollis, Pelham, Strafford, Grafton, Monson (defunct), and Claremont NH honored family connections. He curried favor with the influential Augustus Henry Fitzroy, third duke of Grafton, fourth earl of Arlington, fourth viscount of Thetford, and Baron Sudbury, by naming, respectively, a New Hampshire town and three Vermont towns after his four titular seats.

Ultimately, five New Hampshire towns honored prime ministers: Walpole, Grenville (now Newport), Pelham, Chatham, and Pittsfield (dual honors for William Pitt, earl of Chatham),[47] while six more recalled men who were ministers in the Pelham administration (1746–1754). In Massachusetts, another five who served under Bute-Newcastle (1761–1762) were honored. American sympathizers might be doubly honored: Foxborough and Holland MA for Charles James Fox, Lord Holland; and the adjacent Berkshire towns of Lenox and Richmond MA, for Charles Lennox, duke of Lennox and Richmond. While still perhaps to modern eyes "All the King's Men," to anyone sensitive to parliamentary nuances, they had become increasingly more Whig than Tory by the decade of the Revolution. Indeed, some Tory-named towns were unnamed: Hutchinson and Gageborough, honoring the last two royal governors of Massachusetts, became Barre (1776) and Windsor (1778). About 1780, the town clerk of Towns(h)end MA began to omit the silent *h* from an honorific brought into obloquy by the infamous Townshend Acts, which, among other things, had levied the tax on tea. (The town had in fact been named in 1732 not for this Charles Townshend, but his grandfather.) This new orthography became standard after 1800. Was this a simplification, as in Fram(i)ingham and Billerica(y), or was it a calculated unnaming?[48]

The custom of honorific town names continued unabated into the republican era. Massachusetts governors and lieutenant governors (Hancock, Bowdoin; Gerry (obs.), Gill, Phillips) were honored, much as their royal predecessors, up until around 1810. However, over the first two centuries of settlement, the quality of the persons honored slowly changed. By the time of the Revolution, patriots supplanted peers, and soon afterward, local worthies such as doctors and manufacturers were selected for immortality. The linguistic character of the names themselves also evolved.

At first, towns had been largely named directly only for royalty (Charlestown, Edgartown MA). Witness the refusal of the General Court ca. 1653 to name what is now Lancaster MA "Prescott" in honor of a local blacksmith. "Whereas no town of the Colonies had as yet been named for any Governor; and whereas it were unseemly that a blacksmith be honored ahead of his betters, the name Prescott could not be permitted."[49] Honors might be conferred obliquely, by naming the town after an associated English place, as the seat of a peer, the birthplace or home village of a first settler, the English parish of a minister. Before the Revolution, town names were required to sound agreeably townlike to English ears. Some habitative surnames (Bellingham MA 1719) were felt able to stand alone. Otherwise, the name had to be amplified with a suffix, as were Holliston, Belchertown, and Fitchburg MA. However, by 1780, unadorned, and untownlike surnames, often honorific of patriots, broke the ice of English toponymic decorum, and the nineteenth century saw such names as Sherman (1802), Chaplin (1809), and Seymour CT (1850), all very untownlike in sound. (This was the wave of the future for American names in general: note, for example, that only 2% of Montana towns have traditional habitative suffixes.) Some honorifics might achieve a certain dual validity by re-rationalizing the significance of a borrowed name. Woodbury VT assumed the form of a bona fide Connecticut town name, but was apparently meant to honor Colonel Ebenezer Wood. More preposterously, Cambridgeport VT was named ca. 1825–1834 for a local mill-owner, J. T. Cambridge, yet it stands on an unnavigable stream, is no port, and it is scarcely conceivable that it was not meant as a pun on Cambridgeport MA.[50]

The custom of honorifics is not well calculated to create map-patterns of geographical interest. However, in some cases geographic influences clearly came into play in the choice of honorific placenames, as in five western Massachusetts towns prefixed or suffixed in -mont. Charlemont, Richmond (originally spelled Richmont), Montague, Egremont, and Montgomery, all ostensibly honorific of persons, were named in a spurt between 1765 and 1788, and all are set among hills or mountains. Was Bolton MA wholly honorific, knowing that Bolton was carved out of Lancaster, and that their like-named Lancashire counterparts lie thirty-five miles apart?

A mercenary etiquette grew around the bestowal of names in honor of living persons, although at this late date it is not always possible to sort out fact and legend. In the eighteenth century, the customary response for those so honored was to make a gift of a bell to the town, as was the case of Littleton, Russell, Shelburne, and, at least in intention, Acton MA. Colrain MA is said to have been named for Lord Coleraine in Ireland "who was so well pleased with the honor done him that he sent the inhabitants a fine bell; but, through the unfaithfulness of the agent to whom it was intrusted, it never reached them. It is believed to be still in existence, and used in one of the churches in Boston." Another legend relates that the duke of Lunenberg (son of King George II) visited Lunenburg MA, bestowed his name, and later dispatched a bell, which never got beyond Boston because the shipping went unpaid. A similar fate befell a church bell sent by the heir of Lord John Somers to Somers CT: it stopped short at Longmeadow MA for want of funds. Likewise, a bell given by Colonel John Hill never reached his namesake Hillsborough NH "on account of threatened molestations from the Indians"; it found its way instead to Groton MA "where it did long and faithful service." In Medfield MA, tradition has it that the "town once received the gift of a bell from Medfield in Old England. There is no record, however, in any way relating to it." (This is not surprising, as there appear to be no Medfields, however spelled, in Britain.) Clearly, fabulation is at work.[51]

Sometimes the gift took another form, as for example, money to build or glaze the meetinghouse (Royalston and Hubbardston MA); Berkley MA (1735) was favored by its namesake, sometime resident Bishop Berkeley of Cloyne, with the gift of a church organ. Benjamin Franklin departed from custom and acknowledged the honor done him by Franklin MA (1778) by a gift of 116 books, mostly religious, still displayed in the town library. The town is said to have specifically requested a meetinghouse bell, but Franklin riposted that those who preferred sound to sense would be better served by a gift of books.[52] This gesture, while not wholly novel (Harvard and Yale both honored book donors), struck a chord that resounded well into the more secular nineteenth century. Theodore Foster gave his namesake Rhode Island town (1781) a bookcase of thirty volumes and a set of still extant town record books. Woodbridge CT (1784) was named for its minister, who returned thanks with a work of theology. Sterling CT (1794) was named for a Dr. John Sterling, who pledged but failed to give a library. Millis and Hudson MA both honor library benefactors. South Newmarket NH was renamed Newfields under the terms of a gift of books and $10,000.[53] Ayer, Douglas, Holbrook, Wales, and Winchester MA honor monetary or other gifts. Sometimes the gift preceded the honor; sometimes it followed. In any case, it was often a venal quid pro quo. Belmont NH is said to have been deliberately named to butter up millionaire August Belmont, but his untimely demise foiled the scheme.

Biblical-Auspicial

Only a small number of town names captured the religious mindset of the first settlers: Salem (= Shalom), Enon (now Wenham), Rehoboth, Concord, Contentment (now Dedham) MA, and Providence RI. In theocratic New England, they were surprisingly few, given the Puritan custom of piously instilling virtues in their offspring with hortatory and biblical names such as Charity, Hopestill, Prudence, Rachel, and Solomon. The fact that Contentment, the popular choice for the name of Dedham, was overridden by the General Court, and that all the above-mentioned names arose between 1629 and 1645, indicates a phenomenon that was curiously short-lived, as though the settlers' "errand into the wilderness" took a sudden turn after their arrival. One late example, Newark NJ—the New Ark of the Covenant—settled in 1666 from Connecticut perhaps reflects a resurgence of come-outer zeal, but even this wears the camouflage of a veritable English placename.[54] Such names, when officialized, appear to have been a breach of good taste, perhaps from fear of offending London nostrils with their Puritan enthusiasm.

Biblical names implicate not simply the influence of the ministry but also the universality of Bible knowledge which made clever, obscure references meaningful to the laity. Enon (Wenham) came from the text of a sermon preached by the Reverend Hugh Peters at Wenham Lake, 1636: "At Enon, near Salem, because there was much water there," a geographically apt and erudite pun.[55] The Hebrew name Rehoboth ("Enlargement," or "The Lord hath made room for us") was chosen in 1645 by the Reverend Samuel Newman who removed there from Weymouth. Lebanon CT (1697) was named for its stands of cedar. While it gives pause that three biblically named Connecticut townships—Goshen, Canaan, and Sharon—were all settled about 1738, at the height of the Great Awakening, such names, connoting lands of plenty, had become by then secularized clichés that testified to far less religiosity than Enon and Rehoboth of the century before.

Biblical names fared best at a local level, in the unincorporated names of villages and districts. The Bible is a fertile source of placenames, and in the hands of a biblically literate society offered a more nuanced and universally understood nomenclature for the New World than did transplanted English names. Drawn from a single, timeless source, their allusive power is more succinct: how the fishing quarter of Newburyport came to be called Joppa (Jonah's ill-fated port of departure) is more readily explicable than why a Deerfield farm neighborhood should take the name of the London dockside village of Wapping. Chapter-and-verse quoters could name a trout brook Ezbon because "thine eyes [are] like the fishpools in Heshbon" (Song of Solomon 7:4) and a mill village Jerusalem because of the "boys and girls playing in the streets thereof" (Zechariah 8:5). Egypt almost always referred in some way to the story of how "Joseph's ten bretheren went down to buy corn in Egypt" (Genesis 42:3). Sometimes the parallel was exact, as when "Pharaoh" Robbins of Egypt, Wethersfield CT, sold seedcorn to all comers after the disastrous "Year without a Summer" of 1816.[56] Of course, not all biblical borrowings were wielded with such originality or precision.

Same-Suffixed Names

Town names can often be most fruitfully examined in terms of the suffixes which convey the notion of town or place, such as *-ton, -field, -ham, -ster,* or *-bury*. These five endings account for 31% of all Massachusetts town names, and while most do not show any blatant localization, a certain amount of echonymic influence is statistically detectable. In six Massachusetts counties, Norfolk (*-ham*), Bristol (*-ton*), Essex (*-bury[port]*), Franklin (*-field*), Hampshire (*-ton*), and Hampden (*-field*), the most common suffix is the same as that of the shire (or half-shire) town, reflecting the prestige of the county seat or early settlement core. In the case of Worcester County, *-ster* is second only to *-ton* in frequency: 10% of all towns end in *-ster*, as against 3% in the state at large. This *-ster* group is an amalgam of several historically distinct suffixes (*-caster, -minster,* even the *-ster* of occupational surnames) and does not constitute a true linguistic category. However, as a factor in name choice, *-ster* is highly valid, as the sound of a name carried far greater weight than its actual etymology. Connecticut yields similar statistics, where 38% of all towns end in *-ton, -ford, -field,* and *-bury,* and another 14% end in *-on* (Sharon, Lisbon) or some like-sounding combination (Goshen, Darien). None of this is irrelevant. The study of New England placenames is not so strictly the realm of the historical linguist as is that of British placenames. It involves principles of name choice and mechanisms of word coinage quite unlike those of pre-modern times.

One "natural" suffix, *-town,* typically appended to a grantee's or settler's name, was historically more prevalent than today, due to renamings. But how natural is natural? Diffusionism, as opposed to inventionism, suggests that New Englanders did not reinvent the wheel. This group, perhaps rifest in New Hampshire (Salisbury NH was previously known as Baker's Town, Stevenstown, and Gerrishtown), strongly resembles Irish placenames such as Mitchelstown and Thomastown. Anglo-Irish toponymy particularly favored this form because many of the some sixty thousand administrative "townlands" in Ireland are prefixed by the Gaelic Baile (often spelled Bally), meaning farmstead, homestead, or the catchall translation, town. The homestead, Baile Liam, for example, would be rendered Williamstown by the English conquerors. Against a background of such anglicizations (perhaps dating to the thirteenth century), purely English names such as Cookstown and Clementstown, which honored settlers and landowners, emerged naturally. (The suffix -borough was also occasionally used.) Thus John Bull's other colonial adventure provided likely models for New England placenames much, as we shall later see, as it did for townplans. (Jamestown VA was, of course part of this same phenomenon. Indeed, its etymological twin, the Irish village of Jamestown—gaelicized as Baile Shéamais—was royally chartered in 1621.)[57]

The most famous town suffix-group is that of the -Field Towns of the Connecticut Valley. Originally, field typically denoted saltmarshes or fresh meadows, but in time it became simply a customary suffix designating a township or outlying parish, without real topographic meaning. The actual word histories of the -field names are highly varied. Some are straightforward English transplants, such as Litchfield CT or Hatfield MA; others, such as Springfield MA, are also proper English placenames, but of an obvious topographic origin that invited imitation. Most are names of local coinage, and are naturally descriptive, or honorific of persons, as Deerfield and Pittsfield MA. A few are surnames that themselves ended in -field, as Chesterfield and Mansfield MA, the individuals so honored being chosen largely perhaps for the felicitous form of their names. The first examples appeared in Connecticut, with Wethersfield (1637) on the River, followed by Fairfield (1639) on the Sound, probably the first such name of local coinage, although reference to some forgotten English locality may have been intended. In Massachusetts, the usage arose virtually simultaneously on the coast at Marshfield (1642) and inland at Springfield (1641), but never caught fire in the east as it did in the west. Only 4.2% of eastern and central Massachusetts towns end in -field, while 14.9% of western do. Topsfield (1650), known at first as the Village of the New Meadows, Medfield (1651) with its fresh marshes on the Charles River, and Byfield, a parish of Newbury on the Parker River, well illustrate the early toponymic equation of -*field* with meadow.

In western Massachusetts, after 1641 when Springfield was named for the Essex home of leader William Pynchon, the -*field* towns spread slowly up and down the Connecticut and its tributaries, trunk and branch, for almost three hundred years. This spread was disrupted by a near century of frontier unrest (1675–1759); when the pent-up surge of settlement was released after the Fall of Quebec, the suffix -*field* had ceased to carry any topographic significance. Ashfield and Chesterfield were upland towns without meadows. Sandisfield and Murrayfield (now Chester) were merely surnames augmented to give them a more townlike ring. Partridgefield (now Peru) honored one of the proprietors, Oliver Partridge, while chiming etymologically with Deerfield. Chesterfield honored the fourth earl of Chesterfield, statesman and man of letters, but clearly his appeal as a titular saint lay as much in the last syllable of his name as in any qualities of his person. (Earlier we questioned whether honorific -*mont* names were not similarly ambiguous.) The Revolution broke the momentum of this trend, with only a moribund revival afterwards. Longmeadow, set off from Springfield in 1783, was a baroque variation on the theme; Lynnfield (1814), Enfield (1816), and Wakefield, a surname (1868), were the last wholly new examples in Massachusetts, although the breakup of the Brookfields spawned others until 1920 when the last, East Brookfield, was set off as a separate town.

Connecticut was more secure within its borders during the French and Indian Wars and lacked the slow, phased chronology of the Massachusetts -*field* towns. Indeed, what is most remarkable is that while the suffix originated in Connecticut, it had largely fallen from fashion there by the mid-eighteenth century. (Names with the suffix -*ington* afforded stiff competition exactly where -*field* names would be most expected.) Bloomfield (1835), named either for its orchards or for a supposed Hartford family whose existence has never been firmly established, is a last and belated example. That -*field* names are characteristic of the Connecticut hearth is amply validated by the mapped loop that encircles 74% of the pre-1780 examples (see fig. 4). However, the overall current distribution is more scattershot. After 1780, -*field* names became such an established template that they quickly lost their subregional distinctiveness. Rarely are particular artifacts the visible monopoly of one single area; rather, it is the degree (often best expressed statistically) to which they are locally embraced that sets them off as placemarks. The noise in the data—here the post-1779 examples—must be filtered out or statistically explained away to reveal the "true" pattern. But how far can one go in enhancing a pattern without creating a pattern?

A later, nineteenth-century suffix-group also exhibits a statistically significant if visually less than clear-cut cluster. The *-Dale* Loop on the Worcester-Providence axis encircles the Blackstone Valley, the early cradle of the American Industrial Revolution (see fig. 4). Here are to be found 38% of New England towns and localities with *-dale* names, and more impressively, 70% of all with locally coined *-dale* names, not the ubiquitous chestnuts River-, Oak-, and Glendale. We know *-dale* to have both romantic Wordsworthian echoes as well as practical associations with early industrial England (as Coalbrookdale), so as a poetic name for a millstream valley it came well recommended. In 1800, Paul Revere bought a house and trip-hammer shop on the Neponset River in Canton MA for his country seat-cum-copperworks, which he called Canton Dale, the charms of which he later extolled in a copper-engraved poem: "Not distant from the Taunton road / In Canton Dale is my abode," clearly showing how the romantic and the industrial were melded.[58] The use of *-dale* has early, utopian-industrialist connotations as perhaps most perfectly exemplified by Peace Dale RI (outside our core area), as well as strong affinities to paternalistic mill-villages of the Rhode Island type. Often *-dale* was suffixed to the name of the mill-owner or even the article of manufacture (as Lensdale, Southbridge MA, seat of the American Optical works). When shopworn *-dale* names are winnowed out, this concentration of original coinages in the Blackstone Valley, augmented by scattered examples elsewhere (as Risingdale, Stockbridge MA, seat of the Rising Paper mill), produces a map-pattern of genuine historical interest.

Stylistic Zonation

In the *-field* towns, the suffix serves as a vernacular "placemark" that subregionally distinguishes the Connecticut Valley. But might suffixes serve as "timemarks" as well? Specifically, if name-fads in suffixes occurred in succinct sucession, and if settlement advanced inexorably in well-defined stages like a tide, and if new towns were only incorporated and named on the frontier, then the town name-cover would exhibit an obvious stylistic zonation. However, when a composite graph for frequencies of five common New England town name-types was drawn, it quickly laid to rest any notion that these occurred in "succinct succession" (graph not shown).[59] While the popularity peaks are clearly phased, the overlap is so great that no mutually exclusive name-types are found, and all five were in use ca. 1800, at the very acme of town-namings. Furthermore, we know that the idea of a "settlement tide" is itself too simplistic: early towns were continually subdivided within the increasingly populous and industrialized old-settled areas and given new-style names; established towns were randomly renamed; huge territories (as much of Vermont) were granted en bloc and autocratically named in a relatively short period of time, well ahead of any real advance of population.

The most, then, that we might expect to find is a series of intergraded bands of higher and lower frequencies of certain town names. The Burg-Boro line as mapped treads the most reasonable line between a northerly zone of many -burgs and few -boros (11:4) and a southerly of few -burgs and many -boros (3:32) (see fig. 4). While based on official town names (variously spelled), it applies as well to similar, unofficial, and often jocular names of localities such as Chickenboro or Johnsonboro NH. The Burg-Boro line is also a timemark, as it closely approximates the trend of the 1760 limit of settlement, suggesting that stylistic zonation, albeit subtle, does occur.

Compass-Point Names

The most straightforward way of naming a new town (or village within a town) is simply to affix north, south, east, or west to the name of the adjacent town from which it was derived. When the mother town was a large, early grant, the result was often a cluster of like-named towns colloquially lumped together as "the Bridgewaters," "the Brookfields," "the Andovers," a practice widened to encompass sister villages, such as "the Eastons," that in fact constitute only one town. This highlights a drawback of compass-point names: widespread confusion as to whether such places as West Springfield, North Reading MA, or East Providence RI are municipalities in their own right. Many compass-point names were replaced in the nineteenth century to enhance civic identity. East Sudbury became Wayland in 1835 and North Bridgewater became Brockton in 1874. Compass-point names are generally written as two words, although some (discussed further below under Blend-names) are written as one. Northampton gave rise to the only Massachusetts cluster to box the compass. The unmodified town name cannot always be taken to indicate the nucleus

of settlement; West Bridgewater, not Bridgewater, for example, now includes the original seat of settlement. Despite the adage that "it is an ill
wind that blows no one good," all winds, and consequently all compass
points, are not equal. In regions subject to the Prevailing Westerlies, fashionable suburbs, be they in London, Paris, or Boston, were located to the
west, upwind of the smoky city. Even in the names of fictitious "no-
wheresvilles" such as East Overshoe and East Oblivion this pattern is
negatively corroborated. North was also disproportionately favored.

Deep directional biases, often originally based on geographic advantages, but reinforced by the cachet carried by well-known fashionable
names, have skewed the expected even distribution of north (31%), east
(15%), south (17%), and west (37%) town names in Massachusetts. Furthermore, recorded name changes corroborate this particular prejudice
against south-named towns. As a control, the compass-named villages and
localities listed by the 1915 state census were also tallied, and showed a
much more even distribution: north (26%), east (24%), south (28%), and
west (22%). Beyond perhaps a name on a postmark, these names are unofficial, more spontaneously descriptive, and less subject to the boosterist
soul-searching of legislated town names. In consequence, directional biases
do not manifest themselves.

This remarkable pattern in favor of town names qualified by north and
west is explicable in two ways. The snobbery thesis, just outlined, supposes
that in the vicissitudes of town division, all cardinal points are equally
likely, but that in the course of three centuries of settlement, an innate
prejudice against living in a town demeaned by the words *east* or *south*
asserted itself and skewed the percentages. An alternative, and perhaps
ancillary explanation, is that in the general westward and northward march
of settlement, the mother town that served as the point of reference for
subsequent namings most often lay to the east or south, yielding west and
north daughter settlements. If we examine the data to see how many town
names, in total, are now, or once were, modified by a compass-point, a
rough equivalency can be descried in north (15), south (12), and west (16),
while east (6) is dramatically underrepresented. Exactly this would be expected, if the frequencies were influenced by the overall east-to-west
course of inland migration. The pattern of Connecticut towns tends to
corroborate this by counterexample: north (28%), east (36%), south (16%),
west (20%). East here is surprisingly in the ascendancy, given impulse by
the fact that most original town settlement was on the west bank of the
Connecticut, which gave rise to eastern daughters.[60]

Blend-names

Blend-words combine characteristic elements of two or more words to form new ones, as, for example, motel, from motor hotel. Informal blends of town names are rife today, particularly in the titles of businesses such as Arlex Oil (Arlington-Lexington), or Box-Top Realty (Boxford-Topsfield). While such names may seem to be a modern barbarism, the practice in New England town-naming is at least as old as Saybrook CT (1635), which blends the titles of its two patentees, Lord Say and Sele and Lord Brooke. Connecticut coined such names in abundance because it routinely established intertown parishes (ecclesiastical societies) such as Hadlyme (Hadley-Lyme, 1742), Stanwich (Stamford-Greenwich), and Torringford (Torrington–New Hartford, 1763) which were not an invariable prelude to separate township status. Winsted, from Winchester-Barkhamsted, was chartered as an ecclesiatical society in 1788 and has become a "city" within the town of Winchester.[61] Connecticut town names were boldly broken up and cast in the toponymic crucible, and even tripartite blends were not unknown. The parish of Wintonbury (now Bloomfield), was established by settlers from Windsor, Farmington, and Simsbury in 1736. Harwinton was settled from Hartford, Windsor, and Farmington. For these and other reasons nearly one-third of all Connecticut towns have names that bear some resemblance to an adjacent town.

Massachusetts was equally fertile in blend-names; however, partly because parishes rarely straddled town-lines, these names were far less outlandish than their Connecticut counterparts. Most common were simple compass-point blend-names. Sturbridge was divided and South Sturbridge became Southbridge; Uxbridge yielded Northbridge; Chelmsford gives rise to Westford; Leominster to Westminster; Marlborough fragments into a cluster of boroughs, North-, South-, and West-. Many of these blend-names make plausible claim to English originals. Westminster is "officially" believed to have been named for the London borough and seat of Parliament, although the town historian notes with perplexity that none of the first settlers is known to have had any connections to that place. Obviously, the dual validity of the name, locally apt and yet seemly to colonial officials in Boston, would make it a strong choice. In New Hampshire, two town blend-names commemorate first settlers. Barnstead NH was settled from Barnstable MA and Hampstead NY. The more outlandish Gilsum NH, named for its two chief grantees, Samuel Gilbert and Benjamin Sumner, was founded by Connecticuters undoubtedly well familiar with the practice of blend-names.[62]

Echonyms

By now it is clear that locally chosen town names were felt to be more seemly when they harmonized in some way with those of surrounding towns. Blend-names did this overtly, echonyms more subtly and more debatably. Two such examples are Ashburnham (1765) and Ashby MA (1767); Plymouth (1620) and Plympton MA (1707), where the echoic element is the currently meaningless initial syllable of each name. Both pairs comprise bona fide English placenames, but the choice of name for the later, nearby town seems to have been influenced by a desire for euphony. Granby CT (1786) borders Granville MA (1754), so echonyms might even straddle state-lines. In its simplest and commonest form, echonymy is achieved by alliteration. In Massachusetts, the five commonest initial letters (W, S, M, B, H) account for 54.7% of town names. If we assume that every town borders six other towns, then the numbers of alliterative contiguous pairs of towns (as Mattapoisett-Marion) beginning with these five top letters is generally about three to ten times greater than random probability; only S, with but one pair (Sturbridge-Southbridge), conforms to the theoretical norm. In all, there are 21 alliterative doublets, 6 triplets, and 1 quintuplet in Massachusetts. This fivesome, Watertown-Waltham-Weston-Wellesley-Wayland, is at the crux of the often confused "W-suburbs" of Boston: the prominence of the latter three as "country club" towns has enhanced the mystique of the W-suburbs as seats of privilege.

Terminally echoic elements also occur. Often these are true, habitative suffixes, such as the five contiguous Connecticut -burys that sprang from Waterbury. Other times, as with the -sters of Worcester County MA, the resemblance is one of sound, not sense. One town-name ending that has proven prolific in echonymic clusters in both Massachusetts and Connecticut is -ington. While no longer viable in word formation, its original Anglo-Saxon etymology is well illustrated by Lex-ing-ton: the town of the followers of Lex. In Greater Boston, Lexington (1713), Wilmington (1730), Burlington (1799), and Arlington (1867) form a continuous belt; while all these names have perfect English pedigrees, the overall distribution of -ington names suggests that this name-string could not have arisen other than from mutual influence. Arlington, renamed from West

Cambridge, rounded off this group, and was said to have been championed by an influential citizen, the Honorable Joseph S. Potter, who stemmed from Arlington VA. There is no reason to doubt this story, but the explanation for the name choice runs deeper. Arlington in 1867 was a sonorous and fashionable placename with ambiguous appeal to both North and South. Arlington Street in the Back Bay of Boston was laid out in 1859; the Arlington Mills in Lawrence were established in 1865. Arlington, the confiscated estate of Robert E. Lee outside of Washington, became a Union cemetery in 1864. Garden suburbs and garden cemeteries have an entwined, romantic aesthetic, both in layout and in names, and the use of a cemetery name for a suburb (and vice versa) was not unusual. We have seen how in the nineteenth century compass-named towns like West Cambridge sought to get out of the shadow of their mother towns by rechristening themselves. Arlington was a nationally approved name, and also an *-ington* name with local echonymic appeal. This suffix did not constitute a self-sufficient rationale for the adoption of the name, but would make the final choice seem less arbitrary.

In western Massachusetts, Worthington (1768), Cummington (1799), and Huntington (1855) are contiguous, as was nearby Washington (1777) before Middlefield intervened. Great Barrington (1761) is only separated from this group by Tyringham (1762), which notably contains the *-ing* infix. Nor is the town of Mount Washington far distant. By even a conservative interpretation, 66% of all *-ington* town names in Massachusetts occur in one of these two clusters, a percentage far beyond random probability, which suggests that echonymy was at work. In Connecticut, the main *-ington* cluster around Farmington by 1806 had grown into a formidable gerrymander of six contiguous towns and three parishes: some purely English, as Burlington; some English with dual local validity, as Southington; some odd blend-names, as Harwin(g)ton, as it was formerly spelled.

Name-Changes

So far we have approached the New England town name-cover as though it were a thousand-box crossword in which we read up, down, and all around in search of significant patterns. But this is only its synchronic surface; it also possesses a third, diachronic dimension of depth or time. Many towns changed their names in the course of their history; Weare NH did so an extraordinary five times. These changes often marked char-

acteristic stages in settlement history: a plantation, casually named Lambs-
town (now Hardwick MA) after a principal grantee, might upon incorporation be renamed after someone nearer the governor's heart, in this case,
the earl of Hardwick. Failed plantations might be regranted and renamed;
ill-faring townships that felt jinxed by a name might change it. Bromley
VT (1761) became Peru (1804); similarly, Random VT (1791) became Brighton (1832).[63] Daughter-towns outgrew their apron strings, as when South
Reading MA became Wakefield (1868) and South Scituate MA became Nor-
well (1888).

Name-changes can exhibit the same onomastic principles as collocations and echonyms; but rather than being contiguous in space they are
consecutive in time. Saltash VT (1761) changed its name to Plymouth (1797);
the originals of these names all but adjoin each other in the west of England.[64] Medway VT (1781) after some uncertainty finally fixed on the name
Mendon (1827). Medway and Mendon lie eight miles apart in Massachusetts and were both hometowns of the early settlers. Hamilton MA (1793)
was originally a part of Ipswich called the Hamlet; Chester NH (1722) lay
in the Chestnut Country. Echonymy surely had a hand in both these
changes. Sometimes the rationale for the old name was simply reworked
in the new, as when a Tory honorific yielded to a Whig (Hutchinson to
Barre MA) or a Federalist to a Democrat (Adams to Jackson NH).[65]

Before turning to street names, we shall examine five broad patterns or
"namefields" that well illustrate the dynamic complexities that lie beneath
the static calm of the New England name-cover. Both folk and town
names are considered; while some have certainly caught the eye of travelers, most of these patterns are as invisible as they are vast, and to my
knowledge have never been systematically discussed before.

The Great-Big Line

"Great" was a ubiquitous modifier in colonial New England toponymy:
Great Brooks, Great Ponds, Great Roads everywhere abound. Paired with
"Little," doublets such as Great Brewster and Little Brewster (islands in
Boston Harbor) conveniently killed two toponymic birds with one stone,
facilitating place-naming in a nameless land. On an outline map were
plotted the approximate locations (as compiled by Stanley Attwood) of
Maine geographic names modified by the adjectives Great or Big, as in
Great Spruce Head, or Big Heath. Virtually all were mapped—78 Greats
and 99 Bigs—except those judged to be tediously reduplicative, as Great

Spruce Ledges which adjoin Great Spruce Island. (Please note that a certain necessary editorial judgment and approximation has entered into the compilation of all the arcane statistics I shall now cite.) It was found that 95% of all Greats lay south of a wavering east-west line drawn through the latitude of Bangor, while 82% of all Bigs lay north of it. This "Great-Big Line" tightly entwines the 1800 settlement line in Maine[66] (see fig. 6).

What could account for this odd toponymic dichotomy? The simplest explanation is that slowly between 1620 and 1800 an across-the-board linguistic shift occurred in New England, in which colloquial use of the word Big supplanted Great in the sense of "large." Because settlement in Maine crept slowly inland from south to north, and downcoast from west to east, the transition from Great to Big was etched prominently in the toponymic cover. The virtual absence of Greats north of the Great-Big Line and the respectable minority (18%) of Bigs south of it are quite as expected. Bigs south of the line may represent early harbingers of the general shift, or Greats colloquially mutated to Bigs in afteryears, or relatively recent namings. Looked upon archeologically, while we would not expect to find eighteenth-century artifacts beyond the eighteenth-century limit of settlement, we would certainly expect to find post-eighteenth-century artifacts within it: life, after all, goes on. Not impossibly, a dialectal factor was also involved, based on the cultural groups that peopled these areas.

Whoever named northern Maine—conceivably they were surveyors, loggers, trappers, and guides, and not true settlers—may have spoken a different dialect from the old-stock New Englanders who settled the south. They may have spoken Northern English (the Scotch-Irish?), or a lower-class English, or the broken English of French-Canadian loggers or Abenaki guides. The reasonable correspondence between the Great-Big Line and the shift from irregularly to square-surveyed townships—a virtual fault line in the Maine town-line grid—might specifically implicate surveyors, if the two are not simply coincidental, nineteenth-century phenomena. Geographically, this use of Big roughly correlates with the almost exclusively Maine use of Stream in the proper names of rivers as mapped by Wilbur Zelinsky, a usage also suspected to be of nineteenth-century origin.[67] Whoever named rivers Streams (and no regional English dialect appears to do so) was likely also responsible for the proliferation of Bigs. Once Big-Little as opposed to Great-Little became an established place-naming template, Bigs would beget Bigs and quickly reproduce themselves over the land.

The Great-Big Shift was not a homegrown phenomenon, but one that

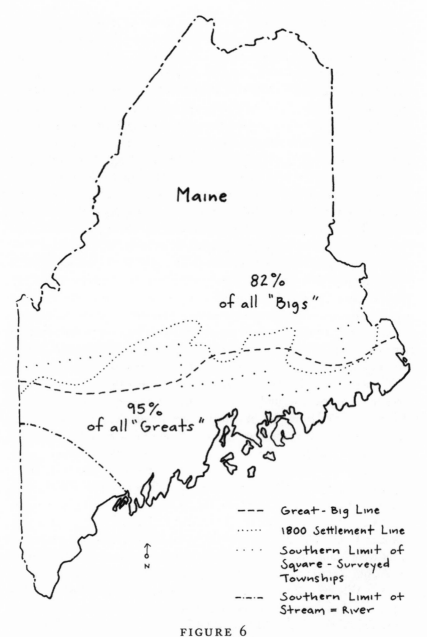

Maine

82%
of all "Bigs"

95%
of all "Greats"

---   Great - Big Line
......   1800 Settlement Line
· · · ·   Southern Limit of
       Square - Surveyed
       Townships
-·-·-   Southern Limit of
       Stream = River

FIGURE 6

The Great–Big Line

The distribution of placenames modified by "Great" or "Big" reflects a
sharp divide in Maine settlement geography.

had been working itself out for centuries in British English and that left its mark on the toponymy of the entire English-speaking world. Great is the Old English word; Big is a possibly Norse newcomer that spread slowly southward from Lincolnshire and Northumbria after about A.D. 1300. The meaning of Big slowly evolved and infiltrated the semantic domain of Great, apparently supplanting it in the basic sense of "large" in northerly regions of England and among lower classes of speakers. Unfortunately, this sort of semantic infiltration is a complex spatiotemporal question only sketchily traceable in the *Oxford English Dictionary* and the *English Dialect Dictionary*. However, the complete absence of the word Big in any of its senses from the Authorized Bible of 1611 (contrasted with some 1050 Greats) provides a telling benchmark in the history of its slow assimilation into the "King's English."[68]

In terms of the overall British toponymic record, the *Ordnance Survey Gazetteer of Great Britain* gives a 13:1 ratio of Greats to Bigs. While Big is in the clear minority, its distribution is nonetheless instructive. Fully 83% of all Bigs occur north of a slant line drawn from the Severn to the Wash; 75% occur north of the Humber if one ignores the one repetitious term, Big Wood. However one cares to draw the Great-Big Line in Britain, the early settlers of New England came principally from the south, or Great side of it. We do know from the *Dictionary of American English* that in 1815 John A. Pickering (himself a New Englander) observed of Big: "This adjective is used by the people of the Southern States where a New Englander would use great or large." The Great-Big ratios for the New England states are also indicative, where the higher the ratio the earlier the settlement, and probably also the "purer" the South of England stock of the settlers:

Massachusetts	37:1 (*that is, 37 Greats to 1 Big*)
Connecticut	11:1
New Hampshire	4:1
Rhode Island	4:1
Maine	2:3
Vermont	1:2

Presumably, the dominance of Big in the American South (and by later diffusion, the West) was due to the North Country and possibly lower-class dialect of the settlers, and their more recent, post-seventeenth-century emigration. While not so dramatic as that of Maine, the Great-Big pattern in North Carolina is not dissimiliar, although marked by its own historical peculiarities of settlement and peopling.[69] Great is mainly a mid-to-northern Tidewater term, crowded densely around Albemarle and Pamlico Sounds and their inland headwaters in the northeast third of the state. Roughly 80% of all Greats in North Carolina lie east of Raleigh. The westernmost mountain counties are the high citadel of Bigdom, where 63% of the Bigs occur in an area comprising only 15% of the state. Thus, the Greats lie in an area settled in the late seventeenth century from coastal Virginia, and Bigs in an area settled in the late eighteenth century by an inland flanking movement of Pennsylvanians, many of Scotch-Irish descent. In the United States as a whole, 68% of all Greats are located within the thirteen original colonies, despite their relatively small land area. Many of the Western Greats occur in physiographic terms such as Great Plains, in the names of mines, as Great Wanamingo Mine, and in stock phrases such as Great Western, which derive from flamboyant nineteenth-century commercialese.

Examination of placenames elsewhere in the English-settled world throws further light on both the chronology and causes of these Great-Big patterns. In Nova Scotia, Greats tend to lie west (or very slightly east) of a north-south axis bisecting the province through Truro and Halifax, a line customarily taken to demarcate westerly Yankee from easterly Scotch-Irish settlements.[70] Indeed, it would almost seem that a geographer could age the settlement of British colonies by their Great-Big ratios. On the two extremes of the spectrum we find the 13:1 ratio for Britain itself, as dramatically opposed to the 1:8 and 1:11 ratios of Australia and Tasmania, which were not settled until after 1789, when the Great-Big Shift was well consummated. Thus even ordinary words, such as Great and Big, without specific topographical content can, through accident of linguistic history, prove excellent spatiotemporal differentiators, or toponymic tracers, of settlement patterns.

Reversed Center Arc

This namefield (see fig. 7) runs in a 135-mile inland band from Strafford NH to Belmont ME, lying roughly 20 to 35 miles back from the coast. It is marked by a curious adjectival use of the word *center*, as in Center Sandwich NH, to designate the central village of a township. The Arc swings conspicuously wide of Portland ME, so the usage likely did not emanate from there, and it probably postdates the settlement of the coast generally. Nor does the Arc trace any colonial inland artery by which it might have spread. Of the thirty-three known current or former U.S. examples, fifteen lie within this Arc, another sixteen elsewhere in New England or New York, with a final two (Center Moreland PA and Center Belfre OH) in New England–settled areas further west. A few related forms, such as Central Nyack NY and Centerlisle NY appear on the peripheries, and presumably represent later corruptions or rationalizations of placenames increasingly felt to be at variance with standard English. For this construction is clearly backwards—or is it?

Consult a British gazetteer in vain for precedent, as placenames employing the word *center* are not historically British. (Central Wingland, Sussex, probably a modern civil parish name, is the closest known example.) Yet center stage, center field, even the center ice of the hockey rink are all accepted as idiomatic English, so why balk at Center Sandwich? Furthermore, metathetic compounds, as the Yankee sidehill for hillside, and the Southern peckerwood for woodpecker, are staple American dialect features, so isolated speech communities clearly evolved contrary standards of order in word formation. Center Sandwich is then not necessarily unidiomatic English; indeed, it may have arisen when the toponymic usage of center was entirely novel, and predate the establishment of any idiom at all. It may have even been consciously stylish. In any event, there is safety in numbers, and the fifteen-town arc of reversed centers in New Hampshire and Maine afforded this variant usage a bulwark of mutual support.

Conversely, isolated examples on the peripheries might undergo rationalization or disappear (Center Marshfield and Center Grafton MA). Center Conway NH is known to the railroad as Conway Center, which serves to remind that the indulgence of the post office has undoubtedly been critical to the survival of the Reversed Center Arc. The official *List of Post Offices in the United States* in 1859 recognized twenty-five of our total and most of the Arc proper. However, recorded usage in the nineteenth century is often nonchalant and self-contradictory,[71] and even in the 1920s, in its own annual excursion programs, the Sandwich Historical Society blithely flip-flopped between Center (or Centre) Sandwich and Sandwich Center, so defense of their toponymic heritage seems to have been of little moment, even among local historians. These center villages are often precisely that: a village or a hamlet, at the geographic center of the township, not necessarily the primary populated place. Indeed, the reversed centers generally have obscurity in common (as anyone attempting an authoritative tally quickly learns) and their survival may have in

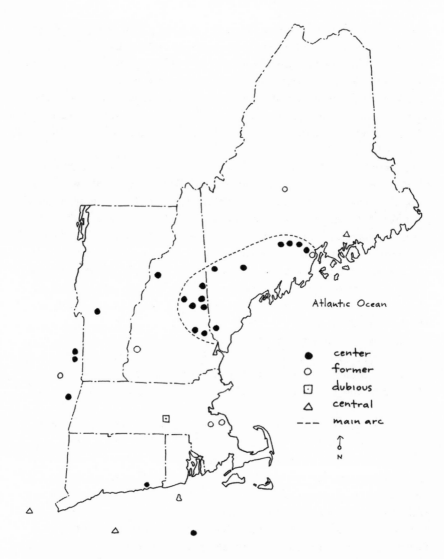

Atlantic Ocean

- ● center
- ○ former
- ⊡ dubious
- △ central
- --- main arc

↑
o
N

FIGURE 7

Reversed Center Arc

Locations of villages and towns with "backwards" names like "Center
Sandwich," New Hampshire (see appendix for full list of towns).

some way depended on the semantic evasion that Center Lovell did not quite mean Lovell Center.

While the dating of these village names is critical to their interpretation, this is no easy matter; villages, unlike towns, are not legally incorporated. An early, if dubious, precedent may be Centerbrook, a village of Essex CT, a presumed contraction of Center Saybrook that came "gradually into use after 1723."[72] Township incorporations within the Arc point to the period of 1770–1800 as the most probable time of origin, but post offices were few and early postal lists (1804 and 1808) hold no confirmation. Centre Sandwich was not established as a postal village until 1828, although the name itself may have come into use shortly after 1765. Centre Rutland VT was earlier Meads or Gookins Falls before it received its own post office in 1850; Drakes Corner became Centre Effingham NH in 1866.[73] Intriguing collateral evidence for the heyday of this usage is furnished by Center Harbor NH, named in 1798 both from its location between Moultonborough and Meredith Harbors on Lake Winnipesaukee, and in punning allusion to the locally prominent Senter family. While not a true reversed center placename, the calculated choice of a name in this particular form in the midst of the Arc suggests that reversed centers were by then so well established as to invite parody.

The dearth of U.S. examples outside of New England and New York may reflect the fact that the very concept of a center village of a namesake township is peculiar to the history and governance of New England and has little application elsewhere. Wilbur Zelinsky, the dean of American toponymists, who mapped Center placenames in their more usual, unreversed form, regarded the usage as a New Englandism, one likely older than the earliest citation (1791) he could find. He makes no mention of reversed centers, but his overall distribution, while much ampler, resembles ours, and likewise fades westward from New York into Ohio.[74] The usage of reversed centers in particular would not have been widely transplanted to the West if its burst of popularity either predated or postdated the general New England exodus (1790–1840).

Evidence in support of an earlier flowering comes from eastern Canada, where the only other significant concentration of reversed center (or variant Central) villages, is found, mostly in Nova Scotia and New Brunswick. Two of the New Brunswick examples are actually in French (Centre-Acadie and Centre-Saint-Simon), but these are typical of the amusing "franglais" to be noted in many Canadian village names.[75] As no plausible British prototypes exist, this disjunct American-Canadian distribution is not simply a case of two localized imports. If we discount independent invention (which would imply an idiomatic usage), then the history of emigration makes it far more likely that the Canadian usage derived from that of New England than the other way around. The highest Canadian concentrations are found precisely in the Maritime Provinces, where the cultural impact of Loyalist emigration after the American Revolution was most profound. Nova Scotia, moreover, knew New England emigration as early as the 1760s, and the Canadian reversed centers are a toponymic tracer of this overall influx.

The sum of the chronological evidence focuses on 1760 to 1800 as the era in which reversed center placenames first flowered, probably in a vacuum of established idiom. And indeed what the overall U.S. distribution of reversed centers most resembles in shape is a late eighteenth-century frontier-line, a tidemark of settlement traced across northern New England and western New York. What we have identified as the Arc proper is the most conspicuous relic of this line, where the custom took deepest hold on the Maine–New Hampshire frontier, where it became officialized on maps and in postal directories and was held up for local imitation. Elsewhere where it lacked the contextual support of like-named villages, it was "corrected" out of existence. Interestingly enough, we shall see later that the Arc also outlines the same cultural subregion identified by Thomas Hubka as the source of the Connected Farmstead.

Of Balls and Cobbles

Two terms for hill or mountain, *ball* and *cobble*, figure prominently in the toponymy of western New England and eastern New York. Both pose difficulties of origin and identification. Cobble, formerly derived from the German and assigned a Hudson Valley origin[76] is now by many authorities regarded as probably English and akin to the *cobble-* of cobblestone. Cob(b) is a widespread English dialect word for the summit of a hill. Ball, of which there are far fewer examples, has gone collectively unremarked, and is easily confounded with the surname Ball and the common adjectival use of Bald in the names of summits. Ball has also been cryptically compounded in names like Saddle Ball and Buckball, which makes recognition in its basic sense of rounded mass difficult.

The distribution of twenty or so indisputable examples of Cobble in the dialect sense of rounded hill is largely confined to a zone between the Hudson and Connecticut Rivers. By indisputable I mean examples in which Cobble stands alone as a noun, without confusing pleonasms, as in Cobble Hill or Cobble Mountain, where an adjectival sense of "cobble-stone" may have been meant. At least seven of these employ the definite article, as The Cobble or The Fox Cobble, and while this occurs in obscure folk names for mountains, as The Ballyhack or The Barrack (both CT), it also occurs specifically in eastern New York with names of patent German origin, as The Gipfel quite literally, the Peak. (This last lies on Kipple Road, Sand Lake NY, which illustrates how such Germanisms might become locally corrupted by English-speakers.) Most are singular in number, but some are plural, as Shaftsbury Cobbles, or the Cheshire Cobbles; a few may be both, as Bartholomew's Cobble(s). This uncertainty may be due to folk-etymologizing meant to second-guess the origins of an obscure term. The name Bartholomew's Cobbles has been attributed to the "out-cropping of queerly shaped boulders" found there,[77] while the sugary lumps of quartzite everywhere underfoot on the Cheshire Cobbles surely must strike climbers as justification enough for the name. Alongside these twenty or so sure examples are to be found another fourteen disputable, possibly pleonastic ones, such as Cobble Hill; but even here, one, The Cobble Mountain (Wilmington NY), retains a definite article, which suggests an original in the form of The Cobble. Most importantly, the few Cobble toponyms in eastern New England all occur in the pleonastic form of Cobble Hill, with no evidence that Cobble ever held anything other than its common meaning.

While discounted by lexicographers, this concentration of Cobbles between the Hudson and the Connecticut Rivers makes it difficult to dismiss the idea that it was an anglicized German (or Dutch) toponym carried eastward into New England, where acceptance was facilitated by its resemblance to the English word for rounded stone. "Kofel" in Bavaria is applied to Alpine summits.[78] One might also suggest the German Kuppel, or cognate Dutch Koepel, meaning "dome" as possible origins. (The sometimes instanced German Koble for "rock" is not found in standard dictionaries.) Certainly, the compound name Cobbleskill NY, unless it is a hybrid, suggests that the first element, like the second (kill = stream), may be Dutch. Cobble, regardless of its etymology, manifests a distribution similar to that of the Hudson Valley words *stoop*, *bellygut(ter)*, and *pot cheese* used by Kurath as index features of this dialect subregion.[79] Thus it seems not unreasonable to regard Cobble as a toponymic tracer highlighting the area of New York infiltration into western New England.

Ball, a comparable, albeit minor term, was applied to round summits.

It is not a striking metaphor, and it did not replicate itself over the land-scape like Haystack and Sugarloaf did; in fact, Ball as a word-image is entirely colorless and lacks the force and clarity that the synonymous use of Knob has. It has no history in English as a descriptive of summits, yet so simple a metaphor might be coined by any speaker, anywhere, at any time. (In French, *Ballon*, and in German, *Bölchen* are so used.)[80] The five principal examples lie along the Taconic-Green Mountain axis, from Round Ball Mountain (NY) northward to Tom Ball, Saddle Ball (MA), and The Ball, then eastward to Buckball Peak (VT). Examined together as a namefield, they illuminate each other's obscurities. Round Ball Mountain prosaically explains the metaphor and The Ball vindicates it as a term strong enough to stand alone. Tom Ball and Buckball remain puzzlements until correlated with more familiar names such as Mount Tom and Buck Hill, and likely refer to the plenitude of game (turkey and deer) to be found there. Buckball Peak and South Buckball Peak presumably arose from later misunderstanding or a desire to eschew a vulgar construal of the Buck Balls. The weight of these examples makes it easy to render Saddle Ball Mountain as the Ball of Saddleback (Mount Greylock), and not some obscure word for pommel or saddlehorn. While Ball Hills en-countered elsewhere suggest that the metaphor may have had some cur-rency in folk speech, the five Taconic Balls stand out like beacons along the north-south Housatonic-Hoosick valleys, and may toponymically trace settlement migration into early Vermont.

The China Syndrome

A cluster of some thirty Maine townships named between 1787 and 1849 bear the names of exotic cities and countries, such as China, Poland, Den-mark, Palermo, Belgrade, Rome. The exact count depends on how "exotic" is defined; it seems reasonable to concentrate on truly novel names, and to exclude those that partake of other or earlier name-fads: places in En-gland proper; biblical places; German names clearly meant to flatter the House of Hanover (specifically, Brunswick in 1737). These latter all of course broke the ground for the general flowering of exotic names in the early republic. The United States was a new nation that had lately assumed its station among the powers of the earth; perhaps in token of this, the names of its towns began to scintillate with the brilliance of the firmament in which it was the newest star. Wilbur Zelinsky believed that "these exotica document the extroverted buoyancy and expansiveness of spirit that many observers identify as American."[81] Some may be referrable to the revolutionary tide of self-determination that America heralded and inspired. Peru was liberated in 1821, the year Peru ME was incorporated: yet there were already Perus in VT (1804) and MA (1806), so the plausible incidental occasion of the name is almost irrelevant. Holland MA (1783), Wales MA (1828), and Milan NH (1824) all honored people, but how could it be denied that by dual validity these names do not clearly anticipate or participate in the exotic placename fad?

When plotted, the dates of 47 exotics identified throughout New England clearly reveal the "battleship-shaped" curve characteristic of style trends in all manner of artifacts from pottery to gravestones. The three earliest (and remember, I exclude names best explained otherwise) were all the work of Benning Wentworth: Alexandria NH (1753), Candia NH (1763), and Corinth VT (1764), and struck an innovative spark amidst the mass of his sycophantic namings. Alexandria was in fact named for Alexandria VA, seat of an important colonial governors' conference held in 1753, but was a clear exotic prototype. Candia is an obscure Cretan city; Corinth was a masterstroke much ahead of its time. Almost two-thirds of New England exotics, however, are found in Maine, principally because elsewhere most townships had already been granted and named by 1800, when the trend reached its height. Only on the Maine frontier could it flourish unhindered.

The weight of the evidence when plotted strongly argues for Limerick (1787) as pivotal to the namefield known here as the China Syndrome (see fig. 8). If we identify and date the Maine exotics and then connect them judiciously with isochronic lines, this "China" cluster shows a definite outward growth pattern. Intervals were chosen to accentuate this trend, but it is virtually impossible to plot the data to some other conclusion. While curves of this kind cannot properly be interpolated or "smoothed," their crinkle-crankles plainly show a chronological concentrism, though not of the simplistic bull's-eye sort. Rather, they better approximate three tangent, nested ovoids whose long axes run northeast with the flow of settlement, and whose broad ends lie toward the frontier. This pattern is simply the theoretical bull's-eye as it exists in real life, one distorted by the tides of settlement, and by a southerly and easterly barrier of already named towns that made equal outward growth in all directions impossible. If we accept the numbered Maine highway network as a legacy of nineteenth-century travel-ways, then the interior northeast axis of the ovoids (rather than some inland trend from the coast) is as should be expected. No single

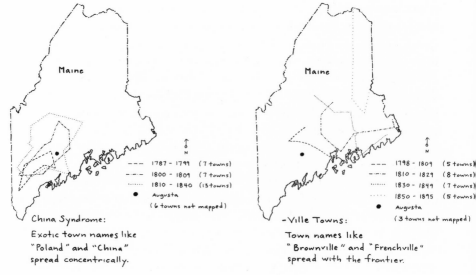

FIGURES 8 and 9
Maine Town Names:
Diffusing Differently

road could serve as a cultural corridor to so vast an area, but the route now designated as Maine 11 does link the extremes of the cluster, at Limerick and Corinth. The China Syndrome paradoxically embodies a national name-fad that nonetheless spread by concentric diffusion; that is to say, the names are popular in type but vernacular in their mode of dispersal.

An instructive contrast is offered by Maine township names in *-ville*, a popular name-type with a more popular pattern of diffusion. The raw isochrones are far from neat, but can be idealized as in figure 9. The lines limn out "tidemarks" of settlement, perhaps not so much frontier-lines as population density contours that more closely reflect when a township or plantation was ripe to be formally organized. Unlike the China Syndrome, there is no closure to these isochronic lines, nor any true nucleus of outward diffusion. As with most popular style trends, the fad for *-ville* names did not emanate from anywhere locally; improved communications made of it a national phenomenon that was, so to speak, "in the air." In Maine, the trend made its debut to the north and west of the exotic names, started later (1798), and held steady and even in popularity without overt peaks. Incorporated *-ville* names are of course only a fraction of *-ville* placenames; nonetheless, could all be dated and mapped, they would undoubt-edly shed further light on the spatiotemporal progress of the style trend.

Street Names

What the street name-cover lacks in historical depth, it makes good in density and accessibility. Full, map-indexed street lists afford a rich testing ground for analytical techniques that would falter for lack of data when applied to thinner layers of the toponymy. The study of street names is also not without intrinsic interest. A proto-suburb such as Arlington MA did not name its streets until 1846 and a rural town such as Plympton MA not until about 1900. The chief motive given for the selectmen of Medfield MA to name the town streets in 1855 was cadastral: "the convenience of bounding lands in conveyances."[82] Even in cities where names were im-posed much earlier (Boston first officialized them in 1708),[83] the bulk of street names will be found to be the product of mid-nineteenth- to twentieth-century residential subdivision. Most of these are not site-specific in the way often characteristic of pre-urban and some industrial street names such as Poor Farm Road (Harvard), Lime Kiln Road (Shef-field), Farmers Row (Groton), or even Incinerator Road (Dedham), Waste Disposal Road (Westford) and T.V. Place (Newton MA). Yet while they may lack immediate geographic significance, their potential for spatiotem-poral patterning as artifacts must not be underestimated.

By 1800 urban street names had become articles of fashion, and as such, were compounded of one part innovation and nine parts conformity. If a street was not named for the place to which it led, or a person connected with that place, then it came geographically unstuck and spun with the winds of fashion. The fashionable spread of street names was facilitated at first by rote imitation of regional style centers (paragons) and later, as the body of names built up, by self-perpetuating templates (paradigms). Cities such as Boston, New York, and Philadelphia saw distinctive ensem-bles of their street names emulated by lesser, often upstart towns within their regions. Newburyport has twenty notable historic Boston street names (among them, Congress, Olive, Franklin, Temple, Milk, High, Court, Summer, Winter, Orange, Charter), while Lynn has at least four-teen. Salem has only six, reflecting perhaps its age and a surety in its own identity lacked by the later-risen Newburyport and Lynn. Certain of these borrowed names, as Tremont and Beacon, are virtual Boston trademarks; indeed a few, as Batterymarch, proved too idiosyncratic for export. A notable absence of more recent names like Commonwealth Avenue (1862) indicates an early nineteenth-century flourishing for the Boston paragon. Many of these borrowed names are so obvious and bland that only their regional pervasiveness and repetition en bloc argues for their Boston origin. Athol MA, eighty miles inland, has both a Park and a Tremont in the vicinity of the Uptown Common, an obvious double borrowing made more understandable as the town was long a stop on the stage road to Boston. Emulation also occurred more locally.

The most influential nineteenth-century paragon did not in fact emanate from any one style center, but rather was composed of an amalgam of Boston, New York, Philadelphia, and common descriptive street names that were banal enough to fit anywhere. They were not recognizably borrowed, politically charged, or laden with baggage as other big-city names such as State, Federal, or Wall Street would have been. This standard paragon recurs throughout New England in the names of the chief (often radial) nineteenth-century thoroughfares of towns and includes Main, Pleasant, Washington, Center (Central), Prospect, Pearl, the various tree names, such as Chestnut or Elm, the four cardinal directions, and simple descriptives as School, Short, or Cross. Without doubt Main and Pleasant top the list; exactly what comes next and in what order only an inventory of street names could determine. George Stewart rattles off State, Federal, Congress, and secondarily Summer, Winter, Spring, Pleasant, and Commercial as the New England standard, but one detects an urban bias in this ensemble of names.[84] A reasonable complement of such standards is universally expected; its absence worthy of remark and conjecture. Plympton MA with only forty-one streets, possesses seven standard names. Yet Rochester MA with fifty-five streets possesses none. Why? At a guess, the latter may have been tardily named, or else lacked a strong nineteenth-century village identity and civic pretensions.

Rapid urbanization in the mid-to-late nineteenth-century exhausted the potential of paragons, and paradigms, or templates, began to coalesce. Like bred like. The poetics of a name, the word-images it evoked, its number of syllables, its cadence, its alliterative possibilities—all were studied and copied. Names composed of agreeable prefixes and suffixes were disassembled and reassembled into plausible pastiches. Once tree names were sanctioned, tree names blossomed; once a few officially named streets bore forenames, the field became crowded with forenames. Where paragons held up only a dozen or two trademark names to be copied verbatim, either singly or as a group, paradigms offered a limitless license to manufacture new names of certain fashionable types. Indeed, many twentieth-century subdivisions are virtually mono-paradigmatic. Below, we shall examine the chronology and evolution of a half-dozen of the chief paradigms. However, our look at street names best begins back-end first, with the datable style trends that refashioned streets into roads, and courts into terraces.

Generics

Scarcely a dozen generic terms for town ways have enjoyed widespread favor since the early eighteenth century, and these constitute the clearest (if overbroad) timemark for a street name. The historic record for Boston and its suburbs affords insight into three centuries of such cycles. By traditional definition, a street was house-lined and urban; a road was rural and often named for the town where it ran (the Concord Road); a lane was a narrow way; an alley a passage between buildings. In the late eighteenth century this simple scheme began to be elaborated with newer, fashionable terms. In Boston, Franklin Place became the first recorded "place" in 1792; the first avenue, Franklin Avenue, followed in 1817, although this term was often initially borne by alleys and only attained its majestic French sense with the development of the Back Bay, as in Commonwealth Avenue (1862). (Thoreau spoke of the "back avenues" of Concord in *Walden*.) Waverly Terrace (ca. 1859) was a block of brick buildings and a terrace in the "true" English sense; Austin Terrace, opened by Arthur W. Austin over his land in 1868, may be its first use for a mere sidestreet or cul-de-sac. But by far the most significant upheaval in street-name generics began quietly with Audubon Road (1884), the first identified use of Road for a residential street in Boston. This is clearly a London import, as Roads were the staple of Victorian villadom. By 1896 an impressive web of Roads, with British names (Selkirk, Kilsyth, Sutherland, Chiswick) had appeared in the Chestnut Hill section of Brighton: prototypical, curvilinear, affluent, suburban roads.

The slow rise of Road over Street is better charted in the suburbs, where Street peaked about 1930 and was overtaken by Road in the postwar development boom. Overall, a definite age-grain in street-name generics is discernible; something no one is likely to calculate, but that leaves nonetheless a cumulative impression. If we examine the street lists of Cambridge, Arlington, and Lexington, we find, in that order, Streets represent 53.3%, 34.9%, and 27.4% of the name-cover. Conversely, Roads constitute respectively 6.0%, 31.8%, and 38.4%. The chronological order in which the three towns were built up is patent. These towns lie along a northwest outbound axis from Boston and their name-cover represents a radial transect of upwards of a century and a half of metropolitan growth (ca. 1835–1990). Although a relatively minor term, Lane more than any other generic epitomizes postwar suburbia in the popular mind. In 1929 there were only four Lanes in Lexington, of which Hayes Lane appears to be an unaffected nineteenth-century holdover, while Larchmont Lane is clearly of the new order. By 1993 there were thirty-two Lanes or 5.9% of the total streets. If one progresses further out to a later suburbanized town such as Sudbury, one finds forty-nine Lanes, representing 13.5% of the street-name cover, indicating that Lexington boomed too early to ride the post-1960 crest of the Lane names.

Colonial

For eighteenth-century street names we naturally turn to the old port towns, which early achieved a degree of congestion that made street-name codification necessary. Unfortunately, there are few such towns, and several factors make intact survival of the colonial name-cover unlikely. Chief streets bore royalist names that, with few exceptions, were extirpated in the 1780s. In Boston, King and Queen Streets became State and Court in 1784; George became Hancock in 1788. Only the relatively innocuous Hanover Street was suffered to stand. The English custom of according distinct names to prominent stretches of what was in effect the same road yielded, partly under the influence of house-numbering, to a single name throughout.[85] The new name Washington Street in 1824 strung together the Orange, Newbury, Marlborough Streets and Cornhill of 1722. Tavern names, which do much to enliven the street names of Old London (and the hamlets of Pennsylvania and the Tidewater), in Boston began to lose sanction with the first street-list drawn up by the selectmen in 1708: Black Horse and Green Dragon officially disappeared; the bland Union and curious Salutation (from its tavern sign: a figure doffing its hat) survived. The custom withered. Many old streets were mere alleys that were later closed and built over. Crooked, Turnagain, Elbow Alleys and Blind Lane have all been obliterated. Private ways were renamed at the whim of the owners. As tastes in street names changed, roads improved, and civic respectability asserted itself, old folknames fell from grace. Cow Lane became High Street about 1798; Frog Lane became Boylston Street in 1809. The few

London transplants, except for Fleet Street, disappeared by the mid-nineteenth century: Inner Temple, Cornhill, Cheapside, signaling an end to the colonial mentality.

The introduction of New York street names, notably Pearl Street (1800), and Broadway (1804), and the Philadelphia pattern of tree names[86] beginning with Chestnut Street (1800) betokened the rise of American style centers and naming trends. Of the 124 names that figured in the Boston street lists of 1708 and 1788, 56% were obsolete by 1834, among them the most flagrantly picturesque. The decolonialization of street names witnessed in the first half of the nineteenth century was offset by a rise in Anglophile names in the latter half, as heralded in the Back Bay, where a discarded colonial name, Marlborough Street, was resurrected and reapplied in 1858. Most ironic was the upsurge in Saint-named streets in the Puritan capital. Among these, Saint Paul's Row (1826–30) was perhaps inevitable in that it took its name from the Episcopal church; the two Saint James Streets (1848; 1860) may have been surnames. Saint Charles Street (1871) was surely unwitting as this was the ultraroyalist name for the beheaded Charles I! Saint Botolph Street (1880), honored the titular saint of Boston, England. To add insult to injury, Cromwell Street was renamed Saint Germain in 1894. Conceivably, some of these are surnames; however, by the early twentieth century the custom of naming streets for Catholic parish churches, as Saint Gregory's Court (ca. 1896), Saint Ann Street (1910), or Saint Mark's Road (1925) had been clearly established, and the puritanical taboo shattered.

Floral–Arboreal

In 1682, William Penn established north-south tree-named and east-west numbered streets in Philadelphia, a paradigm consistent with Quakerly plainness and disdain for honorifics.[87] From there tree names spread nationally, becoming fashionable very early in nineteenth-century Boston: Olive (1796; now Mount Vernon); Chestnut, Elm, Poplar (1800); Oak (1805); Pine (1822); Maple (1822), changed to Willow (1823); Beech (1825); Mulberry (1834); Cherry (1837); Birch (1879). This chronology is in part explained by changing tastes in trees themselves. Poplar enjoyed a brief vogue in the early nineteenth century and was planted on Boston and Salem Commons; Mulberry mania swept Massachusetts about 1830 as an adjunct to a short-lived bubble in native silk-culture. Olive may betoken the post-Revolutionary peace and Willow perhaps reflects the romantic vogue for melancholy. The eclipse of Maple by Willow in 1823 underscores that more than a century (and the elm blight) would elapse before the maple emerged as the arboreal emblem of New England.

Jacob Bigelow's floral-arboreal nomenclature for Mount Auburn Cemetery (1831) was widely admired. Established both as a place of burial and the arboretum of the Massachusetts Horticultural Society, names such as Cypress Avenue and Amaranth Path served both sentiment and science. Rural cemeteries in the image of Mount Auburn burgeoned throughout eastern Massachusetts by the 1850s, bringing with them curvilinear streets, floral-arboreal nomenclature, and even wrought-iron signposts, which were often the earliest avatars of suburbanity in still-countrified towns. However, the exotic florals such as Honeysuckle, Gentian, Wisteria, Verbena, and Eglantine, favored by Dearborn for Mount Auburn path names never caught on; while no such named streets are found in Boston, all exist in London. This may be simply because English Victorians were more horticulturally minded, or because the English cottage garden was the outward symbol of domestic bliss. Or, it may be that in Boston the street was felt to be a man's realm and ladylike dalliance in flowers was best kept in the parlor. Even Dearborn tacitly acknowledged this, naming his avenues for trees and only paths for flowers. By the late nineteenth century the fund of simplex names like Oak Street was exhausted and Anglophile augmentations, such as Oakcrest, Oakdale, Oakhill, Oakhurst, Oakland, Oakley, Oakmere, Oakridge, Oakview, and Oakwood (all Boston) became common. In the 1950s and 1960s a debatably redundant *-tree* was added to form such names as Appletree, Peartree and Maple Tree Lanes (Lexington), or Oak Tree Road (Dedham), and Pine Tree Lane (Abington). Later still, names such as Shagbark Road (Concord), Scotch Pine Road (Weston), and Jack Pine Drive (Sudbury) broke new ground, with novel botany and cadences.

Honorific

Honorific, commemorative and literary names are often highly dated as they tended to celebrate topical figures, events, and works at the hour of their glory. Identifying these amidst the Boston street lists amounts to a trivia contest for the date-minded sightseeker. Some likely conjectures, concisely stated, will illustrate this potential for historical association:

Lafayette Avenue, 1825 (Marquis de Lafayette toured United
     States 1824–1825)
Wilberforce Place, 1843 (English abolitionist, died 1833)
Ashburton Place, 1847 (Webster-Ashburton Treaty 1843)
Limerick Place, 1847 (Irish Potato Famine 1846)
Fillmore Court, 1857 (U.S. President 1850–1853)
Alaska Street, 1868 (Alaska Purchase 1867)

The triumphal 1850–52 tour of the "Swedish Nightingale" was commemorated in Jenny Lind Streets in North Easton, Taunton, and New
Bedford MA. Similarly, in Rockport MA, Penzance Road and other
Cornish-named streets were platted for the Land's End cottage colony in
1889, ten years after *The Pirates of Penzance* took the world by storm.[88]
Priscilla Road is a fad name that spread widely through at least thirty-two
towns in eastern Massachusetts, where the memory of Priscilla Alden and
her poet-celebrant Henry Wadsworth Longfellow burned brightest, and
may perhaps be correlated with the Plymouth Tercentenary (1920). As a
cautionary note, the popularity of the Waverley names (Ivanhoe, Mon-
trose, Abbotsford, Kenilworth, Marmion) long outlived the novelist Walter Scott.

Personal Names

The largest category of street names derive from personal names; while
some are honorific (see above), most simply record the name of an old
local family, a landowner, builder, or speculator. For instance, when John
P. Wyman of Arlington MA subdivided part of his farm ca. 1888–1895 for
houselots, he laid out two streets, Wyman and Palmer, the latter undoubtedly his middle name.[89] His example was endlessly repeated. Prominent
and prolific old Yankee families deeply entrenched in certain quarters of a
county might achieve toponymic immortality in the names of natural features and streets that, when plotted on a map, trace kinship areas or "kin-
doms" revelatory of early settlement history. Only distinctive names work
well as tracers. Ten Belcher street names in a tight belt from Quincy to
Norfolk MA appear to define such an area and are corroborated by two
historic houses, as well as by modern telephone listings. The Leonards
were a dynasty of colonial ironmasters whose first foothold was gained at
the Saugus Iron Works before 1650. An old adage opined that where there

were ironworks, there were Leonards,[90] and the distribution of Leonard Streets still approximates an index of the early iron-working towns of southeastern Massachusetts. Thoughtful attention to old gravestones and war memorials can supplement the street-name record for anyone interested in the geo-genealogical uses of placenames. Before World War II, English Yankee surnames predominated, but to one familiar with ethnic patterns in the agricultural resettlement of Massachusetts in the early twentieth century, the significance of Chmura Road in Amherst, or Arena Terrace in Concord is abundantly clear.

Surnamed streets clearly dominated in the nineteenth century, but forenamed streets were not unknown. Oliver Street (ca. 1850, North Easton MA), named for industrialist Oliver Ames, may have implied either the master-and-man familiarity typical of early industrialism, or the classless towny leveling that long kept nearby Langwater Pond (an estate name) simply Fred's Pond, after Frederick L. Ames. Exotic first names began to figure commonly in the Boston street lists in the 1880s and 1890s: Carlos, Caspar, Eulalie, Josephine, Oscar, Sydney, and Sylvia. Since this first forename wave was perhaps more flamboyant than actual child-naming practices of the time, many may have honored no one in particular, but simply cashed in on the exotic possibilities once a license to coin forenamed streets had been granted. On the other hand, names such as Edna, Eugene, Evelyn, Harold, and Kenneth tally so convincingly with known names of the period—whether of famous authors or one's own great-uncles and -aunts— that the conventional wisdom that they were shaken from the developer's family tree is entirely creditable. Tract streets, both grids and the more recent curvilinear webs, can be found with fashion-dated forenames that appear to commemorate whole families in asphalt. The dearth of conspicuously Irish surnamed streets in Boston (e.g., only eight O' names, such as O'Brien Court, in the city) suggests that forenames may have been a means by which immigrant groups memorialized themselves without disturbing the Yankee status quo. A twinge of towny modesty may also have spurred the practice: surnamed streets may have struck self-memorializing developers as too grand.

Full-named streets have existed since the mid-nineteenth century when an unambiguous honorific was required, as in John A. Andrew Street (1871) in West Roxbury MA, named for the popular Civil War governor. In rural districts, the desire to single out one among many kindred bearers of a surname might result in names such as Frank Schnopps Road and George Schnopps Road in Hinsdale MA. Avenue Louis Pasteur (1906), the axial approach to the Harvard Medical School, is an obvious gallicism in the style of Parisian honorific avenues. However, where nineteenth-century honorifics like (William Tecumseh) Sherman Street or (Philip Henry) Sheridan Street saw no need to belabor their names, for various reasons by the mid-twentieth century the trend began to change. The names of Irish politicians began to be fully and reverently attached by legislative act to massive public works that made jobs and got votes. Full names may have been in part dictated by the extreme commonness of many Irish names, which resulted even in daily life in a heavy reliance on nicknames and initialisms. Gold Star streets and hero squares from the World War II era carried rank and full name. Simple surnamed honorifics began to be regarded as too ambiguous (as King versus Martin Luther King), or too understated. About the time of the Bicentennial, a final trend in full-named streets arose in reaction to sprawl and the blanket of unhistorical street names that overspread once-small towns. Early town worthies, often with archaic names or titles, began to be very posthumously honored by a new type of street name with little historic precedent, such as Solomon Pierce Road (Lexington), Spooner Cornish Drive (Plymouth), or Deacon Hunt Road and Captain Forbush Lane (Acton MA). This trend coincided with skyrocketing real estate values in many towns, and such names often connote a new sort of exclusive suburbia.

Bithematics

Bithematics are artificial two-element compounds: Thorncroft and Crest-view are two examples that epitomize the British and American extremes of the type. Gillian Bebbington traces this paradigm to mid-Victorian Hampstead, London, where building societies wanted " 'pretty' names to attract buyers" to their mushrooming streets of semidetached villas and terraces. "Their only aim in naming streets was to give an impression of genteel, vaguely rural, desirable residences. Hence the number of coun-trified suffixes and prefixes found in Hampstead streets. Croft is the most popular; Ferncroft, Hollycroft, Rosecroft, Greencroft, and Lyncroft. End-ings like 'wood,' 'grove,' 'bourne,' 'hurst,' 'leigh,' 'ridge,' and 'dale,' are fruit-ful basic elements."[91] While some of these compounds may be names of actual, if obscure, British places (or people), it is likely not some real Rav-enscroft that the nine London streets so-named commemorate, but rather the power of the name itself to dreamily evoke the rural depths of prein-dustrial England. Collectively, these names give every impression of having been formulated from highly stylized, archaic, and poetic elements chosen to avoid etymological incongruities and yield plausible pastiches of genuine ancient names of rural cottages and manorial halls. Much of this vocabulary, as *furze-, alder-, hazel-,* or *-croft, -hurst,* and *-wold,* is intrinsically English and without real resonance for Americans.

The American bithematic paradigm lacks the toponymic tradition and codified propriety of the British, and the vocabulary is more intelligibly modern (*-view, -wood*), more American in its word choice (as *rock-* for *stone-*) and in its taste in tree and flower names (as *maple-, pine-, cedar-*). The clipped meter of Stoneleigh gives way to Mountainview and Meadowbrook, names with the ring of turn-of-the-century dairy farms. (The custom of naming common farms has been dated to ca. 1880; while such pretensions often irked the neighbors, they appealed to summer boarders.)[92] American bithematics never took on the hothouse fulsomeness of their English cousins, and are not so cloying or tediously hybridized. Nor are they so rife.

The tendency of London streets to emulate the name fashions in suburban villas, manorial halls, and rural cottages, raises the question of whether in New England parallels might exist between the names of country estates and streets. Unfortunately, only a very small proportion (perhaps less than 5%) of addresses in the Boston *Social Register* for 1899, 1915, and 1930 include estate names, although the custom was clearly more prevalent. For example, while all of the numerous Ames family estates in North Easton bore names, none were listed. (Such names may have been a pretentious indulgence largely confined to letterheads and gate piers.) Yet even if the actual extent of estate-naming cannot be gauged, there is no reason to suppose that the estate names listed varied in character from those that were not, and the names listed are not forbiddingly aristocratic. Some suggest summer camps or cottages as Tween Waters, Keewaydin, Doneroving; some play on husbands' and wives' family names, as Bird's Nest (Mr. George K. Bird), Frosalt (Frothingham-Sault), or Ricefields (Mrs. Arthur T. Lyman, née Margaret P. Rice). However, the horse or dairy farm and not the English manorial hall was the beau ideal of the bulk of pre-1930 estate names, which reflect or anticipate a broad range of twentieth-century street-naming trends; typical are Edgefield, Meadowmere, Glen Ridge, Brookwood, Deerfoot, Four Winds, Greystone, Fox Ridge, Stoneleigh, Windy Knob, Bushcliff, Clearfield and Hillcrest. The suffixes *-hurst, -wold,* and *-mere,* are applied only sparingly to confer distinction; on the whole, the estate names are only a slight degree snobbier than street names. Dreamwold, the Scituate mansion of Copper King Thomas Lawson, and Wistariahurst, the home of Holyoke silk magnate William Skinner are flamboyant, and perhaps nouveau riche, exceptions known from other sources.

In Boston, a few bithematics began to appear in the mid-nineteenth century: Woodside Avenue (1849), Oakland Street (1853), Glenwood Avenue (1856), Greenwood Avenue (1857). In 1868 the board of Aldermen adopted a highly indicative set of name changes for Roxbury streets: Beech Street to Beech-Glen Avenue, Myrtle Street to Glenwood Street, Orange Street to Elmwood Street, Oak Street to Oakland Street, Park Street to Edgewood Street. Here we note the shedding of duplicate or passé names (as Myrtle, Orange), the ambiguity as to hyphenation, the popularity of the Walter Scott–inspired Glen, and the use of an augmentative Suffix to render redundant names distinctive: Oak to Oakland. The rich combinatorial possibilities of bithematic names answered well in the postbellum rise of the city, and the expansion of the street-grid. What might be regarded as cracklingly British bithematics did not appear until later: Stonehurst (1893), Lyndhurst (1894), Heathcote (1894), Kingsdale (1895), Athelwold (1896). All but the first and last have London counterparts.

Despite their relatively recent date, British bithematics pose the same classic problems of diffusion as do all dispersed artifacts and, being linguistic phenomena, they are particularly slippery to trace. How to account, even in the most general way, for Applecroft Road (ca. 1920) in Welwyn Garden City, England, and Applecroft Lane (ca. 1980) in Conway NH, particularly as the word is unhistorical and there are no English villages, or even London streets by that name? Do the two examples represent independent invention from a common bithematic paradigm, or a complex chain of borrowings from the well-known British new town to northern New Hampshire? Were such transfers haphazard, or facilitated by street atlases or even possible master lists in trade magazines, analogous to the pattern-book diffusion of architectural styles? The busy traffic in street names is the more puzzling: only a few people ever name streets; no one person is likely to name very many; yet all of these random, amateur acts of naming participate in widespread patterns and trends.

Sound-alikes

There is ample reason to believe that fad names ushered in other likesounding names in their train, and that alliteration boosted the popularity of whole sets of nineteenth-century names. Among the more familiar are Pleasant-Pearl-Prospect, Albion-Avon-Auburn-Alba, and Kenmore-Kenwood-Kimball-Kensington. While the earliest, the P names, all have distinct individual meanings and histories, the later A names are but obscurely allusive, and to some degree interchangeable. The K names, last of all, are even more so, and their popularity coincides with a mania for K names generally, as witnessed in old trademarks (Kodak, Kleenex, even Kenmore itself). Studied over time, street names often resemble not so much fixed landmarks with discrete if meager histories, but rather fashionable, protean sound-blobs like the Albion variations above.

Nowhere is this unsettling suggestion stronger than in the records of name changes adopted by the Boston Street Laying-Out Department, beginning with those made in 1868 when Roxbury was annexed to the city. Indeed, the principal motive in such changes seems to have been entirely bureaucratic and practical: to eliminate duplication, a common enough problem clearly aggravated in the late nineteenth century by Boston's appetite for suburban towns. The method chosen, which became highly routinized, was meant to maintain a semblance of sound, with the transformations so subtle as to go, so to speak, unnoticed (compare Esso to Exxon, 1972). Names were not so much changed as updated and, as with architectural renovations, the specific alterations—both in terms of what is added, and what is removed—dramatically highlight the fashionable dos and don'ts of the time. Augmentation by some trendy suffix, or outright substitution of an alphabetically adjacent name plucked from who knows where, were the standard techniques. Such name-changes were regularly made when a street was "laid out," that is, surveyed and accepted as a public way. While even the smallest towns have problems with street-name duplication, the scale and method of the Boston solution must surely be unique to older metropolitan cities. The far-reaching influence of such cities as style centers, however, meant that new trends adopted there under the force of necessity (exhaustion of the conventional name-pool) would be emulated elsewhere in towns with relatively few streets as marks of fashion.

In Boston, the extirpation of duplicates was most vengefully pursued in the late 1920s, but continued for decades thereafter. The fruits of three years of this labor, excerpted below, are eye-opening. From a toponymic standpoint, the most alarming realization is that it is entirely unlikely, given the 1927 examples listed here, that anyone named Wardman, Col-

borne, or Adamson ever existed. That is, that even surnamed streets may
be sound-blobs, devoid of even the slenderest historical significance.

> 1926: Linwood Place > Lawnwood Place; Foss Road >
> Fossdale Road; Rosemont Road > Glenrose Road

> 1927: Westminster Road > Wardman Road; Linden Street
> > Leniston Street; Colonial Road > Colbourne Road;
> Adams Street > Adamson Street

> 1928: Parkvale Road > Parklawn Road; Woodbine Road >
> Woodbriar Road; Bates Road > Bateswell Road; Oak
> Avenue > Oakton Avenue; Walnut Park Road >
> Waldren Road

Artifactuals

A whole, novel street-name vocabulary of colonial cultural artifacts burgeoned after World War II. What I take to be the prototypical example,
Lantern Lane, dates to the interwar years; and while not without descriptive logic, it might no more have lanterns than Maple Street, maples. In
itself highly prolific (over forty in Greater Boston), as a model for literal
imitation, possibilities were limited to the vocabulary of pre-Edisonian
illumination. It inspired Candlewick Close, Gaslight Lane, Lamplighters
Lane and Lanthorne Lane: even Bayberry and Candleberry were used.
Conceptually, however, Lantern Lane broke new ground; soon any decontextualized object evocative of hearth or farmstead was fair game in the
artifactual paradigm: Hearthstone Way, Teakettle Lane, Whiffletree Lane,
Wagonwheel Road, Haystack Lane, Weathervane Road, even Sledgeham-
mer Lane. The imagery widened. Firearms were taken down (Flintlock,
Gunstock), ships broken up (Keel, Rudder, Halyard, Mizzenmast, Spin-
naker). Anywhere within five miles of the shore, the full fury of the nautical lexicon might be turned loose. Ultimately, all Colonial pretense could
be dropped, as at The Greens Condominium, with Brassie, Mashie, and
Niblick Ways (North Reading MA). The artifactuals are the most radical
and stylized departure in twentieth-century street-naming, comparable
only to the Bithematics in the breadth of their influence. While the bithematics were grounded in Anglo-Saxon place-naming, the artifactuals in

their evocation of decontextualized images owe more to the modern, manipulative uses of language perfected by Madison Avenue.

Test Names

Some fad names, by their frequency, significance, and rough datability, single themselves out on street lists as possible test names by which to gauge both the age and perhaps even the specific character of the built-scapes they mark. Once such a correlation is suspected, the armchair toponymist can investigate it through street-indexed maps, such as those of the 136-town *Metro Boston Street Atlas* (Arrow ca. 1993) used in the three inquiries below. Dry statistics, however, cannot convey the undeniable excitement of the chase as one flips madly through atlas pages to establish the topographical affinities of one's quarry. While Pleasant Street and Main Street belong to the common legacy of nineteenth-century towns regardless of size, other, more obscure, names, such as Albion, are more typical of the mid-to-late nineteenth-century densely settled core. Albion, the ancient and poetic name of England, became a fashionable, patriotic name in England ca. 1760–1836 during the reigns of the unpopular Hanoverian Georges, for whom little was named.[93] It was quickly eclipsed in Britain on the accession of Victoria, but the name bloomed late in the United States, in part, one suspects, due to the obscurity of its patriotic significance (unlike, say, Britannia). Albion IL was established in 1818 by Englishmen; Albion MI in 1833; various of the six recorded Albion streets, places, and courts in Boston date from 1848 to 1870. Of the seventeen Albions listed in the *Metro Atlas*, most (76%) are found in cities that roughly trebled to septupled their population between 1860 and 1900, such as Chelsea, Brockton, and Fall River. Typically they are short (71%) grid streets (82%) near railroads (76%). (A clearly late example, Albion Road in Wellesley MA, least fits this profile, lying in the midst of curvilinear motor suburbia.) The specific toponymic and geographic context in which a street name is found may shed light on its rationale. Albion lies near Alfred and Marion in Medford, but near Scotia in Salem: euphony was a factor in the former, root sense in the latter choice. The situation of Albion Street in Millville in the Blackstone Valley is so unusual as to suggest that it took its name from a textile mill or was inhabited by emigrant English workers.

Cottage Streets catch the eye because cottage was both a novel term and concept in the mid-nineteenth-century New England townscape. The word was applied equally to small, picturesque pattern-book houses, to simple, depot-clustered "homes of one's own" built for a new class of blue- and white-collar railroad commuter, and as a euphemism for the trackside shanties of Irish railroad laborers. (In a letter to Thoreau, Margaret Fuller refers to the makeshift houses of Concord railroad workers as "quaint cottages"; Emerson, in his journal, calls them shanties, a verdict sustained by Shantytown Path at Walden Pond.) Whatever the sense intended, as a rampant mid-to-late nineteenth-century street name, proximity to rail lines and depots would be expected as these were the growth points of the time. Only later did cottage become chiefly associated with summer houses, large and small. Unlike villa, another term dear to Victorian taste- makers, cottage has become fully naturalized in American English and has lost all trace of novelty. Of the 83 Cottages in the *Metro Boston Atlas*, 17% cross or lie perpendicular or parallel to railroads, while a total of 56% lie within about a half-mile of railroads. Only about 19% lie near the ocean or lakes, the locales with which we most closely associate cottages today.

A term of similar vintage to cottage, but without its picturesque over- tones, is mechanic, as in Mechanic Street, or Mechanics Court or Lane. Outside of garages, the term is almost quaint now, but it once denoted a new and presumably respectable class of workers skilled in machinery, who stood sufficiently apart from the population at large to justify naming a street after them. Predictably, of the twenty-two such streets in the *Metro Boston Atlas*, 36% cross or are perpendicular or parallel to railroads, and a total of 64% are near railroads. Most are back streets. More interestingly, 64% are also near ponds, so-called reservoirs (inferred to be millponds or power reservoirs), cross millpond outlets, or are near dammed rivers. In Attleboro, Canton, and North Easton, the onetime significance of these bodies of water is explicit in their names: Mechanic's, Forge, and Shovel Shop Ponds. This very high correlation with water-powers implies that mechanics as a distinct class flourished before the rise of the large, coal- fired factory. Most of the towns in question were small; many may have been mill-villages, although the exceptions are glaring. The actual distri- bution by towns of Mechanic streets is not entirely random. Half occur in contiguous, often rail-linked clusters of five, three, and two towns each, suggesting emulation as an important means of dissemination. Further- more, with the exception of Mattapoisett, no Mechanic Streets are to be found in Plymouth County, consistent with the dearth of nineteenth- century industry there, particularly between the collapse of the colonial ironworks and postbellum rise of the central-shop shoe industry.

One last, testable, nineteenth-century name-type is well represented in Gardner MA by Emerald, Dublin, and Limerick Streets, in a neighborhood of tightly lotted streets near both chair factories and railroad. An 1878 village map gives a profusion of Irish-surnamed householders such as Doyle, Murphy, O'Neill, confirming the obvious. Such ethnic imprints seem rarely to survive (Dublin Street in Cambridge MA is now Sherman Street), perhaps because such names were regarded as inappropriate, perjorative, or were never officially accepted.

Affinities

Mapwork is clearly but a start. Fieldwork would be required to establish the true age, architectural character, and social class of these streets. Are there cottages on Cottage Street? Company-owned houses on Mechanic Street? Ultimately, the great challenge is to recognize in every street-atlas grid-square a swatch sample of the urban fabric, to appreciate it not as a mishmash of randomly named streets, but as a composite artifact, marked by street names, street shapes (layout, lot size), house types and styles. Such a grid square in Chelsea is far different from one in Wellesley MA, and the open question is, how well can differences so visible in the field be decoded from the interwoven lines and names on the map? The name-cover should not be seen as an interconnected but self-contained layer of the landscape. Its patterns need to be correlated with those of other artifacts. Much of the pleasure of landscape study derives from an appreciation of the tightly woven texture of the "obvious." The Reversed Center Arc in NH and ME strangely outlines the likely source area of the connected farmstead; what could possibly be the connection? More bizarrely, the Tophet name-string overlies a postcolonial zone of yellow-painted churches.[94] Yet "strange" and "bizarre" serve only to describe our reaction to the incongruity of the combination: most combinations of cultural indicators are in fact mere happenstance: berets and baguettes, for example.

Placenames and house types often go hand-in-hand in the settlement landscape, a classic example being found in the New York–settled southeast corner of Wisconsin. There a New York name-scatter (Rochester, Clinton, Troy, East Troy, Palmyra) coincides with small but conspicuous numbers of cobblestone houses, typically of upright-and-wing design. Such unusual houses originated in upstate New York, and are most densely concentrated within a sixty-mile radius of Rochester.[95] In Massachusetts, the East Anglian timber-framed Fairbanks House (1636) stands in the East Anglian–named town of Dedham. Settlers from southeastern Massachusetts brought the Cape Cod house with its eccentric attic gable windows to Franklin County MA; did they bring as well a propensity to name Leyden MA after the Dutch city in which the Plymouth Pilgrims first took refuge? In the popular era, Arlington Street (Cambridge MA) with its many mansards brought together postbellum style trends in architecture and toponymy, while Harold Court (Gloucester MA) combined a then fashionable forename with the Stick-Style and Queen Anne. A Stick-Style house on the corner of Sherman and Sheridan Streets in Lexington MA is almost too good to be true. Lantern Lane and Flintlock Road are not unrelated to the postwar popularity of the Cape Cod Cottage and the Garrison Colonial. These correlations are less remarked upon but not less remarkable because their implications are "obvious."

Town name and town shape can likewise be correlated. In the mid-nineteenth century, compact manufacturing enclaves, often centered on valuable water privileges, might be carved from adjoining towns. Such new townships might be named for the industrialist who promoted them. Sometimes the two trends in shape and name coincide, as in Lowell, Lawrence, and Maynard MA; Sprague and Ansonia (for Anson Phelps) CT; and Proctor VT. Contrast the crazy quilt of town shapes and motley of town names in southern New Hampshire with the squarish grid of English-named towns of southern Vermont. These latter, were they encountered on the Great Plains, would seem to be township blocks named by a bored Anglophile railroad director. In fact, they were largely the work of Governor Benning Wentworth of New Hampshire, who hastily disposed of a vast tract beyond the Connecticut River over which he exercised only a questionable jurisdiction. Thus the regularity of the layout and the predictablity of the names are products of an arbitrary expeditiousness not unlike that of our fancied railroad magnate.

In the preceding pages, we have experimented with a range of techniques for analyzing New England placenames, not individually, as a ragbag of picturesque word origins, but collectively, as an artifactual record of settlement history. These techniques have met with considerable success, despite their reliance on simple diffusionist models and simplistic notions of emigrant groups with a single or predominant place of origin. For town names, we have proposed plausible rationales that run counter to local tradition, but it is in the nature of "scientific" etymology to yield such disturbing contradictions. Even blandly modern street names have revealed an unexpected richness. Most importantly, we have discovered sufficient numbers of map-patterns to conclude that the New England name-cover is by no means a random scatter of static points, but an interrelated and dynamic array. Many questions still remain. The placename palimpsest will well reward further study.

{ *Three* }

BOUNDARIES AND TOWNPLANS

Places can be much more precisely defined by boundary lines than by names; and while the New England landscape may be under-named, it certainly has no dearth of bounded, if effectively nameless, places. An individualistic society based on fee-simple tenure and founded as scientific surveying came of age devoted much energy to boundary-making. (Good fences make good neighbors.) It also expended equal energy in untangling the confusion that imperfectly made boundaries created. Reading the landscape significance of placenames comes naturally. They are, after all, words pregnant with meaning and susceptible to the varied analytic techniques of the linguist and cultural geographer. But boundaries are geometric lines and figures whose true form and vast networks lie largely invisible on the ground, and even when visible, how does one read lines?

Cadastre

A cadastre is an official register of the area, ownership, and value of land kept for taxational purposes, and a cadastral map is one of sufficient scale to show boundaries, field-lines, buildings, and other fine detail. Typically European, often ancillary to the Napoleonic Code, there is no New England equivalent, but perhaps if one rolled the county deed books, the tax assessors', town engineers', fire insurance, and topographic maps all into one, we would have something of the sort. (Computerized mapping can do just that.) Here I use the term *cadastre* to refer to the full network of boundary lines, whether marked by fences or not, that private property, agriculture, and civil administration have imposed on the land: lot-lines, field-lines, town-lines, and so forth. Thus, the seventeenth-century cadastre of a town is this boundary network as it appeared at that time, and which constituted a vast and often invisible cultural artifact that would control the configuration of all future development. As Edward Price has perceptively remarked, "the identities of particular societies are often encoded in the geometry of their landownership."[1] Like all artifacts, the cadastre bears witness to the attitudes, social structure, values, and technical capacity of the people who fashioned it. In the agrarian era it underlay the peculiar mosaic of tillage, mowing, pasture, and woodlot; in the nineteenth century it proved refractory enough to break up the urban gridiron; even in late twentieth-century suburbia it can still give subtle shape to sprawl. The cadastre is the immensely conservative legal bedrock that underlies the landscape as we see it today.

Metes-and-bounds surveying prevailed in the older-settled eastern seaboard prior to the Ordinance of 1785. Like many such medieval law terms, it is etymologically a doublet; that is, it redundantly expresses the same idea in both French (mete) and in English (bound). Lawyers and surveyors invariably construe "mete" wrongly, yet felicitously, as the noun form of "to mete" or measure, and the essence of a metes-and-bounds survey is thus one of measured distances between landmarks. Technically, it describes a circuit of a tract that starts (and ends) at a "point of beginning" and that visits each monumented corner and angle in turn, giving all distances, compass bearings, and abutters' names. By custom, this is all recorded in a single, run-on, Faulknerian sentence. Many early deed descriptions lack certain of these elements. If, when plotted, the distances and angles form a closed figure, the survey is said to close. Failure to close (misclosure) is the sign of a faulty survey, and is historically common. Because of the earth's curvature, were a person to walk north a mile, then east, south, and west a mile, the resultant square would not close![2] The accurate determination of compass bearings and the measurement of landlines on slopes was beyond the competence of country surveyors and certainly beyond the willing purses of those that hired them. The quality of a survey rides on the value of the land: when land was cheap, surveys were very cheap. The earliest generation of New England township surveyors such as Thomas Graves and Simon Willard were often paid in grants of land, while private surveys were undertaken at a fee of two or three pence per acre.[3]

The inability to define and locate the parcel rigorously by the immutable laws of geometry accounts for the wealth of redundant and often contradictory information written into deeds. To remedy this, the law establishes a "hierarchy of calls" to guide surveyors who must rerun the lines. This ranking from highest to lowest typically runs: natural landmarks, artificial landmarks, distances, angles, and finally area, which is deemed the least reliable. "Points on the ground hold," is the surveyor's axiom. Fence-lines are among the strongest evidence in the eyes of the law, and it has been seriously stated that the compass bearings given on deeds are merely meant to assist the surveyor in his choice of which fence to follow.[4] In colonial times, in raw country, boundary monuments were often simply marked trees or stakes and stones. Indeed, beyond the necessary mathematics, a thorough knowledge of common trees was requisite to the surveyor's work, and the species recorded in early surveys have been mapped and analyzed as a random-sample census of the old-growth forest.

Oak trees, plum trees, even stakes and stones proved far from permanent, and the periodic perambulation of the bounds was essential to their maintenance. Human memory—at times reinforced—was also important. Jacob Farmer of Concord MA was called upon as a boy to witness the setting of the bounds of his mother's woodlot. One of his elders calmly whittled a birch switch with Farmer's own knife, then all but dealt the youngster "a blow which would have made him remember that bound as long as he lived." And remember he did: years later in 1857 his testimony clinched a dispute Henry David Thoreau was hired to resolve. Another source tells of head-bumping, with small cakes or cookies as a reward. (This custom is at least as old as marking farm and vineyard boundaries among the Franks. Salic Law ca. 622 prescribes: "Let him give a box on the ear of each of the little ones, and let him twist their ears in order that they can give testimony.")[5]

In the popular mind, metes and bounds connotes crazy-quilt boundaries, but in actual fact, rectangular parcels from the earliest settlement were favored by New England lot-layers, if only to facilitate acreage calculations and allow for a more impartial and methodical distribution of the land. Regular lot lines were easier to locate, remember, and recover if boundary marks were lost, and deeds less open to ambiguity and lawsuits. (They were also more amenable to being recorded verbally without plats in deed books.)[6] This custom of common-sense rectangularity enabled an Essex jury in 1678 to find for the defendant, John Winthrop Jr., in the oft-cited Kites Tayle suit, in which willful misconstrual of the title deed would have skewed the rectangular bounds of his Salem farm, making it "like unto a kites tayle, the like president not being known in the Cuntry."[7] On Nantucket, an ancient parcel is known as the Crooked Record, precisely because of its odd angles, which were presumably unusual enough to be descriptive.[8] Tiers of lots were often laid perpendicular to a road or river, bending with it fanwise at some loss of rectangularity. Perfect squares did not enjoy particular favor for tracts smaller than a township or farm. Fields were traditionally oblong, as in England: long fields gave everyone a fairer cross section of slopes and soils, afforded equal road or river frontage at the narrow end, and could be tilled with fewer turnings of the plow.

Orderly land division gave the metes-and-bounds cadastre in the New England colonies a vastly different character than elsewhere. Direct grants by the colony to individuals were relatively few; rather, land was granted through the intermediate agency of the town proprietaries, who allotted it—quite literally—in numbered lots drawn by lot. Various formulas prevailed, and while the size of one's lot might vary according to one's station in life, still the process was open, equitable, and orderly within the terms of its time. It was also slow: a series of four or more divisions of the commons was typically made, and a century might elapse before the last of a township's twenty thousand acres was doled out. Even in Connecticut, which allowed more individual choice in the location of grants or "pitches," the law specified that lots be in one piece and "in a comely form."[9]

This process was vastly different from the abusive warrant system that prevailed from Pennsylvania southward. Under this system, a warrant for a certain acreage was issued by the surveyor general of the colony and forwarded to the deputy surveyor of the county in which the grantee intended to settle. The deputy surveyor would lay out the lot of the grantee's choosing, certifying it (generally perjuriously) to be of the stipulated acreage, and that the land was unoccupied and did not overlap anyone else's grant. (Prior claimants could enter a caveat to void the grant.) Quite predictably, the choicest bottomlands were occupied first, with grotesquely gerrymandered boundaries, but on paper at least, of the proper acreage. Less desirable land was left to latecomers or as ungranted wasteground: enough leavings to endow the state university in Ohio and to cause contentious lawsuits in East Texas when oil was found to underlie otherwise worthless scraps of land.[10] The result of the warrant system at its worst was a legal bramblepatch: gores and overlaps were rife, and one modern Virginia surveyor has lamented that each parcel is a bit of a jigsaw puzzle in which the pieces do not fit together.[11] North Carolina, with only 55,000 square miles of territory, granted more than 100,000 square miles in land, while the grants made in twenty-four counties of Georgia exceeded by three and one-third times their actual area.[12] The extreme contrast between the settlement patterns governed by town allotment and by individual warrant cannot be doubted to have had an effect on the regional character of the people and the communities they formed.

Toolmarks

Toolmarks are of keen archeological interest. An adze mark on a floor joist or the kerf marks of a crosscut or circular saw tell us how, and often roughly when, the timber was fashioned. The apparent lack of toolmarks, as on a perfectly formed modern nail, is as much a toolmark as the conspicuously clipped edges of an old one. Boundaries are artifacts and, like all artifacts, they bear signs of how they were fashioned. The U.S. Public Lands Survey is an artifact, a tenuous geometric web spun out over millions of square miles, full of telltale flaws that are in fact toolmarks. Among the most notorious are the wildly skewed section lines in the iron-ore belt of Michigan, due to magnetic deflection of the compass.[13] Toolmarks in the federal grid are easy to spot, as they are departures from an absolute standard, and because modern cartography puts their imperfections under the microscope. One prominent toolmark in New England is the Massachusetts–New Hampshire boundary, run to the detriment of the latter by Richard Hazzen with an overcorrected compass in 1741. Flip backwards from page 28 to 23 in the DeLorme *Massachusetts Atlas* and watch the state-line diverge ever wider from the statutory parallel just south of 42° 42'. New Hampshire lost over fifty square miles and only accepted the mis-surveyed Mitchell-Hazzen line in 1895.[14]

The evaluation of toolmarks in the New England cadastre is usually much more problematic, because the intentions of the artisan are harder to divine. It also requires that we understand something of the history of surveying instruments and methods. Probate inventories of colonial surveyor's instruments are rare, and often the court official who itemized them was ignorant of their precise nature, lumping them "as divers mathmatical instruments."[15] Even if he had laid a name to them, the meanings of such terms as dial, circumferentor, and theodolite have proven to be exceedingly slippery over time. References in fieldnotes, on plats, and in deeds to "statute chains," to "chain-bearers" or to distances "by estimation," as well as the refinement of the recorded angles (e.g. Northwestwardly versus N 21° W) shed light on the instrumentation and the sophistication of the surveyor. Perhaps our most vivid image of the colonial surveyor plying his trade comes down to us in the form of marginalia and other embellishments on the plats themselves: doodled houses or scenes of surveyors in the field with tricorn hats and circumferentors on Jacob's staffs. Even doggerel. The affinity of surveyors, deed registrars, and land court lawyers for bad verse is notorious.[16] Richard Candee quotes one ditty (1730) inscribed on a division map of Stoughton MA. It begins: "Upon our NEEDLE we depend / In ye THICK WOODS our COURSE to know / Then after it ye CHAIN Extend / For we must gain our DISTANCE so." It is not only amusing; it also testifies that the use of the Gunter's chain was by then standard practice.[17]

A thumbnail history of some of the more typical surveying tools for angle and distance measurement may be of interest. The earliest compass was the mariner's compass graduated in 32 points (N, NNE, NE, etc.) and references in deeds to actual degrees rather than compass points are often lacking before the 1730s, when the more finely calibrated surveyor's compass came into wider use. The vernier compass of David Rittenhouse (1732–1796) had a six-inch needle and was graduated to one-half degree.[18] In ordinary fieldwork the uncorrected magnetic compass was commonly employed in the eighteenth century; even in 1813 the reputable cartographer J. G. Hales still issued maps, such as his map of Portsmouth NH, with only the magnetic meridian. By 1830, published maps typically showed magnetic and true north with half- and full-barbed arrows. The circumferentor was a considerable improvement on the compass. It was much favored in Ireland and the colonies, but scorned in England as not sufficiently accurate. It consisted of a six-inch diameter compass mounted in a wooden box that was screwed on a tripod or one-legged "Jacob's staff." A rotating arm with front and rear sights (no telescope) enabled angles to be measured quite independent of the compass. The surveyor at point B could backsight to his previous station at point A, then forward to his next station at point C, and the angle ABC could then be "turned," that is, measured, without recourse to the compass. This was the instrument at its most accurate. But the reason the circumferentor was so popular in America was that the angles could also be expressed simply as compass bearings, which eliminated troublesome backsighting in rugged woodland surveys. Point A and point C need not both be visible from point B: the surveyor simply read off the compass bearing of the next line of his traverse. His circuit completed, instead of having a true record of all the internal angles of his polygon, he had instead a set of inherently less accurate compass bearings. In the words of a modern critic: "This was free-style surveying."[19] While the circumferentor was widely used in America by 1800, Abel Flint specifically notes in his 1813 treatise on surveying that New Englanders still preferred the compass (Rittenhouse's was recommended) and chain.[20]

The statute or Gunter's chain, invented by Edmund Gunter in 1621, was 4 rods (66 feet) long and consisted of one hundred 7.92-inch links. It had probably replaced the typically 16½-foot brass-shod "pole" by the early eighteenth century, but as distances continued to be set down in "poles" (alias rods or perches) even when measured by chain, the transition between the two is blurred in deed records. Indeed, many seventeenth-century deeds give distances "by estimation," which suggests they may have been paced. The Gunter's chain remained the standard until after 1900, and was not officially retired from public land work until about 1930. (Country surveyors were still using it in Colebrook NH twenty years later.)[21] The advent of the steel tape in the 1860s was followed by that of the transit in the 1870s, a precision angle–measuring instrument with telescope. By one estimate, more than two-thirds of federal surveyors were using some form of transit in the early 1880s. They were in everyday use by 1920. It may be that certain fancily turned angles (as the ca. 1890 boundaries of the Blue Hills Reservation, Milton MA) are recognizably transit-work to a trained eye. Transit and Tape closed the age of Compass and Chain and ushered in the era of modern professional surveying, although even these "modern" instruments have now been outmoded by electronics.[22]

A board, when cut, leaves an obvious toolmark in the kerf, but it also leaves a toolmark in its length: if it is six feet long, it was laid off with an English rule. However, even the most common measures were not always well standardized. Some Pennsylvania German vernacular houses were built on a 13-inch foot;[23] certain towns in eastern New England, settled by the Scotch-Irish, were surveyed with an 18-foot customary perch or rod (common in Ireland) rather than the statute one of 16½ feet.[24] Benjamin Crane, who acted as surveyor for the proprietors of Dartmouth MA from 1710 to 1721, is believed on the basis of the true, as opposed to the recorded lengths of his lines to have employed a rod of something like 17.2 feet. Since Crane had previously worked in Taunton, where he used the "Taunton Town measure," and since Dartmouth at the time had no standards of weights and measure of its own, it has been suggested that Crane's Dartmouth rod was based on Taunton's. (True to form, Crane's fieldbook also reveals a penchant for doggerel.)[25] Mensuration is a valuable clue, but it applies only to objects amenable to precise measurement, which historically has not been the case with land. ("We're measurin' a fence line, not buildin' a violin," ran an old adage.)[26] Uncertainties even clouded statutory town-lines, as in New Ipswich NH where the town historian in 1852 lamented: "in no two perambulations have the length or direction of the boundary lines been reported the same."[27] In broad terms, errors of 1' in 100' were usual in colonial times, 1' in 500' in 1870, 1' in 1000'–5000' in the early twentieth century, and 1' in 70,000' today (1993).[28]

When all else fails, puzzled surveyors running the lines of an old deed are advised to select one positively identifiable leg of the survey, remeasure angle and distance with modern instruments, and by comparison with the original title description, devise a conversion factor to be applied throughout.[29] While only a professional possesses the facts and skill to evaluate the cadastre, on rare occasions the sightseeker can attempt to field-check the work of the colonial surveyor. One is to measure the width of the right of way (typically two rods or 33 feet) of an old wall-bounded "public highway" long lost in woods. Another is to examine walled fields or lots laid out regularly with predictable variations on the standard acre of 2 by 5 chains. It can be fairly eerie when these measurements work out evenly in rods: as though one had ritually reenacted the surveyor's work of centuries ago. If nothing else, the exercise will prove that even small land measurements accurately made with a modern tape are not child's play. When extrapolated to a heroic scale, and combined with the difficulties of compass-work, one can perhaps understand the awe expressed for one early surveyor, Captain Jonathan Danforth, in this rhymed tribute: "He rode the circuit, chained great towns and farms / To good behavior; and by well-marked stations / He fixed their bounds for many generations."[30]

North by Northwest

The toolmark with the richest potential for landscape study is the magnetic variation of the compass. The conventional compass does not point true north, but magnetic north, which wanders slowly east and west with changes in the earth's magnetic field (called secular variation). True north could not be determined in the field with unfailing accuracy until the introduction of the solar compass in 1835. Before that, compasses had to be corrected by star observations of Polaris at elongation.[31] Henry David Thoreau who, in common with most New England country surveyors, worked with a magnetic, not a solar compass, described the process in his fieldnotes. "True Meridian. Found the direction of the Pole Star at its greatest Western Elongation (1° 58½')? at 9h., 26m. P.M. Feb 7th 1851. It coincides with a line drawn from the s.e. corner of the stone post on the E side of our western small front gate, to the side of the first door on the W side of the depot."[32] Variation troubled Thoreau; to ensure accuracy he compared his compass corrections with those of fellow surveyors. Thoreau was a Harvard man, and one wonders whether all his contemporaries were as fastidious as he. (His Boston-made C. G. King surveyor's compass is on view, tucked in an alcove of the Concord Free Public Library.)

The magnetic declination (printed near the north arrow on topographic maps) indicates the number of degrees magnetic north deviates from true north in a given place and year. Currently, the magnetic declination in the Boston area is 16° W, meaning the compass needle points that many degrees west of true north. This declination varies slightly each year, increasing and then decreasing in centuries-long cycles. No full cycle has yet been recorded for Boston, although extrapolation from scant colonial data suggests the one begun in the early 1600s is due to complete itself. (One source prematurely declared this reversal was already apparent in 1962.)[33] Magnetic declination decreased from about 16° W in 1600 to about 6° W in 1780, and has slowly and reciprocally increased ever since. In theory, a north-south boundary line surveyed by the needle in the mid-seventeenth century would deviate about 14° west of north, while one surveyed in the mid-eighteenth century would bear only 7° west of north. Thus, had colonial New Englanders evinced a scrupulous regard for compass north, the secular swing of the needle would have impressed on the landscape a sort of archeomagnetic grain for dating the cadastre.

The primacy of north even on maps is of no great antiquity, and the true north fetish is a product of the late-eighteenth-century scientific revolution. New Englanders were particularly resistant to a foolish consistency in this regard. The New England delegation to the Continental Congress made repeated efforts to repeal the True Meridian clause of the Northwest Ordinance, believing that magnetic north was sufficient, and further, that terrain and local conditions should be considered in laying out township lines. This attitude was embodied in a Massachusetts man, Rufus Putnam of Sutton, who was appointed surveyor general of the United States (1796–1803) at a time when one's abilities as an Indian fighter weighed as heavily as scientific acumen in filling that post. Putnam was a practical man, an able leader, but weak in mathematics and uncomfortable with the profound geodetic problems inherent in so vast an undertaking. His tenure was marked by a very lax regard for the True Meridian clause.[34]

This late eighteenth-century New England indifference to true north is symptomatic of a general indifference in colonial times to the primacy of almost any kind of north at all, other than, perhaps, what surveyors vaguely call "assumed North." Strictly, any northerly line in eastern Massachusetts that does not fall somewhere between 6° and 16° west of north cannot be readily explained by magnetic declination at all, even though it is frequently done. Any effort at archeomagnetic dating is thus predicated on a scrupulousness for magnetic north that may have been generally lacking. Furthermore, the slow and short swing of the needle in New England means that the entire range of the secular variation operates within limits no greater than what might in prudence be ascribed to instrumental or human error. By contrast, London's variation (1580–1812) was thrice what Boston's is believed to have been over roughly the same time.[35]

In the late nineteenth century magnetic declination was historically better documented in the United States than anywhere else on earth, due to the efforts of L. A. Bauer of the Coast and Geodetic Survey, whose work was motivated by the need to interpret the true value of bearings given in colonial deeds and charters. Enthusiasm for this sort of research would reach its peak in his lifetime, although even Bauer pulled in his horns a little between the first edition of his famous tables in 1902 and the second in 1905. (Mathematically modeled tables of United States Historical Declination are now available on-line from NOAA.) Bauer himself vaunted the utility of such data in his 1902 treatise. Coast and Geodetic surveyors correctly inferred that Chestertown MD (1702) had been settled in the early eighteenth century on a northwest-southwest grid from the alignment of its High Street. Conversely, from the street plan of Oxford MD they surmised the magnetic declination prevailing when the town was laid out.[36] Breed and Hosmer (authors of the surveyor's bible) gave Bauer's methods some prominence in their 1908 edition, but by the 1950s surveyors were warned against a blind reliance on them.[37]

A classic archeomagnetic anecdote, so neat as to arouse skepticism, concerns the Josiah Coffin house on Nantucket. Like most early saltboxes, it was south-fronted, and by measuring the exact bearings of the flanks of the house Professor Henry Mitchell determined they were aligned on a magnetic north-south line in 1723. Knocking on the door, the owner informed him that the house had been finished a year later, in 1724.[38] I hardly doubt that a maritime people might be fastidious devotees of the compass; what does puzzle me is where Professor Mitchell obtained such precise data, as colonial records are far too unreliable for such aplomb. A still more dubious anecdote concerns the Eleven O'Clock Roads in Fairfield CT and its daughter-town of Westport, eleven parallel range roads laid out N 28° W at mile intervals in 1680 when the Long Lots were divided. Their orientation is most readily explained as lying parallel to the old Stratford-

Fairfield boundary: much of this corner of Connecticut has a northwest bias. The more colorful explanation—which defies astronomy—is that the lotlayers calculated that "in this longitude at the time of the year in question the sun's shadow at eleven o'clock in the morning would lie true north."[39] The fact (known to sundial-makers) is that the sun's shadow always and everywhere lies true north at noon local time. The truth in the name may be far plainer. It happens that o'clock to indicate compass bearings has been attested to as early as 1797, and that eleven o'clock is N 30° W. Close enough for government work! (One cannot help adding that the "eleven o'clock" and "four o'clock" were customary hard-cider breaks at barn-raisings and on highway work.)[40]

Vitruvian Hypothesis

The most obvious archeomagnetic artifact would be a street or crossroads laid out by an uncorrected needle to the cardinal points, possible instances of which in New England will be discussed later. However, this idea of a cardinal crossroads presumes that the only proper axial orientation is north-south and east-west in calculated urban design. Particularly troublesome in this respect are alignments that run east of north; in New England magnetic declination has always been westward in the historical period. Two very early townplans single themselves out by their unusual degree of urbanism combined with an apparent indifference to cardinal orientation: those of New Haven CT and Cambridge MA. The plan of New Haven is a tic-tac-toe grid of nine squares believed to have been laid out in 1638 by John Brockett.[41] The orientation here, taking Church Street as the baseline, is N 28° 53' E. Anthony Garvan made an ingenious, but mathematically flawed study of this layout and concluded that not only did the grid derive from a revived interest in the Roman encampment among seventeenth-century military engineers, but also that the street alignment accorded with the recommendations of Vitruvius in the first of his *Ten Books of Architecture*. In the Vitruvian scheme of things, the eight winds proceeded from the cardinal and the intercardinal points, that is from the north, northeast, east, southeast, and so forth. To shelter the city from these winds, he proposed that the axis of the street-grid should be rotated to run to one of the intermediate points, namely north northeast, east northeast, east southeast, and so on, boxing the compass at 45° intervals. As it happens, the actual axis of Church Street lies roughly 6° *east* of north northeast, a discrepancy Garvan sought to explain by invoking an improbably low magnetic declination of 6°–8° *west* of North. Garvan seems to have assumed that these values somehow "cancel" each other out, when in fact they double the discrepancy. Church Street simply does not plausibly lie on a north northeast axis, whether true or magnetic. (Detection of this error is hampered by the fact that the bearings of Church and Chapel Streets have been transposed and corrupted in the text.)

Such a sad end to a lovely theory! However, it seems that the Vitruvian hypothesis, while dead in New Haven, may be alive and well in Cambridge. Old Cambridge was laid out in 1635 on a gridiron aligned true north northeast, within a degree or so of 22½° E as judged from Dunster and Holyoke Streets. Conceivably this was the work of Thomas Graves, an engineer who had previously laid out fortifications and streets (reputedly also on a gridiron) at Charlestown. The Cambridge grid included a market square and gives every evidence of being a highly deliberate exercise in classical urbanism. Its one flaw in modern eyes, that the streets were not cardinally oriented, is thus perhaps no flaw at all. The orientation may be Vitruvian. Such a classically inspired town plan for Cambridge (which was initially projected not simply as the seat of Harvard College but also as the new colonial capital) seems entirely appropriate. John Reps (who otherwise discounts the Vitruvian theory) argues that the layout of New Haven was actually planned in Massachusetts, so the Cambridge grid may have been a directly influential, if less elaborate, model for it.[42] The idea that such an artificial scheme should briefly flower in the first decades of New England settlement is not so absurd as it first may sound. In a vacuum of local, practical experience of town-planning, it would not be unreasonable for learned men to turn to the best European thinking then current on the subject, however abstruse. It is equally obvious that they did not toy with Vitruvius for long.

Through A Glass Darkly

The lack of commonly available cadastral mapping makes the study of the New England boundary network frustratingly difficult. Furthermore, while the passage of time has only enhanced the visibility of the federal grid, a century and a half of agricultural abandonment in New England has obliterated the farmfield-woodlot patchwork that is the strongest cadastral expression on the ground. Contrast the blank, pond-glinted woodlands that greet the air passenger on the final descent from Providence to Boston with the mosaically demarcated metes-and-bounds farmscapes on the approaches to Charlotte or Atlanta. The cadastre is that aspect of landscape history where the sightseeker—who holds visible evidence at a premium—feels most furiously blind. Consequently we must rely on whatever documentation falls serendipitously to hand: town plats, aerial photographs, fire insurance maps, acquisition plans for parks, estates, and public works. Black-and-white air photos often clearly reveal woodland tract boundaries through differences in tree-cover that are lost in the uniform green of topographic maps.

Many town histories contain sketchmaps of the original lots; indeed, as early as 1836 the town historian of Rutland MA thought fit to include a rude plat with his history.[43] A higher standard was achieved in the latter nineteenth century, particularly in Boston, where reestablishing, at least on paper, ancient boundary lines became the genteel avocation of conveyancing lawyer-antiquarians such as Nathaniel I. Bowditch and Uriel H. Crocker. Bowditch, of illustrious family, quarried his fifty-five ledgers of title abstracts on Beacon Hill for chatty articles in the *Evening Transcript*, but as nineteenth-century urbanization made square feet more valuable than acres in colonial times, such concern for the jots and tittles of old deeds must have truly paid. George Lamb, a retired sea captain, mapped the 1643 *Book of Possessions* in a series of plans of Old Boston purchased in 1879 by the city and later published, presumably for more than sentimental reasons.[44] Reconstructed maps have also been published for Waltham, Watertown, Winchester, Hadley, Rowley, and Sudbury MA, to name but a few.

The map for Cornwall CT, completed in 1896, was the work of town clerk–surveyor G. C. Harrison, who also served as a probate judge—eminent qualifications, indeed. This forty-year labor of love has been likened to "creating an entire 600-piece jigsaw puzzle from a written description of each piece." Add that this jigsaw comprised some thirty thousand acres allotted in eleven divisions between 1738 and 1828, and that its pieces changed hands many times, and the dimensions of the task will be appreciated. Nearly a century would pass before this work was truly completed, when through the efforts of a selectman-surveyor and then a local historian, the map was finally cross-indexed with the land records on which it was based.[45] This goal was no mere antiquarian will-o'-the wisp. Charles Lawson, a professional Maine surveyor, employs this same jigsaw metaphor, and elaborates on the practicalities of the problem. A westward exodus of New England surveyors to take up public lands and railroad work after about 1835 meant that deeds back home were increasingly referenced to earlier deeds, and parcels described by abutters only, the dreaded "by land of Jones; thence by land of Smith." Acreages were given as "more or less," on the seller's say-so. Thus was created a jigsaw with pieces of no fixed shape and no fixed size. The answer to this puzzle has come only since World War II when aerial photos used as tax maps could be marked by assessors with known parcels, then amplified by title searches and public review, until "eventually the town [knew] each and every parcel outline and the owner thereof."[46]

While many academic studies have relied on reconstructed colonial parcel maps, it is well to remember that map users are ever at the mercy of their compilers, and that good draftsmanship may not guarantee accuracy. The photocopier at the local library is the greatest ally of anyone eager to check a map's value firsthand in the field. To trace the history of any one parcel involves a paper chase of cross-references; to assemble the histories of a large number of contiguous parcels requires monumental patience and is the realm of the title searcher, and beyond our scope. For the sightseeker, the most accessible method of cadastral study combines Mylar-grid analysis of topographic maps with compass fieldwork, as will be described below.

Mylar Grid

The most rewarding technique for studying the ancient cadastre is mapwork with the Mylar grid: plastic "graph-paper" overlays obtainable at stationery stores. A tenth-inch grid works best (100 squares to the square inch): each square is approximately an acre on a standard USGS 7.5 minute topo sheet. Also essential are a protractor, the topo sheet of one's study area, and copies of as many supplementary maps—current and historical— as one can lay hands on: street maps, town engineer's, zoning board, and conservation commission maps all show different types of cadastral detail. Familiarize yourself with all linear map features that may exhibit cadastral control: park, playfield, and cemetery boundaries, woodland edges, elbows and straightaways on country roads, street grids, "jeep trails" (often colonial cart-paths or highways), even ditched brooks, which were not unknown as boundaries in colonial times. Do not overlook the occasional fence-line recorded on USGS maps by dashed red lines. The jogs and juts, inliers and outliers of large agglomerative tracts like state forests perpetuate old farm and woodlot boundaries and reward study. With one's attention thus focused, the map reduces itself to a mat of randomly oriented lines.

While these are but an infinitesimal fraction of all boundaries, it is hoped they constitute a random sample of the cadastre at large.

The Mylar grid simplifies the task of finding the parallel straws in this cadastral haystack. Lay the grid on the map and, aligning it with some selected feature, note any instances of parallelism or perpendicularity caught in its meshes. Maintaining orientation, slide it carefully over the map to get full coverage. Then rotate the grid slowly over the map to bring out any and all swarms of like-oriented lines, particularly ones that reappear consistently in scattered places as much as a mile or more apart. The compass-bearings of these lines can be determined by protractor to aid in cross-checking other maps, and ultimately in fieldwork. (Circular protractors are more finely calibrated and easier to use.) Recurrent alignment over several square miles is normally attributable to only two causes. Such alignments may be cardinal, that is, north-south (or east-west), to the degree that different surveyors operating in different decades or even centuries with different instruments and standards of accuracy would find the same north. In general, the four compass-points are the only arbitrary master directions that might be expected to assert themselves spontaneously over time in excess of simple random chance. (The chance that two randomly oriented lines would lie parallel or perpendicular within a tolerance of $2°$ is by my reckoning about 2%.)

The alternative explanation, and the only one for noncardinal alignments, is that they represent traces of the original division schemes by which the lands of the town were first allotted, generally within the earliest decades of settlement. These divisions were the only cadastral "masterplan" most New England towns ever knew, although this term must be understood hyperbolically, as one example will show. To the "great perplexity" of its surveyors, New Ipswich NH was laid out on five different schemes, such that "there is not one square or right-angled lot in all the town, all of them being diamond-shaped in different proportions, or wedge-shaped."[47]

Clearly, cadastral history cannot be read from map- and fieldwork alone, but if it is read only in the registry of deeds then our ability to interpret landscape evidence, and with it, the potential to directly enrich our sense of place, will never be fully realized. With a timepiece, you need never note how the sun moves across the sky, or the stars circle through the heavens. Without a timepiece, you become a more curious, a more cunning, and yes, a less punctual person. While it may be theoretically convenient to speak of Boundaries, Townplans, and Road-nets as separate landscape layers, they are in fact inextricably embedded in one another. Boundaries are most meaningfully examined within the historical contexts of the townships, villages, farms, and fields that they serve to etch out, and it is to these contexts that we now turn.

Township

Between 1620 and 1760 New England settlement underwent a steady evolution from the nuclear to the range township. In the nuclear township, a center village marked by a meetinghouse and clustered core of homelots was surrounded by scattered outfields and extensive common-lands. These outer common-lands were slowly allotted in as many as a dozen "after divisions,"[48] often with evermore generous acreages while the land lasted. Larger lots outside the village proper were coveted for farmsteads as enthusiasm for the official formula of village homelots and tilled outfields waned. The creation of farms, and not of villages, asserted itself as the end-all of the townplan. The range township began to take form about 1720 and was from the start one of dispersed single farmsteads. The totality of the town was regularly and rectilinearly divided, all at once, in advance of settlement, into ranges of farm-sized lots. A small-lotted center village might exist as a legal fiction to satisfy royal charter requirements. Both nuclear and range townplans typically included a center meetinghouse and also not infrequently a common, either set aside from the first, slowly developed over time, or inserted long afterward for reasons of prestige. New England townplans lie on a historical continuum and most exist only as hybrids between these ideal extremes. Needless to say, "Townplan" here is broadly understood, and encompasses slowly evolved arrangements of elements as much as conscious prior designs. This short overview will enable us to keep the larger historical picture ever in mind as we take up the various townplan elements in turn, in an order more often thematic than chronological.

Townsite Criteria

The earliest scheme of town-planning was to make grants of a specified size, as six miles square (Concord, 1635) or eight thousand acres (Deerfield, 1673), to existing towns or to groups of forty to sixty freemen, who would then select and survey the site and have its location confirmed by the legislature. (Such group migration was characteristic until the mid-eighteenth century and underpins the Kurathian Hypothesis.)[49] In 1645, Duxbury MA was granted "lands about Satucket" for a "New Plantation" (Bridgewater) "four miles each way" from a center of its choosing—by tradition, the Center Tree, although by other accounts it was an Indian fish-weir. Concord MA also had such a traditional center, Jethro's Tree, itself at a fish-weir on Mill Brook. Perhaps this was a pattern: Indian trail junctions and river fords made good settlement sites.[50] Open, fertile areas were much sought after, as the English with no experience of wilderness clearance since the Middle Ages still had everything to learn. Most desirable were locations near saltmarshes or fresh meadows, which yielded ready hay for cattle, and fire-cleared Indian "old-fields," often desolated by the epidemics of 1617–1619. (The settlers themselves in imitation soon "burned the woods" to clear pasture for their cattle.)

It must not be forgotten that areas of native occupance, linked by native trails, constituted the Algonquian infrastructure of seventeenth-century New England, one ever more deeply graven even as its visible vestiges were destroyed. Howard Russell has mapped the some sixty towns settled by 1650, and all possessed either hay-marsh or old-fields (and generally both). Every saltmarsh on the coasts of Massachusetts and Narragansett Bays, and along most if not all of the coast from Bridgeport CT to Portland ME saw settlement at this early date. The settlers of Dorchester MA (1630), while intending to locate up the Charles River stayed put instead on the coast because saltmarsh was plentiful and their cattle were starving. Lack of salt-hay at Plymouth (which had only old-fields) early precipitated an exodus to Duxbury and Marshfield, as well as Cape Cod and Narragansett Bay; similarly, the first inland migrations to the Concord and Connecticut valleys were lured by their abundance of fresh meadow. Hundreds upon hundreds of cattle were debarked in the 1630s, and were bred: a contemporary, perhaps too glowingly, put the colonial herds at four thousand head, progenitors of the hardy and once-familiar New England red cow.[51] (The cow on the Vermont state flag is customarily red.) Beyond these general desiderata of settlement location, both transplanted and locally innovated templates guided village site selection.

Nuclear Village

As ordained by the Massachusetts General Court in 1635, all houses were to be not above a half-mile mile from the meetinghouse, unless mill-houses, or farmhouses of those who otherwise had dwellings in town. This was intended in part as a measure of defense, in part to accord with the English village template, and in part as conducive to a certain type of ecclesiastical polity.[52] Homelots or homestalls of as much as four to ten acres were granted "by heads and estates"; that is, proportional to family size or capital ventured in the enterprise. These were laid out in serried ranks of narrow, deep lots. The one-acre (2 × 8 rod) "yard stick lots" at Wescoe on Nantucket (1678) were exceptionally small.[53] Such village homelots, because of their exiguous acreage, might admit but one additional house at densities deemed acceptable by even nineteenth-century standards. Many original blocks were so tiny that to open more streets was impractical if not absurd. The survival rate of these schemes was such that the 1872 county atlas map of Ipswich MA preserves the essential outline of the original seventeenth-century village as reconstructed by John W. Nourse "a skilful surveyor and an enthusiastic antiquarian student" of the town.[54] And it was not merely the lots that survived. In one case, the Lords and Kimballs on High Street, lived side by side on original grants for well over two centuries. The most notable change in this lapse of years was that many lots were divided lengthwise, and the numbers of same-named neighbors attests that this was frequently done to settle an estate. In the generalized English village, these lots were the "tofts," as opposed to the "crofts," or outlying fields.

Infields and Outfields

On the outskirts of the nuclear village, each proprietor was also granted plowland for tillage, meadow for haying, and pasture for grazing in the common fields. He might own his tillage lands outright, or have rotating strips allotted annually to his exclusive use. Amid the English cultural baggage brought over by the first settlers were two distinct traditions of agriculture: the ancient, communalistic, open-field farming of the Midlands, and a more modern one of enclosed fields from East Anglia, the very region whence perhaps one-third of the original New England emigration was drawn. Enclosed field agriculture was both individualistic and entrepreneurial in spirit, and better suited to the expanding New England frontier. It throve in East Anglia, where it took its impetus from the rise in cash sales of crops to adjacent urban markets such as London. (Open-field agriculture was geared to barter and communal self-sufficiency.) Each farmer owned his land (although not necessarily in fee simple), sought to own more, and to consolidate his holdings: to live, in short, in a farmhouse on his own farm.[55] While in New England enclosed-field farming in the long run prevailed, in the short run open-field farming and the village-centered life it fostered held its ultimate tendencies in check and left both a cadastral and often mythical social legacy.

Even in England, open-field farming was a medieval holdover whose days were numbered. In this scheme, villagers worked large outlying common fields ruled off in long, narrow, parallel strips or "lands," tiers of which were called furlongs. The furlongs did not lie all the same way, but in places were butted up at right angles to each other, parquet fashion, to better suit drainage or terrain. Plowland, pasture, and meadow were so divided. Scattered strips were annually assigned each villager according to immemorial custom, and where and what each villager sowed and reaped, as well as the rotation of crop and fallow, was communally decided. Long strips required fewer turnings of the plow and represented a day's work; scattered long strips assured a cross section of slopes and soils that shared out the best and worst land equitably. The costly and cumbersome medieval plow enforced this cheek-by-jowl arrangement, as no villager could own or work one alone. (A similar necessity constrained the first New England settlers, as there was a clear shortage of the heavier implements: for example, only thirty-seven plows in Massachusetts Bay in 1636.)[56]

In some of the earliest-settled New England towns, particularly where the town fathers had emigrated from the English Midlands, records reveal how open-field agriculture was briefly tried and quickly unraveled. No sooner had it begun than it ceased at Plymouth in 1623. While Cambridge, Dorchester, Charlestown, Rowley, and Sudbury were founded as open-field villages, Watertown, Ipswich, and possibly Roxbury MA quickly made clear a desire for enclosed fields. In Dedham MA, open-field farming "began disintegrating almost from the day of its inception." In the 1640s Dedham men were allowed to enclose their fields, and by the 1670s were briskly engaged in enlarging and consolidating their outlots, yet the village center still held and few if any had relocated house and barn outside it.[57] While no definitive study of New England open-field agriculture is available, it seems unlikely that any town settled much after the mid-seventeenth century ever saw it. Yet, communality for other purposes, as in the control of wood- and pasturelands, lasted well into the eighteenth century, and the common herding of cattle persisted, it is said, even into the twentieth century.[58]

Nantucket long maintained its communal customs, but at a price. In 1791, a 675-acre general field (allowing 25 acres for each of the original 27 proprietorial shares) was still fenced and planted to corn alternating with rye and oats. Those with privileges in the field improved their own portion. Although the location of the field was frequently changed, the soils were not replenished, and yields declined such that sometime in the years before 1835 the practice stopped. Even in 1791, the yields on private farms had exceeded those on the common fields, as communal agriculture gave little encouragement to improved husbandry. In 1807 two herdsmen still drove Nantucket's cows one to four miles each day to pasture and back at night-fall. Sheep were likewise kept and fed in common and, as with the general cornfield, communality led to neglect. No shelters were provided and many sheep died in snows on the treeless commons. In Deerfield MA, the pro-prietors of the common fields would meet until the 1850s.[59] Despite the collapse, generally sooner, sometimes later, of communal agriculture, New England rural economies survived on a dutifully recorded and balanced scheme of exchanging work. So while not medievally communal, farm life was highly, if individualistically cooperative. While the open-field system itself was short-lived, the land allotment methods it inspired left their lasting imprint on the colonial cadastre.

In *Puritan Village*, Sumner Powell gives us the classic study of residual open-field agriculture and its formative effect on the plan and cadastre of a New England town. His reconstructed map of Sudbury village in 1650 shows lot-lines "located according to the reasonable estimates of anti-quarians, the descendants of the original grantees . . . , the former Town Clerk . . . and the author."[60] Powell inexplicably omits to note that old Sudbury village now lies in Wayland, slightly northwest of the current center—a source of bafflement to readers who wish to orient themselves in the modern landscape. If we compare this map with one of today, meadow roads have disappeared, new roads have been opened, the bridge relocated and another built. The gravel pit is now the Wayland dump; the Sudbury River may have shifted. Matters are thoroughly confused by the fact that the handsome meetinghouse in Wayland Center today stands almost a mile southwest of the original Sudbury meetinghouse lot, indi-cated by a crossed circle on Powell's map.

This lot is now a burying ground and the footprint of the building itself persists in the shape of a low, level, forty-foot-square earthen platform devoid of graves. The meetinghouse stood midway in a line of twenty-two houselots that faced the extensive Cow Common across the road: much of this river meadow is now conservation land. The narrow, deep character of these houselots still survives, and despite their great age, they strike one as altogether ordinary, because the template for a village lot remained unchanged until at least the 1830s. Of the some 150 parcels mapped—whether house, meadow, or upland lots—most were narrow, decently rectangular, and laid out in serried ranges roughly perpendicular to the roads. Powell specifically located the holdings of one John Goodnow, whose one hundred acres included eleven outlying parcels that averaged eight acres in size; in all Goodnow would need to travel eleven miles to work his scattered fields.[61]

While no single axis dominates the cadastre, under the Mylar grid, only one orientation meets two reasonable tests for magnetic north: namely, that noncontiguous areas should give evidence of the same alignment (north is after all a portable concept), and that perpendicular east-west lines also be present. Three, squarish, isolated lots unconnected with the main ranges are particularly convincing in this regard. From these we can infer a magnetic declination of approximately N 12° W, about what we would expect for 1650 if we work back from Bauer's tables. One other alignment is noteworthy. If we tramp the Cow Commons today, compass in hand, it will be found that paths, lanes, wood-edges, and boundaries run southwest (N 120° W); while we do not know how the commons were specifically allotted, this orientation accords with the 1650 lot-lines as mapped across the river.

Town life ran its troubled course. The settlers who in 1638 had hived off from enclosed field Watertown to form open-field Sudbury were again riven over land matters in 1657, and a group withdrew, establishing Marlborough in 1660. (A number of men participated in both removals.) To quell the general dissatisfaction, Sudbury in 1658 allotted about two dozen 130-acre farms, remote from the village, in the Two-mile Grant on its far western border. Powell's map and description of these lots is rather vague, but if we examine the appropriate USGS quads from as late as the 1950s, it will be noted that regularly and almost exactly every three thousand feet a road, be it paved, dirt, or mere track, intersects the ancient Marlborough-Sudbury line from the old Boston Post Road north into present-day Maynard. There are twelve: enough to demarcate a tier of twenty-four lots if rangeways lay between every two lots. This coincides so well with Powell's description that it is impossible that they do not reflect one and the same thing. Thus at either end of Sudbury-Wayland we find cadastral evidence for a rapid twenty-year shift from a settlement landscape of village and outfields to one of great lots and at least in potential, outfarms.

It might seem that only in rural environs could seventeenth-century field patterns endure, but comparison of the 1638 cadastre of Watertown MA[62] with a modern topographical map shows the layout of the houselots and fields of the original enclosed field village clearly underlies the patchwork of early to mid-twentieth-century house tracts. These fields ranged in shape from the squarish to the more typical oblong, and in size, from six to sixty acres, more or less. Lots ran perpendicular to the roads and to some degree this road-net displays what we shall later recognize as an English blockishness, quite unlike the triangularity that in time came to characterize much of New England. The oblong lots exhibit the medieval English parquet pattern alluded to earlier. This parquet, composed of a multitude of parallel and perpendicular lot-lines, was etched with the indelibility that private property gives, and only a reasonable proportion needed to survive to maintain the effect. When the area around Meeting House Hill was subdivided in the twentieth century, the old field-lines were hardened in bricks, mortar, and macadam. To a degree that seems superficial at first but that grows profounder on closer study, the general effect of the 1630s allotment is preserved in the piecemeal street-grid that shifts directions abruptly at right angles to itself. It takes no great leap of imagination to see the skeleton of the field patterns that lie beneath, and this insight transforms an initially grim street-grid into a wondrous colonial artifact.

Cambridge

Vestiges of the original open-field cadastre are etched in the street-grids of Cambridge. "The outlines of the colonial fields are echoed in the pattern of many suburban divisions of the nineteenth century and the long parallel streets in some cases seem to follow the strip field lots like Mt. Pleasant, Rice, Dover, Fayerweather, May, and Sherman Streets. A fair impression of the size and depth of a medieval long lot can be gained by a walk through Rindge Field and the Catholic Cemetery on Rindge Avenue. Both are partial remains of colonial planting divisions that escaped residential and industrial development in the last century."[63] Well-kept records make it possible to map and chronicle the allotment of the common lands in northwest Cambridge. Between 1630 and 1724 more than two dozen divisions, major and minor, were made within an area of less than two thousand acres. The larger of these divisions bore names like West End Field (1634) and Great Swamp Lots (1659–1662) and were laid out in long lots as small as four acres. As early as 1638 lot-lines were being laid that ran within two or three degrees of true north (the Catholic cemetery is such a one), and altogether perhaps a quarter of the area was surveyed on cardinal lines. Despite this early penchant for northing, the dominant effect is still that of the medieval "parquet."

Long-settled northwest Cambridge was well favored to retain these vestiges. With deeply rooted small holdings and patchy growth around taverns, depots, and brickyards, it offered no tabula rasa for the ambitions of great land syndicates, as in Cambridgeport. Grid-streets could only be quirkily shoehorned into available parcels. Slow development through one of the most volatile eras in American architecture make this history all the more readable on the ground. Still, there is nothing strikingly singular in its street pattern when compared with those of other parts of inner suburban Boston, such as Roxbury. Contrasts, however, are patent with the simpler street patterns of inland cities such as Pittsfield MA (1751) and Burlington VT (1773), settled long after the demise of postmedieval lot-laying. These contrasts are not so much ascribable to the age of the streets themselves, which are largely nineteenth- and early twentieth-century developments, but to a qualitative difference in the colonial cadastre that preconditioned them. The crooked streets of Boston Proper are very old, but in its nineteenth-century inner suburbs, fairly modern crooked streets often result from an ancient quirky cadastre.

Divisions

While the towns about Boston may have been allotted in a patchwork of randomly aligned tracts, the layouts of these tracts grew simpler and the lots inexorably larger over the first century of settlement. If the "small lots" nearest the old village centers were inextricably enmeshed in quirky, piece-meal patterns of lot-laying, the "great lots" in the outlands before the close of the seventeenth century showed signs of being measured off methodically like endless bolts of cloth. The town proprietaries, the increasingly elite joint-stock companies that controlled the outlying "common" lands, were liquidating their stock. This pattern was echoed in the seventeenth-century hinterlands as well. In North Berwick ME, surveyor Charles Lawson has illustrated how the first settled, lower end of town is a "mishmash of lines . . . whereas the outer fringes . . . show a more orderly plan": a layout he regards as typical of many towns in the Northeast. (North Berwick occupies the old Kittery Common of 1652.)[64] The methods and techniques employed clearly anticipate the range township, although rarely employed on a comparable scale, or with the same, single-minded, "one size fits all" philosophy. Such egalitarianism would be particularly unacceptable in the literal backyard of old-established towns with their strict social hierarchy. The rectilinearity of this new geometry makes it increasingly detectable under the Mylar grid. Below we shall explore four division schemes effected between 1636 and 1713 and their cadastral legacy as it is still traceable today. These explorations are speculative, and are meant to be read without maps. The intent is to convey typical histories and cadastral concepts that should presumably apply elsewhere within the 1700 settlement line.

Great Dividends (1636)

In 1636 Watertown MA made a general division of much of its remaining western territory in four "squadrons" known collectively as the Great Dividends, a tract that slightly exceeds the present limits of Waltham. The bounds were marked by John Sherman.[65] In contrast to the earliest Watertown allotments, the Great Dividends were ruled off more methodically in four side-by-side, ladderlike squadrons 160 rods or one-half mile wide. The uneven headline (top rung) of these four squadrons reveals itself in the staggered Waltham-Lincoln line today. The four squadrons were oriented northwest N 67° W (parallel to the present Waltham-Lexington line), and while straight in theory, were bent and offset due to terrain (meadowlands were excluded from the grant). Each squadron was ruled off into transverse lots of variable acreage. To use our metaphor, while the distance between the legs of each "ladder" was fixed at about three thousand feet, the rungs were variably spaced, here wide, there narrow, and ranged anywhere from about three hundred to one thousand feet apart. This made lots in round numbers of twenty to one hundred acres, granted in proportion to the number of cattle a townsman owned. (The lots were individually owned but communally pastured.)[66] Expressed as football fields, such lots were ten fields long and three to ten fields wide: in 1636 these were great dividends indeed!

Thus, a vast tract of some ten square miles was subdivided, on paper at least, by some fifty running miles of boundary lines that, despite irregularities, were aligned on a master axis of N 67° W. Doubtless, many lots were soon traded, sold, consolidated, or went unclaimed; many lot-lines were never run. Still, the dead hand of the law assures that some identifiable components of this scheme will last until the end of time. Indis-

pensable to the sightseeker is Edmund L. Sanderson's carefully reconstructed map of the Great Dividends and adjacent grants, which can be
readily correlated to the modern landscape.[67]

Much of the area has been kept open by market gardens, greenhouses,
rural hospitals, estates, country clubs, scout camps, conservation tracts,
and swamp, but fragmentation is imminent. More than once in search of
ancient surveyors I fell into the footsteps of their latter-day descendants,
whose stakes and orange markers chilled the heart like bloodspots on a
timeless scene. Maps reveal considerable parallelism under the Mylar grid.
Town-lines, woodland edges, thickly built tracts and suburban subdivisions (as at Warrendale), the trend of old country roads (some conceivably
ancient rangeways), and recorded property and park boundaries (notably
Prospect Hill) are found to lie on the square. Otherwise inexplicable short
straights or right-angled bends in country roads (what I call "coaxial
kinks") snap into alignment with original division lines. Four main roads
flirt with the four squadron lines at their eastern ends, sections of which
were prescribed in 1669 as driftways along which the three (later four)
town herds were driven to pasture, their grazing ranges defined to a degree
by the squadrons themselves.[68]

Charlestown Wood Lots

In March 1658, Charlestown made a division of the common woodlands
lying at the north end of its grant, in an area now partly encompassed by
the Middlesex Fells Reservation: "The land was drawn by lot and set off
to the several inhabitants in proportion as they were rated, one half of the
share of each lying in the first division and one half in the second division,
probably for the purposes of equalizing, so far as possible, the value of the
land." The Wood Lots were divided into fifteen east-west ranges, a
quarter-mile wide, with the southerly seven and one-half ranges consti-
tuting the First Division and the northerly the Second Division. Their
orientation (N 73° W) and location can be fixed by reference to the
Medford-Stoneham line, which was statutorily defined as "the North side
of the Fifth Range of the First Division of the Charlestown Wood Lots."[69]
Under the Mylar grid, numerous street-grids in Winchester Highlands
and northwest Medford, segments of main roads, and amusingly enough,
Interstate 95, all show alignments in some way determined by the Wood
Lots of 1658. Playground, hospital, and town watershed boundaries, as well
as town-lines, all fall into register. While not all neighborhoods within the

Wood Lots show such street pattern control, those outside their limits are clearly oriented on other axes. The wood lots, then, are still deeply etched in the urban fabric after almost 350 years, but this is by no means their sole legacy.

The creation of the Middlesex Fells Reservation in 1894 has assured that a good two-thirds of the area is once again rocky woodland. The Cambridge Sports Union (CSU) map of Pine Hill records the ghostly outlines of seven or eight squarish, roughly forty-acre stonewalled wood-lots in the southwest corner of the Fells. As William Stevens noted in 1891, "nearly all the long stretches of wall running easterly and westerly mark these ancient rangelines."[70] The lower boundary of these two par-tially intact ranges lies two and one-quarter miles south of the Medford-Stoneham line, making it the lower limit of the First Division of the Charlestown Wood Lots. This is corroborated by the abrupt shift in the cadastre south of it, where long, neat lots, probably pasture lots, oriented N 10° W prevail (magnetic north ca. 1700?). The Pine Hill map is an orienteering map: one sophisticated enough to be based on photogram-metry, yet naive enough to use magnetic north. Exactly how this ruinous stonewall network was found and plotted is unclear, but its recovery is an impressive feat to anyone who has beat the underbrush in search of rem-nants. So many layers of distortion obscure the surveyor's intent that whether the lots were meant to be chained off evenly in acres cannot be told. In any event, they were sized according to the wealth of the grantee and not all equal. Here they vary between thirty and fifty or so acres. Red Cross Path and Quarry Road can be seen to transect the ranges perpen-dicularly and suggest lanes run between woodlots. While fire lanes bull-dozed in the 1930s by the Civilian Conservation Corps complicate matters, segments of many roads and paths can be plausibly interpreted as colonial relics controlled by the original lot-lines. One can almost see the Fells road-net as a set of wayward vines clambering up the crude trellis of the Wood Lots of 1658. This *trellised vine* pattern is a classic cadastral artifact and one we shall re-encounter many times.

Cambridge Squadrants of 1683

Lexington grew slowly as a seventeenth-century outparish of Cambridge known as Cambridge Farms, and was not set off as a town until 1713. While no town plat as such exists, town histories, entries in the Cambridge Town and Proprietors Records, as well as map- and fieldwork, furnish insights into its ancient cadastre. To understand it, one must review some key events in the history of the Cambridge grant. Very early, in 1636, the General Court enlarged the territory of Cambridge northwestward to the Eight-Mile Line, a boundary just beyond present-day Lexington Green which lay the stipulated mileage from the Cambridge meetinghouse. (Modern maps give this distance as 8.01 miles!) In terms of today, this boundary ran from the Lincoln line on the west to the Burlington line on the east and coincided with segments of Worthen Avenue (once Bound Brook) and Adams Street. The line ran N 40° E, and this northeast bias to the ancient cadastre can be detected in many quarters of the town. Soon after, in 1641, the Shawshine country was also annexed, which embraced all of the territory between the Eight-Mile Line and the original Concord boundary, a line that can be extrapolated today from the trend of the Lexington-Bedford border today.

In 1683 Ensign David Fiske laid out the land beyond the Eight-Mile Line into nine parallel ranges or squadrants, each one-quarter mile wide, with the lots numbered alternately east to west in the first range, then west to east in the second, and so forth. (This customary manner of alternation was later incorporated in the federal grid.) The quarter-mile (eighty-rod) interval, seen also in the Charlestown Wood Lots, was favored because it facilitated acreage measurement. One acre equals 40 rods square, and as most of the 153 lots laid down were in multiples of five acres, the inference is that the typical 5, 10, and 15 acre lots measured 80 × 10, 80 × 20, and 80 × 30 rods respectively. Two-rod rangeways were authorized where needed between the squadrants, with double damages in land awarded to those whose lots were encroached. Acreages in the proprietors records, when tallied, account for only half of the true area of the squadrants, and cannot be relied upon to reconstruct the scheme. The quarter-mile interval between squadrant lines, however, was more accurately respected. Unfortunately, while records indicate Fiske drew a plat of the squadrants, this has not survived. More puzzlingly, Thomas Sileo, who has documented the title histories of the Lexington conservation lands, nowhere mentions the grants or squadrant lines, yet some of the most compelling landscape evidence of their existence lies in these tracts.[71]

While much about the squadrants remains murky, two essentials are clear: their geometric scheme and the location of the Eight-Mile Line to which that scheme was referenced. As we know the number, width, and trend of the Squadrants of 1683, it is possible to plot their probable location and then investigate for traces on the ground. Among the maps consulted were the standard USGS quadrangles, conservation maps, and the local bus map (derived from the town engineer's map) which shows boundaries of all town-owned parcels, although it regrettably gives no scale. The Arrow street map was overlaid with tracing paper ruled in quarter-mile grid squares, and alignments noted. Our initial point of reference is the half-mile stretch of Adams Street that coincides with the Eight-Mile Line before being deflected by a hill. The first squadrant line should lie parallel to this, one-quarter mile to the north, and here we find a cart road, a straight stretch of walled cow-lane in Willard's Woods, which corresponds to the rangeway between the first and second squadrants. This surmise is borne out by Revere Street (historically known as The Rangeway), and other snippets of streets that align with it intermittently to the west. Such intermittent range roads are common cadastral relics.

A quarter-mile north of the cart road we come to the second squadrant line marked by a stretch of stonewall between an old orchard and a wooded swamp. The third squadrant line is attested to by the abrupt elbow of North and Burlington Streets, and the third, as well as the fourth, can also be teased out of lot-lines in the Meagherville conservation area (here the Arrow map erroneously shows and names innumerable "paper" streets, the legacy of a failed 1891 development scheme). Route 128 in the fifth squadrant literally bends into alignment with the Squadrants of 1683, much as Interstate 93 fell under the spell of the Wood Lots of 1658. The cartographic evidence for the squadrants thins here. The sixth and eighth may be streets; the seventh is probably a pipeline-powerline corridor described by Sileo as a "rangeway"[72] where it runs through conservation land. This remote section of town was known as World's End in the 1690s: an apt name for the outermost squadrants. The ninth and final or "headline" is plausibly marked by a monumented corner of the zigzag Lexington-Bedford town-line, that lies the requisite two and one-quarter miles from Adams Street.

This is all highly conjectural, but a wealth of corroborative coaxial lines could also be cited: many street-grids, tract boundaries, a ditched swamp behind Diamond School, the remarkable right-angled bends of Farley Brook, mapped stonewalls on conservation lands, and so forth. Thus the overall cadastral legacy is clear, even if its details are somewhat shaky. Portions of the Squadrants of 1683 also extended into the northwest corner of Lincoln, where most happily, a long swath, traversing the third to the ninth squadrants, falls within Minute Man National Historical Park. This twist of fate has not only preserved its rural landscape, but has also ensured a high degree of scholarly attention to the cadastre of 1775. Of inestimable value to the sightseeker are three maps: the USGS Concord sheet, the Park Service brochure map, and the historic landscape map published in David Fischer's *Paul Revere's Ride*.[73] From these, it can be seen that Mill Street, Bedford Road, and the M-shaped vagaries of the Old Concord Road at Fiske Hill, Parker's Revenge, and the Bloody Curve are all roughly coaxial with the squadrants, as well as a mile of mapped field walls, a number of fence-lines, some wood-edges, and possibly the misaligned zig-zags of Hobbes Brook.

Thus far we have dwelt on the east-west rangeways, laid down by law at quarter-mile intervals. But what of north-south roads or cross-rangeways, for which there seems to have been no specific provision? Among these would number immemorial through roads such as the Concord Road that predate the squadrants, as well as later ones that arose as destination roads to towns upcountry. The landscape sporadically testifies to intervals in multiples of a quarter-mile (often a half-mile) between cross-rangeways or country highways as the desired practice of colonial surveyors. While these roads adhere to no absolute quarter-mile grid as hypothetically mapped, at key points, particularly where they cross squad-rant lines, they still respect quarter-mile intervals relative to each other. Overall, they exemplify the *trellised vine* pattern noted earlier, either by wandering subparallel to gridlines, or by intersecting other roads at or near gridlines or grid corners.

It would seem that while a true grid scheme was too wasteful to execute in its entirety, that a half-mile or quarter-mile grid existed as a template to be invoked where suitable: that is, when a road was needed, and no countervailing "organic" necessity intervened, it was run geometrically, in accordance with the template. Generations of highway surveyors carried such a template in their heads, and the pattern we have noted could have sprung up in 1683 or anytime thereafter. What resulted was a mushy-meshed road-net that, when analyzed geometrically, sporadically yet re-peatedly obeys certain rules of a true grid. It was only in the latter eigh-teenth century that this ideal grid, refined and defined by law, became an integral scheme to be rigidly adhered to, one in which the laying out of lots and roads were one and the same.

Dorchester New Grant

Dorchester in 1637 was enlarged by the grant of a vast tract southwards to the Old Colony Line, the northern limit of the Plymouth patent perpetuated in the Bristol-Plymouth county lines. Something of this ancient territorial sweep is captured in two antipodal placenames, Dorchester Heights to the north, and Dorchester Brook, eighteen miles to the south where it flows across the Old Colony Line and out of the New Grant altogether. However, the cadastral evidence is far stronger and more pervasive than this. Between 1630 and 1713 Old Dorchester underwent a number of divisions, in general laid out rectangularly, but often with no dominant axis. Dorchester proper, north of the Neponset River, displays no consistent orientation, and probably, like Old Cambridge, was allotted in a series of small, piecemeal schemes. However, by 1652, in what is now East Milton, a master axis of N 52° E began to assert itself, based on the orientation of the Old Dorchester–Braintree line, which ran between the summits of Great Blue and Woodcock hills. This axis set its stamp on Milton in the Sixth Division of 1660, best seen in the alignment of Canton Avenue, the four-rod Great Middle Highway which divided the two ranges of the division. Long narrow lots were laid off perpendicular to it, the upper range running to the Neponset, the lower to the Blue Hills; at least some of the side streets along Canton Avenue must have been ancient rangeways under this scheme. Thirty-five years later, between 1695 and 1698, John Butcher was employed to lay out the Twelve Divisions, which allotted much of present Canton, north Sharon, and West Stoughton. This was a crazy-quilt of squarish or compact rectilinear lots, only a small portion of them coaxial with the Old Braintree Line, that appear to be farmsteads deliberately tailored to terrain, watercourses, and the specific acreage to which the grantee was entitled. The lack of overall method in this layout hampers the recovery of traces in the modern landscape.

By contrast, in 1713, the remainder of the Dorchester patent south of the Twelve Divisions was run out by the surveyor James Blake on a strictly methodical basis to form twenty-five long, parallel ranges, about a half-mile wide and two to three miles long, extending south southeast to the Old Colony Line.[74] These Twenty-five Divisions, as they were called, were numbered from east to west and ran perpendicular (about N 145° E) to the Old Braintree Line. As they were authorized in the same year as the Dorchester Proprietors were legally divorced from the town, the likely intention was to wind up the proprietary by an expeditious disposal of all undivided lands. The territory encompassed all or part of four towns. Stoughton exhibits considerable coaxialism in the town center, in the numbered street-grid of South Stoughton, in the straightaways and alignment adjustments of old country roads, in the outlines of three old cemeteries, in the digitated golf-course boundaries, in woodland edges, and in powerline or pipeline swaths (which tend to follow property lines to minimize damages). In Sharon the coaxialism can be read in right-angled road junctions, and in Foxborough, in the alignments of three or four likely range roads. This cadastral bias dies out west of here, signaling the limit of the Dorchester patent and the end of the Twenty-fifth Division.

One tract of Old Dorchester in the Blue Hills Reservation seems at first too rugged and sparsely settled to have received a cadastral imprint. Yet on closer look, it furnishes intriguing indications of method within the madness of its road- and path-net. The most valuable map in this research proved to be the most recent, the reservation Mountain Bicycling Map of 1998, which not only excels its field-plotted predecessors in accuracy, but also sorts out much confusion by graphically distinguishing footpaths and carriage roads. (Ironically, it has required the Global Positioning System to vindicate the ancient surveyors.) Scattered yet compelling evidence suggests that much of this territory was ranged and lotted with lines at quarter-mile intervals, laid parallel and perpendicular to the Old Dorchester–Braintree Line.

While not extensive, stretches of stone wall visible now or as mapped in 1893 run largely perpendicular to this axis. Both the 1690 Ralph Houghton grant and the present-day Brookwood Farm (a likely colonial remnant) are similarly oriented.[75] Other important alignments snap in place when our schematic quarter-mile grid is superimposed on the actual cadastre. The southeasternmost corner of Milton (including Houghton's Pond) coincides precisely with lot-lines in this scheme and many jogs in the reservation boundaries closely shadow them, as if, through survey error, they had slipped slightly out of register. The terrain precludes any dead-straight "Great Middle Highway," but Hillside Street trellises a likely rangeline, and at key intersections, it all but touches the corners of our schematic quarter-mile grid. As a rule, even loose range road-nets dutifully visit such fixed lot corners. Furthermore, other telltale quarter-mile intervals are detectable between roads, paths, and boundaries. One footpath trellises a schematic lot-line for a mile or more, before wandering wide. Finally, and a little disconcertingly, North Skyline Trail closely follows the Old Dorchester–Braintree Line between Great Blue and Hemenway Hills: disconcerting because to all appearances this is a C.C.C.-era footpath. And indeed, geometric evidence points to many other footpaths as relic rangeways and cross-rangeways that long predate the establishment of the reservation in 1893.

The abundance of near misses between the schematic and the actual cadastre argues that this tract was originally ranged and lotted in an orderly if not entirely accurate fashion. (To miss the beat so closely one must first grasp the tune.) Yet there is another layer to our cadastral palimpsest. Superimposed on this faint northwest-southeast scheme is the stronger, true north-south one etched by the reservation carriage roads. While these may yield to the terrain, they were clearly laid down at precise quarter-mile intervals and are intersected midway by the six-mile east-west South Skyline Trail. Here one undoubtedly sees the handiwork of the Metropolitan District Commission overlaid on a far older pattern of lots and rangeways run coaxial to the Old Braintree Line. This recent—and clearly separate—overlay reinforces the conviction that what lies beneath is both significantly older and of different origin.

Range Township

The logical outcome of all these colonial experiments in land allotment was the range township, an ideal six-mile-square, prior-platted township rectilinearly divided into large, equal-sized farmlots served by a grid of range roads, with (or more likely without) any true village center. Many elements of this template were very old, but their collective, single-minded application was new, and would revolutionize New England town-platting and inform the federal grid township under the Ordinance of 1785. As we have just seen, by the 1720s, outlying ends of vast seventeenth-century mother-towns were being set off and their cadastres marked with newer, ampler, and more methodical standards of lotlaying.

New Hampshire in particular became a proving ground for the development of the Range Township, a thirty-year trend that James Garvin traces from the 1719 division of the Two-Mile Streak (Barrington NH) to its ultimate epitome in New Hampshire–chartered Bennington VT in 1750. Bennington was laid out as a six-mile square with east-west and north-south sides and subdivided like a chessboard into sixty-four 360-acre lots— true, generous farms. A regular range-road grid was planned at mile-and-a-half intervals that served all lots. Two of the sixty-four lots were reserved by right to Governor Benning Wentworth, and one each for the support of a minister and a school. Many grantees were "esquires" whose interest was probably only speculative; ten bore the surname Williams, scions of the River God family that long dominated western Massachusetts. At the exact center of the township, 64 one-acre "town lots" were reserved, but such villages were by then largely a legal fiction meant to comply with royal charters that still promulgated settlements on the seventeenth-century Massachusetts model. Laborious clearance of even token village lots was folly when dispersed farmsteads were the going proposition.[76]

The hallmark of the range township is, of course, the range road-net, for which toponymic evidence is usually modest, but ubiquitous. Pembroke NH still has parallel roads specifically named Fourth, Fifth, Sixth, and Eight [sic] Range Roads, and the courses of the others, either renamed or dirt tracks, are still evident, much as they were when laid out in 1729–1736. More commonly, towns with several obvious range roads have but one now explicitly so named; North, Middle, and South Roads often leave the word range to be understood. Cross-range roads that ran perpendicular to the range roads might be fewer and farther between, and survive today in names such as Upper Cross Road, Country Cross Road, or simply Cross Road. Many schemes came to naught. New Ipswich NH was granted as a six-mile square in 1736, only to be redrawn as a rhomboid with range roads every half mile and cross-range roads every mile; yet scarcely one was ever built, and the rights-of-way reverted to the abutters or served to indemnify others for later road-takings.[77]

The degree to which the range road-net survives, or ever existed other than on paper, depends largely on the terrain. With the aid of the De-Lorme atlases, the current status of these road-nets in New Hampshire and Vermont was assessed. Three symbols (#, +, *) were marked on a base-map to record towns with a conspicuous range road–net, crossroads center, or radial road-net. It was judged on immediate impression that 56 of a total of 234 NH towns (23.1%) possessed moderate to strong range road-nets, the strongest being Webster, Pembroke, Gilmanton, Candia, Strafford, and Rochester. Range road-nets did not assert themselves in townships lying higher than 1500 feet above sea-level, and indeed the choicest were all below 650 feet. Difficult terrain and sparse settlement would preclude range roads in northern New Hampshire, even if the township was duly ranged, lotted, and granted in toto.

In Vermont, the same percentage of towns bore conspicuous evidence of range roads (56 of 237, or 23.6%), with the same altitude limits, although much less land in the state lies below 650 feet. The most emphatic were in the Champlain Valley (Orwell, Shoreham, Bridport, and others) as well as the Winooski Valley (East Montpelier, Calais, Barre): nineteenth-century urbanization brought out the cadastre. Overall, the range road-nets of Vermont were weaker, rarely dead-straight, and more likely to abandon their prescribed alignments in order to exploit stream gradients in what I shall call river-twined roads. The intricacy of the drainage was such that a serviceable if organic road-net could be achieved. Thus while Vermont, as manifested in its relatively boxlike town boundaries, often benefited from later and more orderly survey practices than New Hampshire, still, in terms of its range road-nets, it attained no greater regularity.

Our evaluation of range towns on topographic maps will expand upon the critical vocabulary developed in earlier pages (see fig. 10). Range towns were often the last settled and the first abandoned, and while telltale field-edges, fence-lines, and lot-lines may have disappeared, the mapped outlines of larger tracts such as state forests often reveal the essence of the scheme. While few range roads run unbroken the full width of a township, their alignments are often easily traced. Surveyors began work at the borders of the town, and these may display precisely "ruled edges": range roads initiated dutifully at the town-lines exactly to plan but gone almost immediately astray. Some range roads decay to dirt tracks or die out altogether, only to reappear intermittently in perfectly aligned segments some distance away. These distances may prove to be in even multiples of a quarter-mile; such "measured intervals" may also characterize the lengths of right-angled road elbows, or the distance at key points between sub-parallel roads that otherwise freely wander. Crossroads in the middle of nowhere may provide reference points upon which to bring one's Mylar grid into perfect register with the cadastre. Farms may have been precisely ranged and lotted while the roads connecting them belie that order. Edna Scofield perceptively noted that the flat, earlier-settled southern part of Tyringham MA clearly shows the scheme of range roads and farmlots, while the later-settled and rugged northwest does not. Yet if one ignores the roads and concentrates only on the fairly orderly array of houses as mapped in 1904, the lots on both sides of town can be inferred to have been laid down on a regular scheme.[78] The lesson is valuable: range townships with highly compromised road-nets may yet be identifiable by their evenly peppered house-dots on old maps.

We spoke earlier of *trellised vine* roads, roads that appeared systemless but that can be seen to randomly but repeatedly fall into alignment under a grid, like a vine guided by a trellis. What might be called the "peg board"

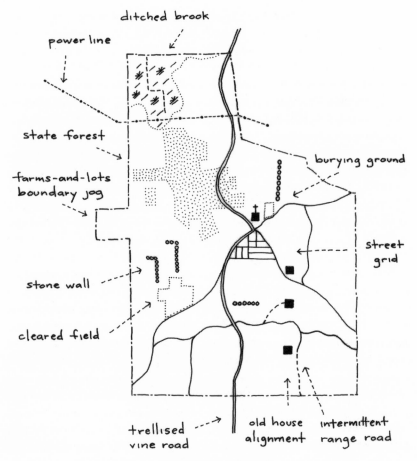

ditched brook

power line

state forest

farms-and-lots
boundary jog

stone wall

cleared field

burying ground

street
grid

trellised
vine road

old house
alignment

intermittent
range road

FIGURE 10

Persistence of the Past

Colonial allotments are the legal bedrock which underlies the
landscapes we see today. They are best revealed in rectilinear
alignments as in these map clues.

theory seeks to explain how technically this might occur. We know that
when a nineteenth-century federal grid township was laid out, the corners
of all sixty-four sections were marked, by stakes, stones, or tree-blazes.
These are the pegs in the board. If this federal survey technique evolved
as did so much else out of the colonial New England experience, then did
even wayward range roads obligatorily visit corner-markers as such, or as

perpetuated in property bounds? For example, Easton MA (1725) is not apparently a range town; yet it can be seen that North Easton is more or less 2 miles north of Easton Center, Pratt's Corner a mile south of Easton Center (though not directly connected by road) and a mile west of Furnace Village, and so forth. The more one studies and measures the more one becomes convinced that while Easton is not a true range town, whoever laid it out carried the template of one in his head, which he ignored and heeded as he saw fit. The year, 1725, and perhaps a land-as-commodity mentality implicit in its history as the east end of the Taunton North Purchase make it reasonable to regard it as a proto-range town.

Half a century later, on the eve of the Revolution, we find the range township universally established, yet universally flawed in its execution. E. N. Torbert well illustrates in seven maps (1767–1930) its imperfect realization in Lebanon NH, a quadrangular Wentworthian township on the Connecticut River measuring roughly 6 by 7 miles.[79] The town is cocked northeast with the trend of the river. The fictitious "town lots" on the first survey occupied a quarter square mile on a ridge-nose where no house was ever built and that was wooded pasture in 1930. Only in the southern half of the town, with its richer beech-maple soils, was a net of range roads seriously envisioned, and perhaps only half were ever built, all before 1830. The last built were the first abandoned (along with the farmhouses they served), upwards of ten road miles by 1860. The maximum road grade in 1830 was 5.7% as against 3.7% in 1930. Because slopes in the township run as high as 13%, it is not surprising that road meshes often enclose the steep contour-whorls of hilltops, and that range-road segments were deviated, early abandoned, or never built due to terrain. At least two road segments ran diagonally from one prescribed grid corner to another, eliminating the longer and hillier L-shaped route expected under the strict scheme. Riverine routes exerted their inevitable dominance, such that Enfield Road, the main west-east axis in 1800, had fallen out of use or repair for more than half its length in 1930. Evidence of a range-lotted countryside is not restricted to roads, however.

House-dot arrays on the early nineteenth-century map of Lebanon reveal rectangular alignments between remote points never interconnected by roads. If houses were customarily seated at the edges or corners of farms (rather than in the interior of the lot), and if the farms were rectangularly laid out, such alignments would arise of their own accord. These early houses tended to stand in clusters of two to four (particularly near road-junctions), or be strung out singly at intervals of one-quarter to one-half mile. By 1930, the lucky clusters survived as single, consolidated farms, while the unlucky were cellar holes in wooded back-pastures on discontinued roads. Even on hard-surfaced roads roughly every other homestead was abandoned (or had gone out of farming). Field-lines mapped in 1930 trace late eighteenth-century range roads that were planned but never built. Croplands keep close to houses, face south, and lie in narrow strips along the main northwest-southeast roads that owe their primacy both to the cadastre and the grain of the land. The largest tracts of the land-use mosaic are woodland and (wooded) pasture, with squiggly edges controlled by the contour-lines where the flattish hilltops are cleared down to the break in slope. Yet enough rectangular wood-edges exist to create an effect of submerged blockiness, the ideal block of which is about a half-mile square (160 acres), although no such perfect pasture or woodland block actually exists. On the whole, Lebanon NH exemplifies remarkably well the utility of much of the analytical vocabulary developed here: template, pegboard, ruled edges, measured intervals, and so on.

Beyond its topographical influences, did the range township imprint itself on the philosophy of the people who inhabited it? Is there a connection between the stereotyped New Hampshire resistance to central planning and government (as epitomized in the motto "Live Free or Die") and a mentality nourished by the dispersed farmsteads of the range townships? Are these to be opposed to the nuclear township and a legacy of greater centralization in Massachusetts? We earlier suggested that the land-granting policies of the southern colonies had obvious and permanent effects on the sociology of the region: were such differential effects also manifested closer to home?

Townplan Templates

The archeological study of townplans asks us to treat what initially seem to be negligible, even nonexistent, landscape patterns as artifacts and to seek precedents for their origins. No layout, however simple, is regarded as so self-evident as to "naturally" recommend itself. Instead, it is presumed that the artificer works from a definite template with a traceable history. This template should not be thought of as an ingeniously technical blueprint; it may be no more than a simple pattern which confers prestige. One such minimal template is the Yale Row (or Old Brick Row), an influential late eighteenth-century campus scheme in which the college buildings were built all in a row, but turned alternately endwise or sidewise to the street. The idea perhaps originated with President Thomas Clap in 1750 and was brought to perfection in a master plan by painter-architect John Trumbull in 1792. Unlike the quadrangle, with its well-known antecedents in the medieval cloister, one might object that a simple college row is intuitive, a virtual nonarrangement, warranting no explanation or commentary, and certainly no scholarly scouring of the landscape for predecessors or successors. However, in the words of Paul Venable Turner, "when the Yale Row was conceived, it was a bold and impressive innovation in collegiate planning."[80]

The true merits of a template can be assessed only once we have classified the full range of artifacts before and after its appearance, and in this way the influence of the Yale Row can be seen to varying degrees in many other New England colleges: Dartmouth ("Dartmouth Row"), Brown, Amherst, Colby, Bowdoin, Washington (now Trinity), Wesleyan, Middlebury, and the University of Vermont. This template also reached schools in New York, Maryland, as well as Ohio, where Ohio Wesleyan, the original campus of Western Reserve, and Marietta testify to early settlement from Connecticut. Even the layout of early factory buildings was not immune. In a pragmatic vein, irrespective of whether the Yale Row template "really" exists, it is more beneficial to posit its existence and thoroughly reevaluate the evidence in light of it, than to vainly await some radical style change or technological breakthrough more agreeable to someone's arbitrary idea of an innovation. A template is also a sieve through which to sift the often unpromising visual rubbish we encounter; a serviceable array of such sieves is both an aid to discovery and an antidote to boredom. Several broad types and phases in the development of village templates can be identified from the way in which certain basic elements are arranged. These include the topography of the site, the layout of the first houselots, the scheme of land allotment, the initial road-net, the placement of meetinghouse and burying ground, and the shape of the common.

Ulster Template

Anthony Garvan contends that as the English had no domestic school of town-planning until after the Great Fire of London (1666), that it is to their medieval bastides in France and their later plantations in Ireland that we must turn for precedents. The three elements of Anglican church, village, and bailey (stronghold) in Ulster translated to meetinghouse, village, and garrison or blockhouse in New England. While the unrealized scheme for Saybrook CT came closest to the Ulster model, the reconstructed Plimoth Plantation makes abundantly clear the importance held by this, the singlemost comprehensive colonial prototype available to the English settlers. Overtly defensive settlement plans continued to figure on the New England frontier until the mid-eighteenth century. In Rutland MA in 1715 we find the sixty-two original houselots laid out "in clusters, in an oblong square, containing thirty acres each, in the most eligible and safe part thereof, and so calculated as to have a Fort, fortified House, or Garrison, in each cluster to flee to, when invaded by the foe."[81] Whether fortified or not, such clustered town lots were long called "citydale" lots or "the citadel," even when laid out as a legal fiction.[82] The defensive advantages of the compact nuclear township delayed the full flowering of the range township until the end of the French and Indian War in 1760.

Brookside Village

A townplan template that might be called the Brookside Village has a long English tradition: Foxton, Cambridgeshire, was settled along a dug brook divergence in the sixth century. "That splendid spade-work fixed the position of Foxton for ever, and determined its shape for the next twelve hundred years. It was not until well after the year 1800 that any house in the village was built further than about eighty yards from the Town Brook, and most of them were within a few yards of it."[83] While this brook is cartographically nameless, and known simply as The Brook, the fact that for narrative purposes it is here called Town Brook is perhaps indicative of what in England such brooks were typically called; they certainly were so here. (Brooks, unfortunately, are not listed in the standard British gazetteers.)

Plymouth MA (1620) was planted where there was a harbor, a commanding hill, cleared corn-ground, and "a very sweet brook," Town Brook, that yielded not only eels, alewives and ultimately water power, but served also as a ceremonial Rubicon when treating with the Wampanoags. In Rowley MA, the original houselots in 1639 were also disposed along the like-named Town Brook, and just as at Foxton: "this brook and its tributaries governed to a large extent the laying out of the town, and nearly every house bordered on, or had access to it, thus assuring all a supply of good water." Bridgewater MA (1651) was laid out in six-acre so-called house or garden lots, each 12 by 80 rods, that stretched back from both banks of the Town River. Concord, for which no seventeenth-century plat exists, was similarly seated along Mill Brook. In Medfield MA, settled in 1650, the original houselots ran back from Main Street to the commonly owned bottomland along the parallel Vine Brook. The generous forty-acre homelots in (West) Tisbury, settled ca. 1671, butted endwise on Mill Brook and measured 40 by 160 rods. Study would doubtless turn up many more.

While brooks can be diverted, culverted, and buried, as village nuclei they may constitute a more ever-fixed mark than the much celebrated and mythicized common. Commons were mutable, movable, removable, and belatedly creatable from a variety of motives, such that when casually viewed as nuclei of original settlement, seeing is definitely not believing. Neither, in truth, are all town brooks so facilely accounted for. The Waterfield of 1638 on the banks of the Aberjona (now roughly Winchester Center MA) stood on a classic brookside site, yet it was originally granted not as a village with serried ranks of houselots but rather in regular squadrons of large outlots, oriented on an arbitrary axis that ignored the trends of watercourses. Close analysis of early villages may reveal other accepted site criteria or imitable templates. Concord, for example, literally dug in against the harsh conditions of its first year, 1635, and house lots faced southwest over Mill Brook Meadow and backed against what is today known as Revolutionary Ridge. Three years later when the adjacent plantation of Sudbury was settled the main rank of houselots was seated against a sunny southwest-facing ridge and looked out over the Sudbury meadows.[84] Lancaster (1653) lies aligned north and south (exactly) along a west-facing ridge-slope looking out over the meadows of the Nashua intervale. Is it unreasonable to suppose that Sudbury copied Concord? Or that Lancaster learned from the two? These under-ridge sites, while totally "logical," may too have had Old World templates, and been familiar as village locations to the first English settlers.

Street Village

This village type, also known as the broad-street village, flourished in western New England, where its form was a natural fit to the flat north-south river terraces of the Connecticut Valley.[85] Its hallmarks are a straight, broad street (or long common) running north-south parallel to the course of the river; narrow, deep houselots running east-west off this spine; and the absence of any significant crossroad, at least originally, due to the absence of bridges spanning the Connecticut. The extreme richness of the soil and the need to build above flood-line would justify compactly arranged homelots long after the nuclear village fell in disfavor elsewhere. While a main road with homelots is discernible in almost any townplan, nuclei of original settlement that meet the formal criteria are much rarer. Classic street villages of the Connecticut Valley type include Middletown (1650), Haddam (1662), Enfield (1680), Durham CT (1707); Springfield (1636), Longmeadow (1644?), Hadley (1659), Deerfield (1673), Northfield (1673), Greenfield (1686), Amherst (1703), Granby (1727), Belchertown (1731), West Springfield MA (1774); Charlestown (1735), and Claremont NH (1762). (Dates are those of settlement or incorporation and not necessarily of the village street itself.)

Necessity was perhaps the mother of invention. Springfield (1636) found itself on the straitened east bank of the Connecticut River rather than, as was hoped, the lusher meadows opposite (troubles over settlers' cattle in their cornfields may have hardened the Agawams against the original scheme). The north-south Town Street was laid out along a neck of dry land, hemmed in between the Connecticut and "Hassocky Marsh." Some forty homelots, eight rods wide and divided by wooden fences ran west from this street down to the river; eastwards, these strips also extended across the marsh to the upland woodlots beyond. In 1660, a marsh brook was ditched parallel to the street and became Town Brook (now a city sewer), which supplied water for almost two hundred years. Court Square so dominates the townplan today that it is uncertain how fully Springfield ever embodied the street village, but its form must have influenced the more clearly prototypical Hadley, settled in 1659. There the street was a generous twenty rods wide and ran north-south across a neck of land formed by an oxbow in the Connecticut River. The narrow homelots to either side were about eighty rods in length; in 1663 these were forty-seven in number. Two lanes ran east and west to meadows and woods, but were secondary to the main scheme. The street was gated at both ends, forming a forty-acre enclosure that "was very convenient for grazing ground, when they had but few fenced pastures." This street village is still remarkably intact, both physically and agriculturally.

The original settlers came from Hartford, Wethersfield, and Windsor CT, and Sylvester Judd speculated in 1905 that "the idea of a street so wide, may have been suggested by the Broad-street at Wethersfield. In forming it, they appear to have regarded both utility and beauty." The Broad Street at Wethersfield lacked parallel edges, and ran northeast-southwest, while Windsor had "a single street running almost due north and south from a central square." Perhaps Hadley blended traits from Springfield and two of its parent towns to help produce a template imitated up and down the middle Connecticut Valley for more than a century.

By 1686 the village-type had reached Greenfield, where Greenfield "Street," was laid out six rods wide (like its mother-town of Deerfield) and ran transversely between the courses of the Connecticut and Green Rivers. The exact orientation N 100° W was likely magnetic west in 1686. Twenty houselots of four acres each were laid out, but the size was doubled to 16 × 80 rods in 1700.[86] In such ways the plan could be modified to suit the terrain or to sate the settlers' land hunger; this flexibility assured its success. The fact that Deerfield Street (1670) is by tradition a measured mile, while the later "Petersham Street" (1733) is three miles, shows the elasticity of the concept in the face of poorer upland soils and the demand for larger, dispersed farmsteads. After 1700, the rationale of the street village began to evolve as it moved upland from the valley. Amherst, Granby, and Belchertown MA are not river towns; while clearly convenient, the street-village plan was no longer rigidly dictated by topography. Instead, custom, prestige, and the potent fact that these were daughter-towns of earlier street-village settlements, seem to have been equally important in its adoption. Similarly, Claremont NH was settled in 1762, four miles east of the Connecticut River. Gracious, grass-verged Broad Street is ten rods wide (165 feet) and runs nearly north-south, perpendicular, not parallel to the Sugar River, along which the town actually lies. Topography and custom were balanced in new ways to yield this choice of layout.

In the days when unmade roads were churned to wallows, rights-of-way were necessarily wide, and the broad village street served as both thoroughfare and common. As late as 1839 John Barber ambivalently described the center of Belchertown as "a wide street or common," and his illustration is equally ambiguous. Only once roads were improved and the common fenced in the nineteenth century did the distinction between public ground and roadway become clear. Exact arrangements varied with the width of available land and local traffic requirements. In West Springfield in 1866 the "single broad thoroughfare" called Broad Street or Broadway was reconfigured as a long east-west common flanked by Park Street and Park Avenue. In Deerfield, where the six-rod right-of-way is too narrow to admit the customary long common, The Street lies between tree-lined verges.[87] In Amherst MA, the common is a broad median mall, while that in Charlestown NH has been shouldered aside by the highway.

Many street villages once bore street proudly in their proper names.

The custom is English, as evidenced by Wickham, Weybread, and Shotley Streets, where the postscript distinguishes sister-villages of like name. (The word derives from the Latin *via strata*, or paved way, which may account for its medieval cachet.) In Massachusetts, Deerfield has been immemorially known as The Street; Hadley Street and Hatfield Street were so recorded in 1709; East Street and West Street denoted the two parallel villages of Amherst in the 1830s. Greenfield was Greenfield Street. Petersham Street (or The Street) is still used locally for the main road through town, while the hill ridge where nearby Athol was first settled is a neighborhood called The Street. In New Hampshire, Keene was originally Keene Street, and we find also Canaan, Conway, and Cornish Streets as village names. In Vermont, we find New Haven Street or simply The Street.[88] While all may not be street villages in our sense, they attest to the prestige attached to the term.

Crossroads Village

In New England, so much emphasis is placed on the spider-web road-net that a significant countertrend is often overlooked: one toward central, even cardinal crossroads, whether expressly laid out, or belatedly pieced together. Indeed, Plymouth, the first New England village, was laid out within a diamond palisade with Governor Bradford's house and four pateros (small cannons) at the central crossroads. But this layout was atypical; more illustrative were the three civic crossroads that arose in the old port capitals of Boston, Portsmouth, and Newport. To this day, we still see the Old Boston State House (1713) islanded in the intersection of Washington, State, and Court Streets; State Street, the best arm of this crude cross, runs to Long Wharf on a line slightly north of east. The first Boston Town House was built here in 1657, and since the 1640s (and likely before) it had been the open-air marketplace where stood meetinghouse, pillory, and stocks. The prominence of this crossroads was underscored by the choice of street names: originally, High and Great Streets, later changed to Cornhill, King, and Queen. In the words of Whitehill: "Public life was centered at the spot." While obviously a civic crossroads, was it meant to be cardinal? Probably not: cardinality was more accurately achieved about this same time in the Beacon Hill pasture lots. While Washington Street looks to have been coaxed into rectilinearity, no one bent heaven and earth to fit an ideal crossroads into the fabric of Boston. (The original effect has been much compromised by Congress Street and Government Center.)[89]

Portsmouth NH, by contrast, has long had a griddy character, since at least 1704, when the glebelands were platted as if by a stubborn child with a shaky ruler. Market Square, on a low hill at the crossroads of present-day Congress, Pleasant, Market, and Daniels Streets, grew steadily in civic prominence with construction of the North Meetinghouse (1712), the Old State House (1758–60), and finally the Brick Market (1803). This intersection is neither right-angled, nor cardinal, although Pleasant Street does run N 8½° W (magnetic north ca. 1775) and the orientation of Congress Street (N 30° W) is re-echoed elsewhere within the town. Newport RI also nourished its civic crossroads (or rather, ankh cross), on the Mall or town square where the Brick Market (1762) rose at the intersection of Queen and King (now Thames) Streets. Both plan and colonial street names ring the changes on Portsmouth and Boston. The extension of Queen Street into Long Wharf (1680) lies along an axis of W 10° S, which likely was the magnetic west of the time. However, not until 1800 did "the town's new square, built after the ravages of fire and war [lend] the first planned look to an unplanned town."[90] Clearly, while cardinal crossroads were an ideal in all three capitals, they could not be inserted retrospectively once dreams of provincial grandeur came within grasp. These crossroads come down to us as crude, and realigned, often jostled awry by building encroachments, unvalidated by adjacent grid streets, and compromised by radial ones. Often the roadways themselves are wedgy-edged, making precise bearings hard to assign. Still, the birth-throes of these misbegotten crossroads affirms the power of a colonial townplan ideal.

Our principal interest is in crossroads villages platted from the outset as such, ones often laid out in compliance with charter requirements that mandated some semblance of compact settlement. Typical attributes of these crossroads are that they are right-angled, cardinally oriented, have directional street names (North, South, and so on), lie coaxial to the larger road-net, stand at the geographic center of a square or diamond township, and are distinguished by a common, a meetinghouse, and the main buildings of the town. Few crossroads villages conform in every particular, and many were but fictions on paper; those that were in earnest might require years to come to fruition. These criteria, while often discernible on maps, are ultimately best evaluated in the field. Dummerston VT, even in the DeLorme atlas, does not suggest a crossroads village, yet when seen firsthand, the hilltop, lazy-X intersection of Middle and East-West Roads, with its Congregational church, tiny green, and half-dozen houses clearly conforms to the type. No attempt will be made here to trace possible Old

World origins of this template, which long exercised a geometric fascination. Suffice it to say that while the crossroads village was too artificial for organic English settlement, unrealized schemes for English "seigniories" in Munster, Ireland (1568) envisioned roughly four and one-third miles square plantations ruled into quadrants by a crossroads, with a square, central church green and laborers' cottages surrounded by a manor house and orderly clusters of farmhouses on rectilinear outlots. A geometrically similar center was platted for Nottingham Square NH in 1724.[91]

Our focus will be on Massachusetts, where the artifice of the crossroads village stands out starkly against a tradition of organic settlement, and where only one-fifth of towns have crossroads centers, and of these, only one-third have crossroads churches. Their distribution is fairly wide, and origins vary. Those in the early settled east often occur in the range-lotted outer ends of mother-towns, or in grid ports (discussed later). Most were clearly not crossroads villages platted as such from the first, although one extremely early example appears to have been. Medfield MA (1650) was laid out at the intersection of four roads now known as North, Main, and Pleasant Streets, "thus forming the four corners which have been from that day to this reckoned as the centre of town."[92] The first houselot, that of Ralph Wheelock, by tradition the "founder" of Medfield and author of its covenant, occupied a place of honor at the crossroads, but ironically not the meetinghouse. Neither was the crossroads cardinal, but laid out on a roughly N 35° W axis. As the Main Street houselots butted on Vine Brook, Medfield in effect blended both Brookside and Crossroads templates. Arlington (Menotomy) was an out-village of Cambridge and its meetinghouse seems to have been opportunistically located at a ready-made crossroads laid out in an orderly allotment of the outer common-lands. Crossroads in Milton and probably many other urbanized towns in Greater Boston, a number of which did not come into their own until the nineteenth century, are explicable in similar terms.

One example well illustrates how difficult it is to interpret the visual and written evidence. At present the Bedford second meetinghouse (1816) stands in a rectangular common in the southwest quadrant of a center crossroads, but this tidy "plan" took more than a century to develop. The first meetinghouse (ca. 1729) was located on substantially the same spot in what then was the roadless center of a new township formed from the easterly parts of Concord and the westerly parts of Billerica. By 1735, at least three arms of a crossroads had appeared: The Great Road (laid out parallel to the old Concord-Billerica line) and Springs Road. Whether the fourth was originally intended or merely "happened" is unclear. In the 1730s South Road was by some evidence still a trodden path; it was a road by 1806, but stopped short at the old meetinghouse and was not realigned and extended to The Great Road until 1847 to complete the crossroads as we know it.[93] The question remains whether Bedford Center is an early eastern example of a well-planned, albeit slowly realized, crossroads village of the 1730s, or merely assumed this form after more than a century. The sightseeker is often at a loss to tell whether the landscape in view results from evolution or special creation.

Crossroads villages in central Massachusetts seem often to be an inevitable outgrowth of the street village and, in many cases, one axis so dominates that the distinction between the two types is moot. The emergence of the crossroads village is obviously one element in the slow formulation of the grid-platted range township; and accordingly, it flourished best in areas settled after 1720. Petersham, settled in 1733, is an early epitome of the type and merits close study. It is a six-mile-square township laid out diamondwise to the four points of the compass with a central cardinal crossroads complete with church and common. (Two other nearby diamond townships, and ten with central cardinal crossroads largely settled between 1713 and 1767 show that while uniquely perfect, Petersham also typified local trends.) The stamp of the compass is plainest at the center of the village, where three arms of the crossroads (East, South, and West Streets) maintain their set courses for a quarter-mile to a mile, oriented on an axis of N 9°5' W. (Magnetic north at Boston in 1730 was N 8°26' W.) The north arm, Athol Road, is not straight at all, although a mile out of town it assumes a N 3° W alignment through open country. Cardinality is favored here and elsewhere in New England by the general north-south grain of the landscape.

The Petersham townplan falls between the two extremes of the nuclear and range township. It was a soldiers' grant, one of many intended to pay off volunteers for service in the Indian wars. Such blatant commodification of land signaled a breakdown of the older, paternalistic norms of Massachusetts town planting and fostered an expeditious orderliness in the plat. Its first division (1733) assigned generous houselots of 55 to 100 acres to each of seventy-two proprietors along the north-south axial ridge of the town where the best farmland was found. Second and third divisions of 100 acres each were voted in 1738 and 1739 and resident proprietors were allowed to "pitch" these lots adjacent to their houselots, rather than draw them randomly by the usual lottery. A smaller fourth (40–60 acres) division was drawn by lot in 1753, and a fifth and all but final of 6 acres in 1770. Thus, even within six years of settlement, an inhabitant might create for himself a 250–300 acre farm, as in a range township. Yet much of the symbolic centrality of the nuclear township was retained.[94]

In western New England, at Williamstown, the east-west Main Street (now Route 2) was laid out in 1750 and the intersecting North and South Streets in 1761. The N 16½° E orientation of this crossroads was probably pure happenstance. Pittsfield, settled about the same time in 1752, centers on a crossroads expressly named North, South, East, and West Streets. Despite the nomenclature, its orientation of N 12° E is clearly not cardinal, lying, as it does, about eighteen degrees east of magnetic north of the time. To the south, in Connecticut, the western lands of Litchfield County are marked by cardinal roads and often crossroads centers.[95] Litchfield, the shire town, settled in 1720, lies at the imperfect intersection of four compass-named streets oriented on an axis of N 13° W, about five degrees west of contemporary magnetic north. North and South Streets are slightly offset and fail to join at the green (it also has a fifth arm). Such "imperfections" are by no means uncommon.

Many Massachusetts towns founded at the right time and place for crossroads villages have centers perversely "spoiled" by a fifth road, topographical diversions, or cockeyed intersections. Still others, most notably Buckland, Windsor, and Middlefield in the Berkshires, exhibit "broken" or offset crossroads in which the two cross-arms fail to meet at the main road by a factor of several hundred feet. It may be that the main road was a middle division line where lot-lines were readjusted, or that the cross-arms were separately surveyed from the town-bounds inward, and never the twain did meet.

Curiously, the term *crossroads* never figures in New England place-names, while prominent in Virginia, the Carolinas, and to some extent Tennessee and Kentucky. Corner (as in Daley's Corner) and specifically Four Corners (as in Phillipston Four Corners) are the local terms, and as a northern dialect term Four Corners was carried westward to Ohio and beyond.[96] In New England, a crossroad if not a cross-range road was merely a side road that "cut across" between main roads. Such Virginian names as Templeton Cross Roads and Turners Cross Roads are without British precedent or New England parallel. They are original to Tidewater toponymy and integral to a very different settlement geography in which backland hamlets sprang up at key traffic points with prosaic names like Millers Tavern, Smoky Ordinary, or Pairs Store. The sociocultural incapacity of New England to generate such placenames is highly telling. Not only did New England lack the term *crossroads*, until the rise of the range township it often lacked the thing itself.

Hilltop Village

As population increased, the frontier moved inland from the coast and upland from the valleys, a shift symbolically echoed in the names Brookfield MA (1664) and Ridgefield CT (1709). A new settlement template took shape, the Hilltop Village, developed at a time when the tight-settled seventeenth-century township was coming undone and when homelots might be small farms of fifty acres. Poorer upland soils alone justified such generosity. The topography of such hilltop sites has been carefully gauged by Michael Bell. Only long, north-south, glacially streamlined hills had the depth of soil and arable acreage needed for farm settlement: drumlins were too small, and bedrock ridges too ledgy. A hilltop meetinghouse and village green stood at the center of a radial road-net and farms spread along the ridge and down the slopes. In Connecticut, settlement dates suggest this plan emerged as early as 1700; exemplars as enumerated by Bell are "Lebanon, Franklin, Sterling Hill, Brooklyn, Hampton, Hebron, Gilead, Columbia, Colchester, Washington, Norfolk, Newtown, Greenfield Hill," "Goshen, West Goshen, and Litchfield."[97] In central Massachusetts, where lands below five hundred feet had all been taken up by 1690, such streamlined hills were fortunately widespread. Towns such as Belchertown, Granby, Dudley, Wendell, Royalston, Templeton, Petersham, and the original settlement of Athol all occupy such sites. While a hub-and-spokes road-net can be found as in Connecticut, the long hill ridges of central Massachusetts were eminently suited to carry on the street village tradition of the lowlands.

Hilltop villages stand out with particular clarity on topographical maps, and portents of their demise are to be read there as well. Belchertown MA, as mapped by the USGS (1949), lies along an elongate, ovoid hill ridge two miles long and three-quarters of a mile wide, moated by swamps and rising 200 feet to a church-crowned summit at 610 feet. (Contour-lines generally vindicate these church sites as the highest buildable spots.) With the aid of the topographical map, one can scan the countryside round about: there is nowhere else to seat a village. The cleared hill slopes contrast with a predominantly wooded countryside, and the village approach roads are lined with houses and barns (indicated by filled and hollow squares). On the summit level a street village and long common, perhaps 1000 by 200 feet, are aligned along the glacially molded north-northwest axis of the hill. In a maneuver pregnant with consequences for hilltop villages generally, the railroad can be seen to seek the lowest-lying route, hugging the swampy hollows along the hill foot, with the station located downslope, two thousand feet from the church. The hilltop village by its very geography divorced the meetinghouse green from the mills and mill-streams in the valleys round about. With the advent of the Industrial Revolution, both wealth and commerce slid down from these hilltops to the factories below, a trend only aggravated by the railroad. In consequence, many towns developed two centers, old and new, agrarian and industrial; if severe, the political polarity could rend the town, as when the cotton and carpet-mill village of Clinton was set off from Lancaster MA in 1850. Even when not, a definite duality still arose, as between Fitzwilliam village and Fitzwilliam Depot NH, where the difference is but a mile's distance and two hundred feet of altitude. While the lack of industry doomed the hilltop village to decline in the mid-nineteenth century, many saw a mild reversal of fortune by the twentieth, as their quiet, rural ambience and staid charm drew colonies of summer people.

Templates Revisited

We have studied four village-types whose recurrence in the landscape suggests that they served as accepted templates for settlement. The templates are by no means mutually exclusive; alert readers will have noted that Petersham MA straddles three. But are these true templates, or a mere classificatory device? Are the resemblances intentional or accidental? Keene NH presents the interesting case of a possibly transplanted, hybrid townplan that highlights these issues. S. G. Griffin's reconstructed 1800 map of "Keene Street," has obvious affinities to the street village, with its broad, north-south axial main road; however, if one focuses on the Y-junction at the head of the street (now called Central Square), one detects an eerie schematic resemblance to the Common (now Monument Square) at Concord MA as it was in the early nineteenth century. Keene had a perilous history: granted as Upper Ashuelot by Massachusetts and settled in 1734, the town was abandoned in 1747 and all but one house was burned by the Indians. Two years later, a number of settlers drifted back, and in 1753 Governor Wentworth of New Hampshire regranted them the town and reconfirmed their ownership. With these facts in mind, let us

revert to the summer of 1734. On June 26, a committee of the Massachusetts General Court, meeting in the house of John Ball, innholder, in
Concord MA "received as proprietors of the Upper Township on Ashuelot
River" sixty-three persons, many of Concord. The next day, the first proprietors' meeting was held, also in Concord, and Samuel Heywood of
Concord was elected clerk. A committee of three to survey "the Whole
Entervail in said Township" was also chosen, and consisted of a Dedham,
a Marlborough, and a Deerfield man. In the earliest days travel from Boston to Keene was made circuitously westward to the Connecticut Valley,
then northward through Deerfield. Later, as the Great Roads developed,
it was via Concord MA and New Ipswich NH.[98]

The bare and perhaps unanswerable question is, did Keene derive its
broad street from the Connecticut Valley and its square from Concord?
Could such early influences slowly work themselves out over a calamitous
three-quarters of a century? Must we look only to the summer of 1734 for
our origins, or could the influence be ongoing, fed by cultural and commercial contacts from two directions? Is it wiser to speak of templates as
though villages were stamped out once and irrevocably by machine, or of
models to which they slowly assimilated themselves over the years? The
ultimate answer lies not in a minute autopsy of Keene Street, but rather
in the unknown totality of analogous cases of possibly transplanted townplans. Only once New England towns have been reexamined with an
open-eyed awareness that a "nonplan" may indeed be a plan, will we have
data enough to speculate about the origin and diffusion of settlement templates. If one can be permitted an architectural analogy, early influences
might indeed take years to bear fruit. When Becket MA built its meetinghouse in 1738, the style, true to an early resolve, was that of the one at
Grafton MA ninety miles away, the town in which the proprietors' first
meetings had been held some fifteen years before.[99]

Commons

Today, the signal feature of the townplan is the common. (The Town
Brook as focus is utterly ignored.) As notable by its presence as by its
absence, the common is the one element of colonial town-planning of
which most people are aware. The village common is held forth as the
prototypical nucleus of the New England town, yet of the scant 15% to
25% that actually possess commons, many were off-center to the original
settlement and many others were only belatedly acquired. The best evi-

dence is that absolutely and proportionately Massachusetts possesses the most of any New England state (149 of 306), yet even there only 42% of towns have them. One should note that while the term *common* is traditional in the Bay Colony, *green* is more usual in the old Plymouth and Connecticut Colonies. (Lexington Green is an egregious, if not entirely official exception, presumably the effect of its apotheosis by nineteenth-century historians: George Bancroft in 1858 correctly called it a common, but preferred green for emotional effect.)[100]

A common, like a clay pot, is an artifact, with identifiable man-made attributes or characteristics. Unfortunately, the common is a maddeningly protean artifact that has changed its size, shape, function, mode of decoration, and its context within the urban fabric, often even its name, over two or three centuries. It is as though our clay pot had been slowly reworked from a redware milk-pan to a china tureen or porcelain punchbowl. Add that only the randomest fragmentary and visually inconclusive record of these transformations exists; that virtually no satisfactory examples of pots arrested in early stages of development are available to trace a sequence; and that ceremonial punchbowls (and the legends that surround them) came so to dominate all thoughts of pots and be so coveted that for centuries to this day "imitations" have been manufactured that are indistinguishable to the eye from the originals. Furthermore, only some three hundred such punchbowls are reported to exist, most of which are only scrappily documented, if at all, and these are gleefully scattered over a territory of perhaps forty thousand square miles. Whether taken as parable or parody, this is the quandary confronting anyone interested in the study of commons.

The data for a comprehensive study of even the most elementary aspects of commons, such as shape, has never been assembled. On standard USGS maps (1:25,000) my little fingernail covers 15 acres, five or ten times the size of many commons. Road widths are exaggerated for legibility, but only at a loss to the true dimensions of street-bounded commons, which are conventionally labeled simply as parks. Churches, town halls, schools and cemeteries may be marked by symbols, but the age and character of the village builtscape itself can only be appraised firsthand. While the early editions of the DeLorme Vermont and New Hampshire atlases classified all extra-urban buildings with half a hundred miniscule symbols, the map-scales used were even smaller than those of the USGS. Only the nineteenth-century Sanborn fire insurance maps report the needed cadastral detail at an adequate scale. Shape is an early, albeit not reliably permanent attribute of commons; it is also mappable. More stylish traits, such as species and layout of trees (whether lombardy poplar, elm, white ash, or maple), as well as fences and monuments, were imitable marks of civic prestige with datable vogues. They also went cartographically unrecorded.

As people traveled to other towns and cities for business and pleasure, there arose a shared vocabulary of fashionable design amid a climate of rivalry and emulation. Frog Ponds, whether fountains or beautified (or euphemized) sloughs, were found on commons or malls as distant as Boston, Lynn, Newburyport, Amherst MA, and Waterbury CT, indicating that the improvers of commons did not dwell in mutual ignorance, and that fashion might make even a blemish into a beauty spot.[101] A graceful meetinghouse, such as that of Pittsfield, could influence church-building for decades in the country round about; if they copied the church, why not the green? Lexington was Cambridge Farms before it was set off as a separate town in 1713 and acquired a common by purchase, the well-known Battle Green. This common, like the larger one of the mother-town of Cambridge, is triangular and lies in a fork in the road. Similarly, Westford MA was set off from Chelmsford in 1729, and both towns have commons made up of wedgy gores where five or six roads converge. Were such similarities accidental or consciously copied? The artifactual study of commons inclines us to entertain the latter possibility. We shall now consider three classic types.

The three-cornered or *wedge common* is probably the most numerous type in New England, and in eastern Massachusetts arguably the most prevalent. They also have precedent in England, particularly the south: defensible square plans being characteristic of the war-torn towns of the Scottish border. Yet there is also an obvious practical justification: because the road-net of a town was a spiderweb centered on the meetinghouse, and not a gridiron, the triangle and not the square is more characteristic of any random road-bounded parcel. Commons that were originally much larger, as the Cambridge cow common, saw themselves ultimately whittled down to road-bounded triangles. Indeed, the chief surviving seventeenth-century commons with certifiable bovine heritage—Boston, Cambridge, and Salem—have all come down to us as of above-average size (despite their urban setting) and more or less triangular in shape. Additionally, because many commons were originally meetinghouse yards, the traditional practice of siting public buildings in the middle of the street played its part. The old statehouses at Boston and Newport, markethouses such as Faneuil Hall, and the old Marblehead townhouse, all equally astound by standing dramatically enisled in the middle of the street. Thus there was nothing rustic in the placement of the first Lexington meetinghouse in the fork of what is today Massachusetts Avenue and Bedford Street; such sites were undoubtedly the genesis of many a wedge common.

Wide turns by a wagon and team meant triangles of wasteground were churned up wherever one road teed into another: these were called heaterpieces, after the shape of the heated core of a flatiron. While generally larger, the outline of the wedge common may have evolved similarly; in fact, the Framingham Common was once known as "the heater." In some cases slivvery commons might be reclaimed from the overample rights-of-way accorded in the days of unmade roads. When commons were fenced and formalized as parks, and the dirt roads that crisscrossed them were closed, their triangularity was reinforced by the requirements of a more restricted traffic circulation. Some commons were never reconfigured: lovely Templeton Common is still splintered into six odd bits of green by converging roads. In view of this, it is hard to imagine that a square or rectangular common was not laid out from the start. For example, the town of Bedford MA prior to incorporation in 1729 acquired some twenty-two acres to accomodate a training field, a meetinghouse yard, and to induce its first minister to settle. While the original two and three-quarters acres allocated to the meetinghouse has shrunk by an acre due to road encroachments, it has always been rectangular.

The incorporation of a common in a crossroads village is a geometric problem with no perfect solution, one in which the integrity and centrality of both features is maintained. Creative compromises were struck in the layout of such *crossroads commons*. Nottingham Square NH was laid out in 1724 as a meetinghouse lot of some six acres (30 rods square) at the theoretical intersection of four roads that did not quite meet.[102] Pittsfield MA (1752) located its square common off-center, in the southwest quadrant of the central crossroads. Other towns resorted to a modified long common that gave primacy to one axis. What was to become the classic American solution, typified by the Midwestern courthouse square, in which the entire central block of a gridiron townplat was laid out as a park, can be seen as early as 1638 in New Haven, but such geometric formality is otherwise notably absent in New England.

Oval commons became something of a rage. As early as 1801, Charles Bulfinch designed an oval park to be called Columbia Square in his plat for Boston "Neck" (belatedly and imperfectly realized in 1849 as Franklin and Blackstone Squares). Again in 1812, his equally visionary "race-track" plan was meant to ring his new University Hall at Harvard with a formal ellipse of some fifty trees. While neither plan was executed, both were symptomatic. The oval or ellipse suited the Federal taste in elegance, and was freely employed in windows, stairwells, bowfronts and other architectural elements. The oval was also ornamental enough to satisfy the later Victorian passion for dressed grounds. Oval parks formed the urbane centerpieces of Boston residential squares such as Louisbourg Square (1834), Union Park and Worcester Square (both 1851), as well as the more ambitious Chester Square (1850), which lay encurved by two townhouse crescents. While the oval might be urbane, rounded corners were all but inevitable under the press of wheeled traffic.

A higher incidence of square and rectangular commons in western New England (itself due to the paper-platting of townships in the eighteenth century) may have contributed to the particular popularity of ovals there in the mid-nineteenth century. An early and presumably influential example at Pittsfield was laid out by 1832. Pittsfield hosted an important agricultural fair and was soon to supplant Lenox as the shire town of Berkshire County, so its common was undoubtedly familiar many miles about. Thus it is of considerable interest that in 1838 John Barber should have encountered something similar, if cruder, at Westfield, thirty-five miles away. There he noted that "a small enclosed common, oval in its form, is in the central part of the area, around which the public buildings are situated; it is new set out with shade trees, and will add to the beauty of the place." His engraving makes clear that this was in fact a rather freakishly formal oval park fenced off at the center of an otherwise unkempt triangular common. The saplings in this oval were aligned much as the mature shade trees at Pittsfield were. The newness of the plantings and the misfit of the oval to the overall site are consistent with a derivative design, quite plausibly in imitation of Pittsfield.[103]

Thirty years later in the spring of 1868, an itinerant photographer captured a similar scene on the triangular common in Rochester VT: a large fence-enclosed oval, its center marked by a stake in readiness for the new soldiers' monument. Ovals were now the rage. In 1872 a special citizens' committee in Temple NH was authorized to improve the common and erect a soldiers' monument. One George Chamberlain offered to pay for a wire or iron fence, preferably oval. An oval granite post and wooden rail fence, enclosing an oval ring of maples, was in fact erected. The next town but one, Milford NH, also had its common fenced and improved in 1872;

perhaps the beginnings of today's Milford Oval. The Litchfield Green (CT) was photographed in its current oval outline in 1867. The Amherst Common (MA), while known locally as the Oval, may never have been more than the round-ended rectangle mapped in 1856. Perhaps the most famous oval common in New England is that of Woodstock VT. Its "sharp-pointed oval green," was laid down, to believe local legend, along the lines of a "ship once commanded by one of the town fathers."[104] Fair Haven VT on the 1869 F. W. Beers map has a much more robust oval, lying in a large rectangular town square called Park Place. Two lesser oval flower gardens adorn its south end, suggestive of fussy Victorian tastes. Some commons, such as those at Newfane and Strafford VT, incorporate elliptical elements that at street-level can frame the scene from key viewpoints as elegantly as a true oval.

Steeple Rows

Lending dignity to a common is often a formal line of public buildings, two or more of which may carry steeples or belfries. Typically these form a harmonious white-clapboard ensemble, slowly built over time, in a range of styles from Federal to Greek to Carpenter Gothic. The bodies of the buildings harmonize, while the steeples are elegantly differentiated, ringing the changes on pattern-book plates, serving to distinguish the civic from the ecclesiastic. A much-photographed example at Washington Common NH consists of a Congregational church (1840), school (1883) and town house (1789) that hold aloft a two-stage Greco-Gothic bell tower, a simple belfry, and an open belfry with octagonal lantern and cupola. In Temple NH, on the west side of the road opposite the common are a square-belfried Greek town hall (1842), a plain chapel used for social gatherings (1887), and a steepled but simple church (1842). As at Washington, the three stand back from the road on a semi-elliptical drive. A nondescript and ignorable brick library (1890) and a wood-frame store (1893) flank the threesome. In Manchester VT ca. 1861, a pinnacle-towered Congregational church (1829) stood alongside a small, elegantly belfried wood-frame school and a brick courthouse with a domed cupola.[105] Other examples are seen in Duxbury, Rochester, and New Salem MA. Often these ecclesio-civic triptychs consist of two landmark buildings flanking a modest centerpiece inserted later. A tremendous impetus to steeple rows was the formal separation of church and state that occurred in Massachusetts in 1833 and earlier elsewhere: two buildings were required where one had stood before. Those examples for which dates are known seem to have been consummated in the 1880s. The steeple rows, like the Yale Row, represent an unacknowledged but pervasive template, but too little is known of them to say whether they demonstrate any subregionality in distribution or architectural character (as in preference for spires or cupolas).

Farms and Fields

The transition from nuclear to range township made for small, scattered farm holdings in the early-settled downcountry, and large, compact, squarish farms in the late-settled upcountry. As this contrast might be misconstrued in purely modern terms as one between market gardens and dairy farms, its colonial roots must be kept in mind. Let us review some mapped examples, circling back counterclockwise from the northwestern periphery to the eastern core. In Vermont, some two hundred hill-town farms surveyed in 1933 were overwhelmingly squarish, compact, and generally of one hundred to two hundred acres. In Litchfield County CT, the 144-acre Ezra Stiles Lower Farm in Cornwall combined two rhomboidal lots and one ell-shaped one, awkwardly pieced together between 1738 and 1753. As was Connecticut practice, Cornwall right holders "pitched" (chose the bounds of) their later lots, disrupting an originally geometric townplat. In Bristol County MA, the 59-acre Old Paul Farm (1811 or earlier) and nearby farms in Dighton and Rehoboth attest to long, narrow, rectangular, range-lotted farms. Finally, in Middlesex County MA, the heartland of the nuclear village, we find complexly bounded, multiangular farms that, while sometimes fairly compact, were neither rectangular nor entirely contiguous. In Bedford, a 1773 map drawn for the estate of James Lane shows a goodly 223-acre farm, but one intricately gerrymandered and barely contiguous, looking like an origami giraffe, complete with a 53-acre appendage called "the neck." The only hint of orderly land division is in the meadow acreage, expressed in multiples of five (and perhaps this merely denoted legal rights to hay). Other Middlesex farm surveys show similar, if less gerrymandered holdings.[106]

Routine trading in small parcels, and more perversely, inheritance, complicated farm boundaries and disordered the cadastre. Otis Dyer sums up the history of one shoestring swamp-lot in Rehoboth MA: "In one case, an original owner owned a long rectangular tract of twenty acres in Manwhague Swamp a little north of Long Neck. After passing through several generations of the family the lot had been divided again and again until some parcels were only 33 feet wide by a half a mile long."[107] Indeed, the crazy-quilt of ownerships on old parcel maps is often best explained by the genealogical relationships among the landowners themselves.

Wall-net

Farm fields will be viewed here from a bird's-eye perspective as a purposeful agricultural network marked on the ground most permanently, but by no means exclusively, by stone walls. Scholarly and popular studies of stone walls focus largely on how and when they were built, their cost per rod, and the incredible miles (over 150,000 in 1871)[108] they run across the New England landscape. Rarely do they treat the network of fields they enclose as an interpretable artifact. Yet if one visited an old mill seat, it would be altogether natural to seek in the ruined stonework the footprint of the mill: the wheel-pit, the head-and tailrace, the footings of the building itself, and so forth, comparing these to surviving mills and diagrams in books. By analogy, a farm was also a vast open-air industrial site, one that processed crops and livestock, and that left its groundplan traced in fields and stone walls. Stone-walled fields, much as clay pots, can be creatively analyzed in terms of their materials, contents, form, and function. While this scheme might seem overelaborate, its virtue is that it compels thoughtful scrutiny of an artifact too often taken for granted.

Materials and Contents

The material of stone walls closely reflects the complex glacial drift and bedrock mosaic of New England. In East Lexington MA, walls tend to be of Salem gabbro-diorite, once known aptly as "greenstone"; in north central Massachusetts, they are often of the native schist. Those in Andover MA combine Andover granite boulders with chinking-stones of Tadmuck Brook schist; dressed walls in North Easton and Scituate MA betray the common bedrock that underlies them both. This variety becomes truly astounding when one begins to correlate local walls with a geological map. More rarely will the construction itself be noteworthy, as in the lace walls of Chilmark MA, where the open stonework perhaps was meant to stymie sheep by the play of light and shadow. Edge-set flat limestone is typical of Windsor County VT.[109] However, most farm walls were of dry-laid random fieldstone with little in their build to distinguish them; mill canals, railroad bridges, and estate walls would be more revelatory of datable or ascribable masonry techniques (as between, say, Irish and Italian workmen).

The "contents" of abandoned fields can be variously construed. On the surface, even in returned woodlands, old furrows can be revealed by a light snow, while underground, tillage left behind plow-scars, discolored soils that record ridge-and-furrow lines, bits of clamshell (added to sweeten the soil), and at times a puzzling assortment of domestic rubbish introduced wherever city nightsoil was used to fertilize fields. After the Civil War, farmers in South Hadley MA experimented with rag waste from local paper mills as fertilizer, and the Button Field on the McElwain farm took its name from the nonorganic residue. "Contents" might equally refer to the tree growth that succeeded abandonment: pines and cedars on lands recently worked, or mature hardwoods after the lapse of a century.[110] Old, broad-crowned maples or oaks crowded in by newer growth may have once been "cow trees" left in midfield to shade grazing herds. Reading earlier land-use patterns from the current forest cover is a provocative challenge on every abandoned farm.

Form and Function

The aspect of New England farm fields most noticed (and criticized) over the years is their small size. Small fields might mark stages in what the English call piecemeal enclosure, as parcel after parcel was won from the wilderness. Smaller acreages eased the burden of stone removal: stones from the middle of even a five-acre square field would need to be carried more than seventy-five yards to the nearest edge. Truly has it been said that it is easier to clear eight one-acre fields than one eight-acre field![111] Varied soils and terrain also conspired to create small fields. Comparison of the mapped field-lines of Hutchinson Farm (Arlington MA, 1898) with the local topography clearly reveals how the terrain mosaic broke up a fifty-acre rectangle into a dozen-piece jigsaw puzzle for agricultural purposes.

By the 1870s, however, such small tillage fields had fallen from fashion, as New Hampshire–born editor and agriculturalist Horace Greeley explained:

> In this section, our minute chequer-work of fences operates to obstruct and impede plowing. Our predecessors wished to clear their fields, at

least superficially, of the loose, troublesome bowlders of granite wherewith they were so thickly sown; they mistakenly fancied they could lighten their own toil by sending cattle to graze, browse, and gnaw, wherever a crop was not actually on the ground; so they fenced their farms into patches of two or ten acres, and thought they had thereby increased their value! That was a sad miscalculation!

Walls harbored weeds, briars, bushes, and vermin; they wasted ground, and cramped cultivation by requiring repeated turnings of the plow. Stone-walled two- and three-acre fields Greeley relegates to "grandfather's day," and asserts that eastern farms have "too many fields and fences," which he recommends be buried or their stone turned to better account in concrete barns. G. F. Warren, a Cornell professor of farm management, beat the same drum in 1914, offering careful bookkeeping to justify hiring two men and two horses to remove 114 wagonloads of stone from a 38-rod fence, which he reckoned cost less than a dollar a rod to dismantle, as well as yielding good stone for a barn cellar. (Wages were 20 cents an hour!)[112]

Does this two- to ten-acre checkerwork still exist, if such fields have been vilified for well over a century? Stewart G. McHenry, working from aerial photographs, has shown that three- to ten-acre fields in Vermont are nothing unusual, and that one- to five-acre fields, found close to town centers, are arguably of eighteenth-century origin. Furthermore, only the sudden availability of surplus bulldozers after World War II made the work of wall removal truly practical. On the Old Paul Farm in Dighton MA the early nineteenth-century stone-walled Grintry (Green Tree) and Peach Tree Lots were cleared of rocks and the wall between them buried by bulldozer in 1946: one of the first local efforts to "tractorize" farmland plowed by horse before the war.[113] Thus if small fields survived even on working farms until World War II, the likelihood of still finding them on gentrified farmsteads and abandoned farmland is surely great; equally great, however, is the likelihood that they would pass unremarked. Indeed, since an acre is roughly the size of a football field, is an urban New Englander likely to regard even a two-acre field as noteworthily small?

Walls served various functions. In colonial times field crops were fenced in while livestock roamed at large, a pragmatic reversal of English custom in a raw land. Later, in the face of increased settlement, tradition slowly reasserted itself, and livestock were penned, with swine usually the last animals suffered to range free by annual vote of the town. (Harvard MA

ceased to do so in 1804; Bolton in 1808.) The golden age of wall-building
has been dated to 1775–1825; barbed wire invented in the Midwest in the
1870s and widely used in New England after 1900 offered cheap and ready
solutions to many fencing problems, and brought to a close the common
era of stone-wall building. One of the basic functions of walls was rock
disposal. Removal and disposal of stone from fields and scythed mowings
was an endless task, for which wall-building was the best-known but not
the sole expedient. Stones were often pitched onto exposed bedrock ledges
in midfield, in piles typically twelve feet apart: six feet being about the
limit one could throw a hardhead. Large outcrops might become field
corners from which a half-dozen stone walls radiated.[114] Stones might also
be dumped into gulleys or on low wet ground, to be reclaimed later as wet
mowings with a layer of soil. Unwieldy boulders were toppled into pits
and buried below plow-depth, or holes might be drilled and charged with
black powder or simply rainwater, which split the rock more slowly but
quite as surely by freeze and thaw. Wall corners became rock bins that
monumented boundary angles for all eternity.

But what of the wall-net as the footprint of an open-air factory? A
common early nineteenth-century farm layout neatly exemplified in Dighton (1811), in Concord, and less perfectly in Petersham consisted of a house
and barn, with a walled cowlane that led from the rear of the barn between
walled tillage fields to the back pastures. (Precisely this layout was condemned in 1916 in USDA *Farmers Bulletin 745* as a waste of land.) Those
fields closest to house and barn one expects to be smaller, more carefully
planned and improved, with more walls and gates, and specialized in purpose. Typical of these were kitchen garden, orchard, cowyard, paddock or
livestock pen, barnyard, tillage field, and pasture. Colonial orchards were
stone-walled to protect the apple (or peach) trees from browsing livestock
and often planted on rocky ground unsuitable for crops. Mapped examples
in Concord, Lincoln, and Dighton were notably square and as small as an
acre, quite unlike the commercial orchards of today. The mapped fields of
the Sanderson Farm in Petersham MA in many places resemble a labyrinth
of livestock chutes and compartments.

Many offshoots of field walls that on the ground appear to run nowhere
and enclose nothing when viewed in conjunction with farm lanes, can be
seen as baffles to funnel stock into the adjacent field. Dyer describes and
maps a stone wall on Great Meadow Hill in Rehoboth that was "built in
such a way that sheep could be herded into an area where converging walls

funneled them into a stone holding-pen." Small, roadside triangular enclosures (as the half-acre "Little Close," Hutchinson Farm, Arlington MA)
that adjoin pastures may have been similarly used. Some walls resemble
guardrails meant to keep cattle from gulleys or swamps. The present network lacks gates, barways, and all wooden elements (surely considerable,
as half of Massachusetts fences in 1871 were of wood),[115] further compli-
cating interpretation. Because these field patterns are the work of traditional and not "scientific" farming, the nineteenth-century agricultural
treatises cannot be expected minutely to detail what they roundly con-
demn.

Tillage and pasture fields are distinguished by "double" versus "single"
walls. Double walls consisted of two parallel stone faces infilled with the
annual crops of plowed, picked, or frost-heaved stones. Wider walls may
indicate intense cultivation; smaller stones suggest root crops or potatoes.
The tillage fields they enclose would be expected to lie closer to dwellings
and roads, and prove smaller, flatter, less stony, and better soiled. Single-
walled pastures were likely remoter, rockier, or hillier, and might comprise
ten or twenty acres rectangularly laid out on compass lines or lengthwise
upslope. The commonplace classification of nineteenth-century farmland
as tillage, mowing, pasture, and woodlot must somehow be correlated with
the wall-net, if at times only negatively. Woodlands shown on town maps
of the 1830s, the era of maximum forest clearance, were undoubtedly
woodlots, and woodlots that had always been woodlots (not early abandoned farmland) are unlikely to be demarcated by stone walls. There are
no stone walls, for instance, in the woods around Walden Pond, which
historically were the woodlots of the Emersons, Heywoods, and other old
Concord families. In Petersham MA, maps of the relic wall-net and of the
1830 woodlands present roughly inverse images of each other. Boundary
walls without original agricultural justification might become embedded
in the field-wall pattern. Boundary walls are typically straight and continuous, and may follow original division lines; field walls skirt wet ground
and do not extend beyond the borders of the field. Until the mid-
nineteenth-century, town roads were often fenced with stone walls except
through woods and swamps.[116] Early railroads also were walled where they
skirted farmlands; it would be of interest to note if the later branch lines
of the 1880s, when barbed wire was available, were also. (Relationships to
roads and railroads of known age furnish some of the most datable contexts
in which to study stone walls.)

Sources of Evidence

Undoubtedly, abandoned farmland that has reverted to forest is the best place to study antique field patterns; too often, however, the ability to comprehend this network as a whole is lost in the undergrowth. By good fortune, the entire 280-mile wall-net in Petersham MA has been mapped. Some of the earliest, if at times dubious, visual documentation of the field network comes from Revolutionary-era maps done in the English carto-graphic tradition, such as those of coastal settlement around Boston in the *Atlantic Neptune* (1780). This shows a precise hedgerowed countryside not readily reconciled with anything to be found in the modern cadastre. In-deed, as some of these "fields" run a half-mile long and a quarter-mile wide, they can only be fanciful infill. On the other hand, the topographic fidelity of the Amos Doolittle engravings of the events of April 19, 1775, is widely acknowledged, and his view (no. 4) of the British retreat through Lexington shows a stone-walled field pattern with remarkable points of similarity to the present-day street-blocks near the old Muzzey School. (The stone cannon on the school grounds pinpoints the British fieldpiece depicted by Doolittle on the knoll.)[117]

Gone are the days when hilltop panoramas revealed a compartmented landscape, such as that shown in three early views of Amherst MA from the Pelham Hills (ca. 1833–1858) assembled by Polly Longsworth. (And indeed ominous for us, the accompanying old photographs confirm with glass-plate clarity that the fences of the nineteenth-century Amherst coun-tryside were perishable split-rail, four-board, zigzag, and not permanent stone!) While the old Coast and Geodetic Survey charts (such as those of Plymouth Harbor MA in 1875 and Edgartown Harbor MA in 1943) show richly detailed field patterns up to a quarter-mile inland, the USGS only rarely maps prominent field-lines with a red dashed line. Deed references can provide sure dates before which walls must have been constructed. Field names as recorded on estate maps and farm-plats shed further light and are often descriptive as to purpose, size, location, soil, or former owner: Ox Pasture, Cow Pasture, Middle Orchard; Nine Acre Lot; House Piece, Barn Piece, Schoolhouse Lot, Meetinghouse Lot; Sidehill Lot, Swamp Lot, Ledge Lot; Nicholls Lot, Wyman Lot (all Abner Marion estate, ca. 1858, Burlington MA). Rock of Ages Lot (Dighton MA) is humorously apt.[118] Old terms such as "piece" and "close" (as in Hubbard's Close, Con-cord) are of linguistic interest.

Finally, a knowledge of local agricultural history is vital to the interpretation of wall patterns. In much of New England, wall-building flourished in the golden age of sheep-raising, which unlike cattle-raising, "requires small fenced fields, so that the sheep may have frequent changes of pasture. With general stock-raising, the fields may be few and large." The dairy farms that succeeded the sheep farms, while heir to miles of stone walls, also benefited greatly from barbed wire introduced in the 1870s. Each era built on that before. Western Middlesex County MA, grew rich on grain, and richer still in cattle-fattening, creating a prosperous countryside of two-story farmhouses and well-tended, well-walled farms. This laid the groundwork for a later landscape of gentlemen's farms, rural estates, and ever grander walls. The gneiss walls that so handsomely bound the old farms of Middletown RI were built by black and Indian labor, a mute reminder of the only corner of New England where plantation slavery took hold. Elsewhere, farm and estate walls amounted to private relief work for men idled by business slumps, as at Winslow Farms, Leicester MA (1873) or the "Great Wall" of Whitinsville MA (1876–1878), which measured six feet high and six feet thick. ("Famine walls" are a commonplace of the Irish landscape.)[119]

Fields as Artifacts

The intriguing question of farm fields as cultural artifacts was explored by Stewart G. McHenry in his article "Eighteenth-Century Field Patterns as Vernacular Art,"[120] which compares the original eighteenth-century plats of thirty-four Vermont towns with current field patterns as seen in aerial photographs. Early Vermont was at the conflux of settlement streams from literally all sides: the Dutch from New York, French-Canadians from Quebec, and Yankees from New Hampshire, Massachusetts, and Connecticut. Each group brought its own ideas of farm layout and habits of cultivation; by selecting towns for which a dominant group of settlers could be established, McHenry sought to identify farm-field traits that would serve as cultural indicators of their presence. Out of some thirty possible traits examined, only those most clearly correlated with particular settlement groups are tabulated here. (Regrettably, he has little to say directly about stone walls, and often treats small fields and whole farms indiscriminately.)

Massachusetts: An acute-angled patchwork quilt of tree-
bounded, hillside fields, strung along irregular roads amid
forests.

Connecticut: A regular tartan plaid of square and rectangular tree-
bounded fields, adhering to original lot-lines, often with
smaller square fields within fields.

New Hampshire: Square or rectangular fields dotted by overgrown circular mounds of rock.

New York Dutch: A patchwork of small irregular or narrow rectangular floodplain fields with few tree-lines.

French Canadian (early): Long, narrow, tree-bounded lots of ten
to twenty acres running perpendicular to the Champlain lake-
shore.

French Canadian (modern): Large, regularly shaped, consolidated fields without tree-lines.

Although McHenry did not himself subsequently trace these traits back
to verify that they were well and truly representative of their presumed
places of origin, at least in the case of Connecticut, soil survey maps (based
on air-photos) for Litchfield County (1970) show the telltale plaid and
squares within squares, while others for Hartford County (1962) are often
shockingly—indeed un–New Englandly—rectangular.

Ditches

Walls and fences were not the only boundary-markers. Ditches were early
dug to the dual ends of drainage and boundary demarcation, and were still
"deemed legal and sufficient fences" by William Herrick in his *"Town
Officer"* in 1870. Ditch-lines as cadastral evidence when encountered singly
are easily ignored but when they occur repeatedly in areas of seventeenth-
and eighteenth-century settlement, oriented coaxially with other man-
made lines, it is well to pause and consider. Like stone walls, ditch-lines
are timeless; once the raw edges have healed, they might have been dug
thirty or three hundred years ago. In Ipswich and Hadley MA ditches
embanked on one side and perhaps made cattle-proof with rails were legal
boundaries within the fenceviewer's purview. At fifty feet a day in light
soil, ditch-digging far outpaced wall-building.[121]

A wealth of mapped examples can be found in the Concord (1979),
Lexington (1971), and Framingham (1950) MA quadrangles. In Bedford, a

straight ditch south of Davis Road through wetlands once known as Sancto Domingo Swamp lies parallel to and twelve hundred feet south of the ancient Concord-Billerica line. Of this and other ditch-lines one local historian remarked: "Nowhere do I know of a place in Concord bounded as in the eastern part of Bedford by ditches a foot or more deep, dug in the 17th Century through woods and field for miles, and surprisingly enough, still to be seen." In Lincoln, a ditched stretch of Iron Mine Brook lies parallel to and within one hundred feet of the ancient Concord-Watertown line and may somehow relate to the Farms Dividend; also in Lincoln, a ditch in Hobbes Brook meadow aligns or nearly aligns with other ditches, wood-edges, roads, and a cemetery boundary in the same neighborhood, much or all within the limits of the Flint's Farm grant made in the 1640s. Edmund Sanderson specifically records that boundaries of the Great Dividends were marked by "stonewalls, fences, or ditches,"[122] and we find evidence of their master axis in six mapped ditch-lines within the Beaver Brook meadows in Waltham and also in adjacent Lexington (strictly outside their scope). In South Sudbury, as mapped in 1950, Landham Road, two unnamed and presumably rechannelled tributaries of Landham Brook, kinks in Raymond and Woodside Roads, and even the New Haven Railroad show parallelism at one thousand– and two thousand–foot intervals, suggesting the effect of ancient rangelines. These intervals are also echoed by benchmarks, boundstones, and woodland blocks. The abundance of evidence from older settled towns argues that such ditch-lines cannot be heedlessly dismissed as of recent origin (although deed research alone can settle this point).

Town Bounds

The township boundary network is on a far grander scale than that of fields and lots, one measured in square miles and not mere acres. Unlike the Midwest, where the quarter-quarter section (the proverbial forty acres) and the township bear a genetic relationship, the lot-lines and town-lines of New England cannot be presumed to be interdependent. Just as field-lines are a cultural artifact, the town boundary network records the three-century sociopolitical evolution of the New England idea of community. The formative years of this evolution, which saw town-planting transformed from a theocratic enterprise into a speculative venture, significantly influenced the character of the six-mile square federal township. The town boundary network also has an age grain reflecting both technical advances in surveying and increasing ambitions for grand design. The Maine township grid is particularly illustrative because it spans roughly two centuries of surveys from 1650 to 1850, and proceeds chronologically from south to north, from "chaos" to order. The southernmost seventeenth- and eighteenth-century town-lines resemble those of Middlesex and Essex Counties in Massachusetts; the northernmost are prefectly oriented grid townships—in outline, at least—on the midwestern model. About Millinocket the grid-lines shift from magnetic (N 11° W) to true north.[123]

While the town boundary network can be examined in detail, mesh by mesh, when we study it as a whole, our focus must be on geometry in relative ignorance of history. Bear also in mind that on a state roadmap scale of 8.3 miles to the inch, 700 feet is less than the width of a fine pencil line, and wobbly lines run true. Smaller-scaled atlases go even further in making the crooked straight. USGS maps, which show not only boundaries, but also boundstones, are an indispensable adjunct to town-line study. At least three types of town-lines can be identified on such maps. There are straight-run compass lines, legally prescribed in advance, which strike across country many miles together. There are meandering metes-and-bounds lines, "tree-to-tree" surveys, which pick their way hesitantly forward, as if traced by an ant with a bump for direction. And there is a third class of artful dodgers remarkable for their abrupt, precise, and often rectilinear jogs and offsets. While the first two types as likely as not reflect original town surveys, this last class of town bounds, common in eastern Massachusetts, as we shall soon see, long postdated settlement.

While the town boundary network in much of New England is highly irregular, it is predictably consistent in one regard. If one counts the number of contiguous towns that each of the sixty towns of Worcester County MA touches, one finds that the average is 6.05, and that fully 42% of these conform to average and touch exactly six towns. This is called the "contact number." The average contact number for Washington County VT is 6.4 and 40% of the towns touch exactly six neighbors. In his influential Central Place Theory, the German economic geographer Walter Christaller postulated that the optimum settlement pattern consisted of town centers surrounded by non-overlapping hexagonal market areas, resulting in a honeycomb where each cell had a contact number of six. By contrast, the contact number, as effectively ordained by law, for a midwestern grid township is eight (just as eight squares adjoin the central square in tic-tac-toe).[124] The New England township, especially in Massachusetts and Connecticut, evolved organically and was freer to approximate the natural optimum. For a fertile period from roughly 1700 to 1900, Massachusetts towns tripled their numbers by a sort of cellular mitosis. Some split in halves, thirds, and not uncommonly quarters; still others ceded corners to new "middle" or "nip" towns that came into being by nibbling off pie-wedges of land from several neighbors.

Until the early nineteenth century, the sacred justification for these divisions, pled before the General Court, was an undue distance of travel to the meetinghouse on the Sabbath, generally reckoned as above three miles. Elaborated with single-minded efficiency, this would create a matrix of hexagonal townships six miles in diameter. Nothing quite so neat occurs, but one does find a honeycomb of multiangular townships that approximates it. Cyclists in Middlesex County, steeplechasing from village to village, are acutely conscious of how they lie about a half-hour's ride apart. This honeycomb would arise in districts sufficiently populous to support more than one minister per town: densities to be expected in longer-settled towns with good agriculture. Perhaps it was more flagrant in monolithically Congregationalist districts, where doctrinally riven parishes rove whole towns without the escape valve of Baptist or Methodist churches. In the nineteenth century, when parochialism shifted its basis from religion to economics, it was often seats of manufacture that were set off from old farming towns; whatever the case, the new enclave would require a definite basis of wealth. Some fifty new Massachusetts towns were set off between 1840 and 1900. Many mill towns that withdrew to enjoy their golden goose were later beggared when the goose ceased to lay (witness Clinton, Maynard, Millville, even Lowell MA). The estate haven of Beverly Farms was foiled in 1890 in its attempt to secede from Beverly to form a "tax-dodgers' paradise," but only after an orgy of rhetoric, pamphleteering, and attempted bribery had shed harsh light on the politics of town division. (The other side of the coin was contemporaneously seen as Boston repeatedly sought to annex the gilded suburb of Brookline.)[125]

By Farms and Lots

Although the largest concentration of these subdivided towns occurs in eastern Massachusetts, it should perhaps not be regarded as a phenomenon of suburbanization so much as one of parochialization. The threshold population that triggered these divisions seems ludicrously small today. Separation was a geographical solution meant to preserve eighteenth-century town meeting government and town polity from the nineteenth-century problems of increased population and manufacturing-agrarian cleavages. It was a move that plunged us ever deeper into the terrible beauty of Massachusetts political parochialism today. The nook-shotten boundaries of newly set-off towns reflected controversies that divided next-door neighbors and even next-of-kin. Lines were drawn to placate border-dwellers cool to the notion of splitting off from the old meetinghouse. By jogging judiciously among the outlying farms, the allegiances of the owners could be respected and sufficient ayes garnered at town meeting. In some cases, curious concessions were made, as that which left the Kibbe Place as a high and dry island of Concord within the limits of the newly set off Carlisle (readjusted in 1903). When a new parish was erected from the corners of Concord, Lexington, and Weston, in what is now Lincoln, an unusual provision allowed dissenters to remain in communion with their original parish and pay their taxes there.[126]

Parochial boundaries were run "by farms and lots," meaning that every effort was made not to split any one farmer's holdings between parishes: a sure prescription for boundary anomalies. Furthermore, records of baptisms and deaths for the parish of Byfield show that "the boundaries appear to have been changed repeatedly for the convenience of various families." Definitive boundaries, "with distances and angles," were not laid out until 1809 and 1816, and it seems that many parishioners felt aggrieved when the bounds were at last legally fixed. (Nineteenth-century town school districts would maintain this same flexibility for similar reasons.)[127] Byfield is a sort of "nip" parish of about a dozen square miles, taking in corners of Newbury, Rowley, and Georgetown. It is a curious fossil in that unlike Lincoln, Rockport, or any number of other one-time parishes, it never became a town. Parochial status did not simply placate the separatists' grievance that the meetinghouse was too far; it also afforded an experiment in self-rule, often transitional to incorporation. Even today, parishes such as the First Parish Concord post warrants for parochial meetings indistinguishable in format from those for annual town meetings. On the eve of the Revolution, towns ceased to be created and a tertium quid was devised: the district, one which excluded legislative representation (this status was essentially a gubernatorial maneuver to limit the size of an already unruly General Court).

When, at any rate, the time came for a parish to be erected into a separate town, and for its boundaries to be fixed, the full fury of parochial politics was unleashed, for town boundaries were immutable and had even greater socioeconomic significance. The resultant boundary jogs—often squarish tabs or notches of a few hundred acres that traced the outlines of border farms to be included or excluded—are a staple feature of Massachusetts town bounds, although often unrecorded in local histories. Sometimes they are treated as unique anomalies (the fate of many a forgotten commonplace); sometimes they are recounted with the air of unverifiable legend, as when in Bedford it is said that the owner of an outlying parcel in the Pickman Woods "seceded" to Billerica. When the southwest corner of Petersham was set off in 1801 by farms and lots to the newly created Dana, the area became known as Pilfershire, a folkname that rings with the contemporary resentment. In 1768 a deviation was made in the Deerfield boundary at its northwest corner, so that John Taylor, who had built there in 1759, might thereafter live in Shelburne. Slightly enlarged in 1896 (undoubtedly by the Massachusetts Topographical Survey), it still exists. In Lincoln, the town-line jogs across North Great Road because members of the Brooks family parted company on the issue of a new town. When Harvard was set off from Groton in 1732, the language of incorporation specifically excluded "Coyacus Farm" and the "dwelling house of James Stone" from the new town. (Nonacoicus Farm was a seventeenth-century land grant to Simon Willard now within Fort Devens; the old boundstones initialed "H-G" now mark the Harvard-Ayer line.) What have been called "crenelated" county lines in Georgia and Alabama resulted from similar options to choose one's jurisdiction.[128]

One notorious instance is the Caleb Wheeler Farm, a 43-acre diamond-shaped outlier of Sudbury MA entirely enclosed by Wayland save for a narrow stub of land. The town-line between Sudbury and Wayland (alias East Sudbury in 1780) was laid out by joint committee and displays the expected "farms and lots" irregularities, breaking the landscape generally along extant natural or cultural lines: the Sudbury River, roads, and for much of the way, property lines, one of which was marked by ditches and others, presumably, by stone walls. While no committeeman was an abutter to the line chosen, family interests seem likely to have been involved, as Jonathan Rice was a member, and Nathaniel Rice an abutter. More significantly, the West Parish protest meeting that cited the unacceptable loss to Sudbury of both its training field and gravel pit was moderated by one Asahel *Wheeler*. Both these places were reserved to Sudbury when the lines were ultimately drawn; the gravel pit location, needed for causeway maintenance, was the diamond-shaped outlier and included the Caleb Wheeler Farm.[129] In the twentieth century, Wayland ingeniously turned the tables. Today this errant exclave of Sudbury, now bisected by U.S. 20, is more familiar to motorists as the site of the Wayland landfill. Thus a relic of colonial parochialism has become a classic case of modern NIMBY-ism (Not In My Backyard), in which nuisance uses are pushed to the edge of, or indeed, out of town.

Such jogs did little to simplify the problem of establishing and maintaining the already tangled town boundaries. Only in the 1880s and 1890s did the Massachusetts Topographical Survey finally wade into this legal bramble patch: statutes and records were researched, hearings held, and the attorney general consulted. As a result, many town-line tangles were cleared up: accepted and perambulated boundaries were ruled in error, duly established corners shown to be omitted, witness monuments found to be mistaken for corners. Some of the more flagrant but legal anomalies were legislated out of existence. In one case, the Waltham-Lexington boundary was clouded by the unknown whereabouts of the farm of one Matthew Bridge, who being "very inconveniently situated for attending public worship of God" had successfully prayed the legislature in 1755 "that he and his estate [might] be set off to the town of Waltham." A conveyancer, retained to look into the facts, concluded that while the Bridge Farm could no longer be located, it certainly had existed within the present limits of Belmont, and could have no bearing on the line in question. Even to this day, chapter 42 of the Massachusetts General Laws provides for selectmen to perambulate their boundaries once every five years, although gaps in the quinquennial dates painted on Concord-Lincoln boundstones suggest this colonial, indeed immemorial, custom (it has English precedents) may be one honored more in the breach. Instead of the colorful assemblage of selectmen stereotypes once witnessed by Thoreau, towns may now act on their own and report to abutters by registered mail. Nor does perambulation guarantee the legality of the lines walked.[130]

Ad Filum Aquae

As the most salient of linear natural landmarks, it might seem that rivers and streams would be among the most common boundaries since colonial times; for example, the Merrimack and Kennebec bounded the grant of the New Hampshire–Maine coast to Mason and Gorges in 1622. Rivers also figure prominently in the language of the original 1629 Massachusetts Bay patent, but not in the way one might expect. Massachusetts was granted the land lying between a line three miles north of the course of the Merrimack River from its headwaters to the sea and a similarly defined line three miles south of the Charles River.[131] Much later, the Massachu-setts–New York line was agreed to lie twenty miles east of the Hudson. River bottoms and intervals offered the richest farmland for crops and fresh marshes for fattening cattle; early grants eagerly took in both banks. Among seventeenth-century towns, Concord and Lancaster still straddle their streams; Sudbury and Springfield lost their cross-river lands to daughter-towns.

While by legal custom riparian boundaries run *ad filum aquae*, to the "thread" or middle of the stream, exceptions occur. As ordained by the privy council in 1764, New Hampshire owns the bed of the Connecticut River to the low-water mark on the Vermont shore. New Hampshire issues the fishing licenses and presumably could collect tolls on the bridges for which it still pays the lion's share of two-thirds or more. Vermont frugally acquiesced until the early twentieth century when the taxable worth of the new hydroelectric plants aroused its cupidity and it filed a bill of complaint: suing and ultimately losing in the Supreme Court in 1933 to have the boundary relocated to the thread of the stream. In 2001, the Court likewise ruled whether the Maine–New Hampshire border runs to the middle of the main navigable channel of the Piscataqua or along its northern shore. "Ad Filum Aquae" was vindicated, and the royal decree of 1740 upheld. Again, money was at issue: whether Portsmouth Naval Shipyard on Seavey Island was truly in Maine, and subject to the Maine payroll tax.[132]

Rivers shift in their beds, and boundaries generally shift with them, as when the Connecticut River broke through the Oxbow one night in 1830, short-cut its length by three miles, and transferred four hundred or more acres from Hadley to Northampton MA. And while the law might fix the Wethersfield-Glastonbury CT boundary as the course of the Connecticut in 1690, the river knows no law. Boundary and river came to crisscross five times before being readjusted to the thread of the stream in 1874. (Through legislative inadvertence Wethersfield still retains 122 acres on the wrong bank of the river.)[133] Acts of God are one thing, man-made changes quite another. Some of the most perversely circuitous boundaries result when waterways are filled. Due to embankments on both sides of the Charles River for Storrow and Memorial Drives, the Boston-Cambridge city-line no longer runs to the thread of the stream as it now is, but wanders lazily back and forth across the breadth of the river, Boston all but touching the Cambridge shore where the Longfellow Marshes were filled. The mean-ders of long vanished salt creeks still dictate segments of the Cambridge-Charlestown and the Everett-Chelsea lines. A two-and-a-half-mile squiggle in the Norwood-Canton town-line was left high and dry when the Neponset River was rechanneled to accommodate Interstate 95. The defunct course of the Swift River still defines town-lines beneath the waters of Quabbin Reservoir.

Ridgeline and watershed boundaries also exist, but are more easily defined in the abstract than located by survey on the ground. Many monumented corners lie in reasonable proximity to hill summits as later mapped by scientific topography. Ridgelines in rough country were virtually unsurveyable, and instead might be approximated, as in the Warwick-Orange MA line, by a stairstep give-and-take line that awarded equal acreage to each town. Richmond and Lenox were similarly divided, a move that "was eminently sensible considering the abrupt mountain range that split the town and presented difficulties of communication." The line was arbitrated by a committee "of indifferent and discreet men" from neighboring towns about 1770; such boundary juries were a customary court of last resort in land disputes. "The final division resulted in a zigzag boundary more or less on top of the mountain which has ever since given rugged exercise to the selectmen of the two towns who are required to walk their boundaries every five years."[34] And indeed heights of land figure often enough as town bounds that uphill pedalers may recall topping off at a town-line marker, only to coast triumphantly down on the farther side.

{ *Four* }

ROADS

T HERE are 1,941 townships in New England, each of which has a
unique fingerprint: its road-net. Without systematic scrutiny, fin-
gerprints are little more than smudges; with systematic scrutiny comes the
touchstone that underpins the whole science of dactylography: each print
is unique. So it is with road-nets. As the "greatest corporate work" of early
settlement, they constitute the largest cultural imprint on the land, their
patterns fully recorded on maps, yet defying interpretation.[1] We lamented
earlier that we could see the cadastre but as through a glass darkly, but the
road-net we can truly see face to face. To study a road-net as an artifact,
we must first identify—and devise terms for—its attributes. These may
describe its basic geometry (straight, crooked, cardinally oriented), the
ways roads are joined (crossroads, star roads, fork roads), the shapes of
road-bounded areas or "road-meshes" (squarish, triangular), and the overall pattern of the road-net itself (gridiron, spiderweb). Its geographic context—that is, its special relationship to the terrain and to cultural features
(houses, churches, mills, allotment schemes)—serves further to define it.
While much of its physical character may be determinable only in the
field, geometric form and geographic context can generally be evaluated
from maps and will be our main concern here.

We have already examined elements of the road-net in townplans, street
villages, and cardinal crossroads where the act of design is incontrovertible,
either because they were plotted on paper in advance, or laid out with a
compass or in accord with some template. Such toolmarks have been taken
as evidence not simply of the hand but the mind of the artificer: proof
that they are indeed conscious artifacts. Yet man-made objects need not
be so obviously contrived. We know that the entire New England road-
net is manmade, that it is an artifact, even though only a small portion
conforms to a simple, arbitrary geometric pattern as do the roads of the
federal grid. Yet the complex variety that makes it maddening to explain
also enhances the likelihood that its patterns may encode something significant about the character and circumstances of the people who made it.

Little more can be deduced from a midwestern road-grid than the technical competence of the surveyor and, when encountered on such a vast scale, the resolve of a strong central authority. To understand the New England road-net, it will be of help to look both backward and forward in space and time: east to England, and west to the Middle West, not simply to establish a thread of historic continuity, but also to elicit comparisons and contrasts. The properties of an orange stand out more tangibly when juxtaposed with an apple and not another orange; such contrastive models will help us to perceive form in a road-net too often dismissed as amorphous, spontaneously generated, or organic.

English Model

Qualifying the road-net as "organic" deftly deflects inquiry into what has heretofore been taken as a fundamental requisite of an artifact: that it have a template. Organic suggests that something occurs spontaneously under given circumstances, without reference to rule or prototype. Did the New England road-net directly follow an English model, or are its attributes the specific outgrowth of the facts of New England settlement? Could the same road-net arise organically in New England and old England? If the scheme of occupance, mode of travel, and facts of terrain were the same, might the natural laws of organic road formation take over and yield the same results? Or might the New England road-net conform to English precedent only insofar as English roads were not planned, but simply "happened?" Was the cultural debt to England merely an indisposition to plan things such as roads (towns, too)?

The New England road-net is unlikely to possess an English template in the same way as a house, a gravestone, or even a village green. Specific segments, whether a village street or civic crossroads or common rangeway, undoubtedly did, especially when laid down by the first planters or others with experience or knowledge of the mother country. The road-net as a whole, however, was a multigenerational effort that took shape long after the deaths of the first comers: it was the work of people who knew little of England or even of the totality of the network they forged. Old and New England had altogether different patterns of property-holding and occupance, and presumably different road-nets in result. New England abandoned nucleated villages and common fields for dispersed farmsteads held in fee simple and thereby worked a radical revolution on traditional English ideas of countryside. Furthermore, the New England road-net was strongly sectionalized by the (roughly) thirty-square-mile township. Each township is a separate windowpane in which a spiderweb has been spun; countries such as England without this characteristic framework, are unlikely to have a road-net that is so regularly compartmentalized, even if the road-nets are demonstrably organic.

Very different mapmaking traditions in Britain and the United States make comparison of their road-nets difficult. The amount and nature of the data recorded, the choice of color, symbols, and typeface, the alien toponymy—all solidly convince one of the differences of the landscapes long before one begins to analyze the road-net. Different scales and standards of inclusion make overall impressions misleading, and raise doubts as to the comparability of coverage. Because British cartographers employ the traditional double-lined road pen, and do not classify roads by color (like Rand McNally), the road-net stands out much more starkly, like the leads of a stained-glass window. Maps of East Anglia and eastern England in the Landranger (1:50,000) series and the Passport (1:300,000) road atlas were examined as typical of the home districts of many emigrants. Overall, the road-net appears to be "stouter," with well-traveled main roads between populated places, while secondary roads seem to be frailer, purely local, and often farm lanes. Such hierarchicalization contrasts with an American expectation of equal and universal accessibility, epitomized by navigating the Cartesian grid of midwestern section roads. English villages and hamlets do not crop up so predictably at main intersections, nor are triangles of various sizes so explicit an element of the road-net over much of the area. In addition to these general contrasts, a stronger variety of localized road patterns than New Englanders are accustomed to was observed. Two may hold lessons for New England.

In northwest Norfolk between King's Lynn and Fakenham an area of about five hundred square miles of presumably large landholdings displays a remarkable triangularity in its road meshes. Triangles are typically from one-half to two and one-half square miles in area, and most of the few villages lie at the center of a spiderweb of five or more radial roads. Much of this district was historically barren, a place where, in the earl of Leicester's memorable conceit, two rabbits might fight over one blade of grass.[2] East and southeast of this area, in more thickly villaged countryside, the triangularity becomes less pronounced, or perhaps covert: concealed by a denser road-net. The suggestion is that triangularity is particularly a feature of less densely settled country with fewer population centers—an idea with obvious bearing on New England. A second area, the Lincolnshire

Fens, from the 1640s onward witnessed massive drainage schemes under the direction of Dutch engineers. It is largely made land, and the road-net is marked by long, straight, parallel lines, in places describing perfect two-mile-square boxes or rectangles a half mile wide by four miles long. Anthony Garvan has pointedly instanced the large-scale allotment methods employed in this reclamation scheme as a contemporaneous English parallel to colonial lot-laying techniques.[3] (Both of which were overscaled versions of medieval common field division patterns.) The fens were fertile, flat, and expensively won farmlands and it is no surprise that the cadastral pattern inscribed on them has proven far more durably visible than that imposed on the New England wilderness. In much of East Anglia, the road-net seems more disposed to quadrangularity than in New England: nothing the least like a grid, but rather road-meshes describable as loosely squarish, whether because right-angled, parallel-sided, or with the look of crushed boxes.

We well know that the true gridded road-nets of the Midwest came about through conscious design and prior survey. How does an even moderately squarish road-net arise that is presumably unplanned, particularly when we have reason to believe an organic road-net is marked by triangularity? The answer comes in two parts, the second of which radically subverts the question itself. The first and fundamental explanation is that medieval roads and paths served and traversed farm fields and that the character of the English road-net was fixed by the rectangular bias of field patterns. They were field lanes. Such perhaps best explains the situation in Essex and Kent, which were untouched by parliamentary enclosures, and which probably retain largely unchanged the local road-net known to the emigrants of the early seventeenth century.

Elsewhere, however, from the seventeenth to the nineteenth century, the enclosure of the common fields would work a revolution that only simplified, rationalized, and magnified this tendency to quadrangularity. This was most marked in the open field counties of the Midlands, with an area centered on Northamptonshire the most profoundly affected. As W. G. Hoskins has observed, here a new Georgian landscape was created, one where "the immemorial landscape of the open fields, with their complex pattern of narrow strips, their winding green balks or cart-roads, their headlands and grassy footpaths" was transformed into "the modern chequer-board pattern of small squarish fields, enclosed by hedgerows of hawthorn, with new roads running more or less straight and wide across the parish in all directions." These roads were laid out by the enclosure commissioners to run nearly dead-straight between villages; indeed, rather than deviate, they might bypass a village altogether, with access relegated to a side road. "There is none of that apparently aimless wandering in short stretches, punctuated by frequent bends, going halfway round the compass, which characterizes the by-roads in country" not affected by modern enclosures.[4]

Thus we undercut cherished assumptions as we gain new insights. W. G. Hoskins's own maxim that the English landscape is older than you think is grossly inapplicable in much of the Midlands and some areas to the east as well. Between 1750 and 1850 more than twenty-five hundred separate acts of Parliament obliterated and remade the ancient cadastre of perhaps ten thousand square miles of England. The revolutionary course of English land reform from open to enclosed fields was foreshadowed in the centurylong transition from nuclear to range township in New England. The heart of England has a new planned landscape, describable as Georgian both literally in time (specifically, George III) but also in its orderly geometric spirit. In New England this same term Georgian (as opposed to postmedieval) has come to be diversely applied by students of material culture to a symmetrical style of architecture, to the way in which household trash began to be buried in pits rather than strewn over the yard (ca. 1750), to the design of Samuel Slater's spinning frame (1790), as well as to the trend towards rectangular land surveys that culminated nationally in the Ordinance of 1785.[5] Thus, while there may be no English model as such for the New England road-net, by studying their points of similarity and difference, together with their root causes, we can begin to discern the distinctive structure of something too commonly dismissed as structureless.

Midwestern Model

The federal grid plan and New England radial plan were fundamental to very different settlement schemes. In the Midwest, the road grid was the basis of the cadastre, as it served to subdivide each township into thirty-six sections, upon which depended all subsequent aliquot property divisions, from the 160-acre quarter-section down to the 40-acre quarter-quarter section. The grand design was to speed the sale of lots, to retire the Revolutionary War debt, promote even settlement, and forestall endless litigation by the elimination of metes-and-bounds. (By the 1770s, the cost of title suits in New Jersey was said to have exceeded the actual value of the land itself.)[6] Into the bargain it accomplished a systematic transect survey of timber, soils, and resources of the new country. It was a farsighted and costly scheme, authorized and underwritten by the federal government.

One way to evaluate alternate road-nets is to identify that with the "least cost to builder" and "least cost to user." In eighteenth- and nineteenth-century New England, the builder *was* the user, because of an annual obligation to work on the roads (or make payment-in-lieu). The basic 84-mile road-net of a grid township was simply not cost-effective in light of true-life New England conditions. It ordained too many roads in sparsely settled areas, too few in thickly settled areas, and dictated their placement irrespective of terrain in an era of axe, pick-and-shovel, and oxcart. In truth, things were not far different in the early Midwest, where farmers worked off their road tax much as in New England, and where nineteenth-century farm-to-market roads were not at all rectilinear. As one Iowan remembered: "The roads were wagon tracks running diagonally from the village to the farms, and in main roads from town to town; but these were gradually crowded by tillage from their antigodlin courses to their present places on the section lines, all running north and south or east and west." While the road-net might be initially organic, in time, the indomitable cadastre reasserted itself and assured that this phase did not last. The power of the postmaster general to withhold rural free delivery (established 1896) wherever routes were impassable would become one of the midwestern farmer's most tangible incentives to keep the road-grid viable.[7]

In New England, the radial road-net, which took terrain into account and approximated the shortest path to the meetinghouse, was better suited to the theocratic nature of the early towns. Central access was more important socially than the maximum interconnectivity of all points. The comparative diagrams make clear how much more efficient and adaptable the radial scheme was (see fig. 11). If the radial roads were allowed to deviate according to terrain and location of houses, an even more commodious system could be bought at the cost of a few extra miles. The radial road-net could also be constructed piecemeal over time, as a multigenerational effort, with each segment added and appreciated for its immediate practical value. As settlement patterns changed, branch roads could be relocated or abandoned, as they served only a finite number of points. As the best lands of a township tended to be taken up first, it is quite probable that primary, secondary, and tertiary roads connected farm

- total miles of roads
- % of township within ½ mile of road
- maximum road distance to town center

Radial

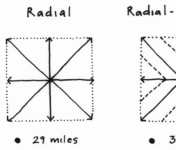

- 29 miles
- 66%
- 4 ½ miles

Radial - Dendritic

- 37 miles
- 100%
- 4 ½ miles

Federal Grid

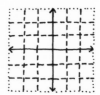

- 84 miles
- 100%
- 6 miles

Star - Road

A - Road

Cross - Road

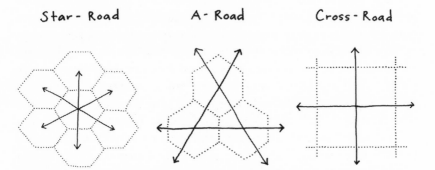

Radial roads favored triangular over rectangular meshes.

FIGURE II
Radial and Grid Roads Compared

sites with first-, second-, and third-class soils. Under this scenario, as unprofitable farms were abandoned, and the branch roads withered, the integrity of the network was not threatened because the trunk roads through first-class farming districts would survive.

Organic Model

In simple terms, an organic road-net is one that evolves slowly over time, guided by natural and cultural factors and features on the ground, and that is free to change as these influences change. Terrain, soils, cadastre, settlement patterns, traffic, mode of travel, road-making costs, population density, type of agriculture—all bear on its formation. It is only free-form in the sense that its form is free to accommodate itself to such geographical forces and lacks a neat, preplanned geometry. The net evolves through trial and error; unused roads wither, others are improved, new ones are opened up. Yet organic is by no means a synonym for random or quirky. In the same way that vernacular architecture can be described as organic even when it follows traditional rules of carpentry, the road-net can be organic and still follow rules and evince a characteristic pattern.

Earlier, we saw how New England towns might be conveniently regarded as hexagonal in shape. In such a honeycomb, if all centers of contiguous towns are interconnected with straight roads, two types of intersection, which I term the *star road* and the *A-road*, dominate (see fig. 11). These are as elemental to the organic New England road-net as is the crossroad to the midwestern grid. The centrality of star roads is implicit in the adage that all roads lead to the meetinghouse,[8] but still others found in godforsaken places, far from church (or mill), belie this wisdom. Star roads need not have six arms; five may in fact be commoner, as attested by placenames such as Five Forks, Corners, or occasionally Points; "Seven Stars" are exceedingly rare. Some arms may not converge on the village center, but rather feed into the main road on the outskirts. While the A-road is rarely encountered with diagrammatic clarity in the landscape, it informs the strong but imperfect triangular lattice of the New England road-net. Typically it is made up of three to five roads, rarely the nexus of six as diagrammed. The triangular road-mesh at the heart of the A may be smaller than an acre, or greater than a square mile, and is more readily recognized on a map than in the field.

Maps enable us to quickly assess the road-net of a town. Thoreau's Concord (ca. 1850) showed seven road forks, two crossroads, two star roads, and twelve A-roads that enclosed triangular road-meshes anywhere from 24 to 350 acres in size. Eleven other road-meshes were loosely shaped polygons; generally larger (125–750 acres) than the triangles, these circumvented natural obstacles such as rivers, swamps, and ponds, and did not represent the conjunctions of true desire lines. The road-net of Bolton MA (1831) curiously had no star road, no central spiderweb, but rather was polarized along an east-west travel-way as old as the old Bay Path, which predated and predisposed settlement. Thus the road-net holds clues to the history of a town, although structural similarities can be deceiving. While Concord MA and Acworth NH may both have a radial road-net, the one lies in a plain and the other on a hilltop and the genesis of each was subject to different topographic controls. The variable nature of New England town road-nets is in marked contrast to that of the ideal midwestern town, which could boast twenty-five crossroads, thirty-six mile-square road-meshes, and nothing in the way of a triangle.

Quo Vadit

So far we have considered the form of the road-net but said little of its genesis. Hodgman's *History of the Town of Westford* (1883), like many nineteenth-century town histories, offers extracts of town records that bear on the laying out of the whole hierarchy of the road-net: county roads, town ways, bridleways through gated fields, "particular" ways (which might grant a limited right-of-way to meetinghouse or meadow), and paths. Such transcripts make dull reading, but they do record bounds or landmarks that defined the course and often the rationale of the road: "a stake and stones on Thomas Heald's land" (1757), "a plumb tree between Daniel Raymond's house and barn" (1764), or "a stake one rod south from Mr. Thomas Coming's house" (1760). Over the years 1719–1778 covered by the extracts, increased settlement is witnessed by the landmarks chosen (such as fields, fences, paths, barns, and dwellings); significantly, crooks in the road often correspond to such cultural features. Over this same time, surveys also grew in precision, with compass bearings for each leg of the line more commonly recorded.

By 1883 much of this colonial road-net had been wiped away, and could not even be located by those knowledgeable persons entrusted to write the town history. Parts vanished almost from the first, as abutters offered "in exchange of way"[9] new routes for segments of town ways they wished discontinued. Before the twentieth century, a more universal network of roads and paths, often redundant, wove through the landscape and choice of route might vary according to one's conveyance, one's load, the weather, or the seasonal state of the ground (frost-bound, snow-covered, March-muddy, summer-dry). Much of this web of ways was never public, and in consequence never mapped. Eugene A. Wright describes the "old cart roads" in the turn-of-the-century Plympton MA of his boyhood. This was the end of the dirt-road era, when the network was "much more extensive than at present" and "took one just about any place in Plympton he cared to visit. Many short roads led into fresh meadows, wood lots, old farms, and fields." The flat, sandy Snappit woodlands, dotted with charcoal kilns, gave rise to cart roads that "wound around all over." If poorly kept up, they were not a tax burden; if infinitely more circuitous than today, they multiplied the ways from here to there.[10] Cars and the advent of asphalt changed all this. Roads became fewer, straighter, better engineered, and more firmly etched into the landscape. Wet-weather runarounds became unnecessary; slow-going cutoffs were no temptation when the longer paved road made better time; unmade ways barely negotiable by oxcart were less so in a tin lizzie, even if they led anywhere anyone in the post-agricultural age cared to go. The road-net shriveled back to its main stems. The rationale for these old roads was essentially the exploitation of resources; as local agriculture and self-sufficiency declined, such reasons began to disappear, and in time the roads themselves.

Contour-keeping

One basic dictate of road location is to maintain the levelest route with the easiest grades, something readily gaugeable on topographic maps by the degree to which a road crosses the fewest feasible contour-lines between two points. Old hill-town road-nets often set this logic at naught, as the original farms (often later abandoned) typically kept to the heights and avoided valleys.[11] Many such roads are too steep by modern standards, and have been discontinued, "pent" (barred), or closed in winter or to heavy tonnage. Another rule of thumb was that good soils made bad roads (the rockier the better!). The sagacity with which old-time selectmen and highway surveyors could marry roads to terrain is all the more remarkable as they read these contours by eyeing the ground. Topographic and historical maps afford an abundance of concrete examples with which to unriddle why old roads go the way they go.

Connect-the-Dots

References to houses and barns as cultural landmarks in early road surveys raise an important question: to what extent did houselots and even houses preexist the roads? Did the houses spring up along the roads (as in the modern experience)? Or rather, did the roads string together houses and farms? One old-time New Hampshire town historian asserted that "the roads crooked . . . around from house to house, without reference to directness, so as to accommodate settlers in the best way."[12] This theme was elaborated upon by a 1928 surveyor's manual: "the early settlers connected their dwellings with paths which have become streets." The location of roads was "from house to house and along private property lines—where each owner gave one-half the right-of-way necessary for the road, or, in order to avoid damage to an individual's farm land, the road location was detoured through less valuable land."[13] The affinity of early houses for bends in roads suggests that house seats determined the location of the road and not the other way around.

The 1830 map of Fitchburg MA shows many houses characteristically positioned on the outside of the turn, some fifty to one hundred feet down from the actual crook, as though the road shifted to skirt fields and farm-yard before the house was reached. Such evidence suggests that the road linked preexisting houses in connect-the-dots fashion. The roads grew out from the center to meet the houses, branching like trees, the branches often crooked as they grew jointedly from outlying house to outlying house —as though the ripened fruit called forth the apple boughs and not vice versa. The loss of original houses due to farm abandonment (noticeable even between the 1830 and 1852 Fitchburg maps) would efface some of this evidence on the ground (unless cellar holes remain) and road relocation might follow. However, even today on country drives the illusion that one is about to plow straight into an old farmhouse before the road veers off is still quite common. Some of the kinkiness of these early-mapped roads is undoubtedly artificial, the result of straight lines drawn between plotted survey points. The Athol MA topo sheet (1970) shows many country roads connecting clearings in the woods, many of them old farmsteads, as indicated by the pepper-and-salt of black and white "house dots," which usually betoken farmhouse and barn. Has the countryside ironically come full circle and reapproximated the conditions under which the road-net was first laid down, a late eighteenth-century wooded landscape of scattered farmhouses on cleared fields strung together by a road?

The practice on mid-nineteenth-century town maps of recording the distance of each house in rods from the center conveniently enables us to determine the distance between houses. On the Fitchburg map of 1830, it is typically between one-fifth and one-third of a mile, often less, rarely more. This means that walking the roads one was likely to go no more than four to seven minutes without passing another farmhouse. Had all pre-1830 houses survived, the bicyclist at eight miles per hour would pass a colonial or postcolonial farmhouse every two minutes, and a motorist proceeding at thirty-five miles per hour would pass more than two a minute. How ghost-ridden these old New England roads must be!

Path-ology

Paths are the ultimate ramification of the road-net that penetrated the roadless places. Accordingly, most paths should begin and end within one road-mesh; cross-country paths that cross roads obliviously must come under suspicion of either predating the roads they cross or being decayed roads themselves. Paths are timeless artifacts, in a negative sense; they appear to possess no datable features and, unless documented as an old stage road or abandoned town way, even the oldest is effectively no older than living memory. In colonial times, paths had no legal character; only when accepted as town ways did they gain mention in the town records. Yet stock phrases such as "as the path now goeth" (1731) and "as the path now runs" (1722), which figure in the official language of road acceptances, indicate that segments of an earlier quasi-public network of paths were continually spliced into the town road-net.[14] As they were never thoroughly mapped, it is not possible to compare the nineteenth- and twentieth-century path-nets of a town to learn of its growth and decay. Path-nets do not appear on the town maps of the 1830s, and even today are beneath contempt to the USGS, unlike in Britain, where footpaths have legal status and are mapped by the Ordnance Survey. Because footpaths in New England lacked legal and cartographic standing, and rarely figured in late nineteenth-century fiction (perhaps because New England rural genre writers were women and their characters often elderly homebodies), remarkably little is recorded about them.

Thoreau's writings form the curious exception, and the cult of Thoreau led directly to a unique undertaking by Herbert W. Gleason, a minister-turned-photographer, to map the path-net of mid-nineteenth-century

Concord, one of the few such records we have for a New England town. Gleason (1906) sketchily rendered some twenty miles of paths frequented by Thoreau, most extensively in Walden and Estabrook Woods. While the toponymy of this map is heavy with Thoreau's own romantic nomenclature (very like that of Jacob Bigelow's Mount Auburn Cemetery, a place undoubtedly familiar to him from his college days), anyone with knowledge of some rural quarter can testify that there is nothing implausible in the path-net itself, other than that it is woefully incomplete. ("How little there is on an ordinary map," as Thoreau himself once said!) Gleason restricted himself to wood roads and true footpaths and not farmer's cart paths over open fields, which Thoreau certainly used but at times spoke of with disdain. Two of the footpaths shown are identifiable on the current Walden Pond trail map (1994) as the Concord and Watertown forks of the Old County Road. These had ceased to be public ways before Thoreau's sojourn and the Old County Road was in fact bisected by the Fitchburg Railroad (1844) without a grade crossing or dry bridge. As currently mapped, the mesh-size of the Walden path-net is generally about one and a half to six acres, which agrees with its longtime use as woodlots by the Emersons and other Concord families, who would need to get around to cut trees and collect firewood. While Walden woods may be one of the most heavily trodden tracts in New England, there is no reason to suppose that it was not well provided with many of these same paths long before the twentieth century. Perhaps "R.W.E.'s wood-path south side Walden" which Thoreau followed on September 12, 1851, is still there.[15] Indeed, the actual shoreline path is believed to have been of aboriginal origin. Open lands in parts of Weston MA as shown on the *Explorer's Map* (1980) also exhibit loosely squarish meshes of one to six acres in size, and overall a path-net that is not radial, but quadrangular, suggesting equal access to field and woodlot was the goal.

If exploitation, not communication, was the principal rationale for the path-net, might not the traditional tax-assessor's fourfold scheme—tillage, mowing, pasture, and woodlot—translate itself into distinct path-nets on the ground? Interpretation is complicated by a changing landscape that has been cleared, farmed, reforested, cut over, and forested again, so much so that Massachusetts (which was three-quarters cleared in 1840) is now almost three-quarters wooded. Yet while the face of the land was altered beyond recognition, the path-net itself might be carried over from phase to phase, and survive. A thoughtful intercomparison of path-nets in present and former tracts of known land-use might reveal how exploitation dictated their form, and how identifiably such forms survive on postagricultural land.

The cowpaths of old pastures are perhaps the most familiar; while not strictly artifacts (being trodden by cows) they do possess characteristic attributes of paths that are man-made. Presumed relic cowpaths in Lexington Great Meadows kept to slight contours on easy terrain, lacked straight lines, had narrow, deep-worn footways, interconnected old pastures and the thickets that succeeded them, passed through stone walls only at gates, barways, or wide gaps; led to miry, trodden brooksides and other watering places, and frayed out in braided sidetracks, or dead-ended in wet meadows where cows fanned out to graze. Lexington Great Meadows has not been grazed since 1872. Thin sandy soils, and frequent groundfires have stunted growth and retarded the return of woodlands. Could a cow path-net under such conditions, and reinforced by human foot traffic, possibly survive for more than a century? Or is the present path-net entirely modern? Comparison with the still grazed Bigelow Field in Concord MA sheds helpful light on such questions. However, each time the path-net is carried over into a new exploitative phase, it renews its rationale and loses much of its obvious historicity. Yesteryear's cowpath becomes indistinguishable from a bike trail that sprang up overnight, in the same way that an old stage road is labeled a jeep trail on a topographic map. Much thought and fieldwork is still needed to unriddle the past history and modern fate of the New England path-net.

Nantucket

Road-nets and path-nets in the pre-motor landscape were so interwoven that they cannot be readily told apart. Are cart paths roads? Are abandoned roads paths? The complexities of these questions are well exampled in the rutted roads over the old Nantucket sheep commons (moors):

> These could hardly be honored by the name of roads, and yet they were the only roads that existed. Their sandy ruts were often worn down to a depth of a foot or so, and once the vehicle was in the channel there was no turning out until one came to what was called a "soft place," an open spot of sand where the carriage might make a turning. If by accident one's chaise or "box wagon" turned into the road to Polpis or Wauwinet, it had to stay there until it reached that destination. When

these ruts became impossible, someone would start another route which in time became as bad as the old one. Thus the commons were streaked in every direction with wheel tracks.[16]

One might conclude that the rutted road-net was entirely organic, and initial impressions, both from maps, and from jouncing over the moors on a bicycle, only confirm this. The two vast tracts where rutted roads can still best be seen are the south plains (my term) in the near southwest and the middle moors in the east. In both areas triangularity is strong. In the south plains dirt tracks fan out from roadheads and run to the south beach. (Curiously similar patterns are also found on the nearby islands of Tuckernuck and Martha's Vineyard.) In the middle moors, a spiderweb of rutted roads loosely converges on Macy's Hill ("Altar Rock").

Yet a closer inspection of maps (that of the Nantucket Conservation Foundation is most helpful) reveals a rectilinear overlay on the boundless commons. In the south plains, a measured interval (likely 200 rods or 3,300 feet) separates Barrett Farm, Ram Pasture, Hummock Pond, Somerset, and very nearly Miacomet Roads, this last deflected by a pond. Some are dirt, some now paved, but all are reasonably straight, parallel, and aligned within two degrees of true southwest (N 137° W), indicating a corrected compass (something Nantucketers would be keen on). This same alignment can be seen on roads further west, as well as the airport runway, which was undoubtedly constrained by lot-lines. This 200-rod interval appears also to have been employed in the south pasture below Milestone Road. The south plains slope gently to the sea, and the surveyor's work was little compromised by terrain. In the middle moors north of Milestone Road however, the swamps and intricate topography play havoc with the rutted road-net, but one can glimpse a familiar pattern in four or five north-trending tracks that run off Barnard Valley Rd. at something like 200-rod intervals.

Clues to the origin of these patterns can be drawn from the history of the Nantucket proprietors. In 1821 most, if not all of the land in question was comprehended in the New Divisions, which consisted of seven major tracts in both the southwest and east of the island. Presumably the rectilinear rutted roads were laid down as rangeways or driftways and date to this time. They must have overlaid a preexistent organic road-net perhaps to be still seen in the Altar Rock spiderweb of the middle moors. Each of these seven tracts was further subdivided into twenty-seven sections, one for each of the original proprietors' shares. As no one person held a full share, and most shares had been infinitesimally divided by inheritance (the fractions calculated in "sheep commons," which were reckoned as 1/720 of a share) each of these twenty-seven sections was held in common by the heirs and assigns of one of the original proprietors. Only by recourse to the courts could any individual have his fractional portion set off to him in severalty; that is, as his own personal property. The land was thus largely held in common, and the extreme vagueness about boundaries and own-erships was of calculated benefit to the raising of sheep, as small, fenced tracts were in no one's interest.[17] Thus the New Divisions did not effect the landscape revolution that attended the common-land divisions on the mainland.

This anachronistic chaos exacted its toll when the sheep boom of the nineteenth century became the land boom of the twentieth. An off-island lawyer, Franklin E. Smith, acting through his Nantucket Cranberry Com-pany, seized his chance and snapped up more than 70% of the "sheep-commons" to gain control of the proprietary in 1909. For the next thirty years he dealt himself remnant tracts of the commons, some as large as four hundred and one thousand acres, often for only a nominal cancellation of a few of his sheep commons.[18] To this day the Nantucket Conservation Foundation lists five Sheep Commons, a legal entitlement without area or location, among its real estate holdings. It would seem that as long as clouded titles lie heavy like a Nantucket fogbank on the land, that such fractional shares have more than sentimental value.

Grid

Grid streets have been laid out since classical times and enjoyed an initial decade of popularity in New England: Charlestown (ca. 1629), Cambridge (1632), New Haven (1638), and Fairfield (1639–1640), were all laid out with regular square blocks of houselots. However, none were expanded upon as the settlements grew, and for about a century, the formal gridiron was largely neglected in New England, after which, potentially at least, it was implicit in the crossroads village and the range township. The earliest gridirons of the 1630s were justified on the grounds of defense (the plan was much like that of the fortified medieval new-town, the bastide), and perhaps also amateur enthusiasm: a tendency to overplan when town-planning was still a novel and momentous exercise. In the eighteenth cen-tury, remote proprietors in their paper plats could still succumb to this. One such, Ezra Stiles (later president of Yale), in 1764 envisioned moun-

tainous Killington VT laid out on the New Haven grid plan: his folly came
to naught. Yale men must have had the New Haven gridiron burned into
their souls, for Timothy Dwight (its next president) on a visit to Boston
in 1796 audaciously reimagined its "narrow, crooked and disagreeable
streets" as a grid of broad, (120 feet!), tree-lined avenues and open squares.
Were this but so, Boston would have been "the most beautiful town the
world has ever seen."[19] The time was now ripe for such apostles.

In the long run, street-grids, and even piecemeal street-blocks, required
denser urban populations to become firmly established. Ports were the first
towns to support a large nonagrarian workforce that gained its livelihood
within a small compass. By the mid-eighteenth century, irregular, compact
street-blocks had slowly sprung up within a walk of Boston and Marblehead Harbors; more regular ones in Portsmouth NH. In 1801, the first
formal grid-plat in Boston was drawn up by chief selectman Charles Bulfinch for subdividing "the Neck," the narrow strip of land traced by Washington Street that, before the mud flats were filled, was the sole road to
Roxbury. As mapped in 1814, it consisted of three dozen almost perfect
square blocks, with axial streets named for counties and cross-streets for
county seats. As it appears today in the vicinity of Franklin Square the
grid is considerably less regular, with the interiors of many blocks opened
up with additional streets. About the same time (1800–1801) in Cambridge,
a few grid streets had begun to appear in Central Square, but the most
ambitious early grid was that of East Cambridge laid out in 1811 for the
Lechmere Point Corporation by Peter Tufts. North-south streets were
named numerically, while certain cross-streets honored prominent inves-
tors. The county court, jail, and registry of deeds were induced to relocate
here in 1813, and unlike many grid schemes that languished as paper-plats,
this one took on a life of its own. The grid is still intact, although extended
south and west over filled marshland to twice its original size.[20]

Street grids spread to the frontiers of settlement with amazing celerity.
Bangor ME began with the log house of one Jacob Buswell in 1769 and a
half century later log houses still outnumbered frame houses. Yet as early
as 1801—the same year as the first street grid was proposed in Boston—
the redoubtable Bulfinch was commissioned to devise a plat for what was
then Conduskeag Point on the Penobscot River, which became the basis
for the Bangor street-grid. The post-Revolutionary upsurge in maritime
trade, ship-building, and the whale fishery, seems to have spawned a discernible class of grid ports. Some like Bangor were newly settled and formally planned; others elaborated upon serried tiers of early houselots laid

out perpendicular to a waterfront street, with perhaps a back street of similar lots behind it. Nantucket[21] and Newport have such a look, and also Provincetown, which, hemmed by dunes, never expanded to the rear. In time rectilinear plats could touch all but forgotten shallow-water ports such as Annisquam (Gloucester) MA. Tight little Chester Square would not raise an eyebrow in Marblehead, but makes heads shake in disbelief in a port that never grew beyond a village. Perhaps by reason of their flat topography, their often belated post-Revolutionary rise as ports, and coast-wise spread by emulation, such grid ports seem particularly common in southeastern New England. (A wider survey of coast towns might modify this view.) Edgartown and Mattapoisett seem to be cut from the same cloth and epitomize the type.

Many port grids have lost their historic focus by virtue of their outward expansion. What may be the crucial significance of the New Bedford grid is easily overlooked because its vast uniformity reads instantly on a map as that of a mid-nineteenth-century boomtown. Yet even at first sight the grid of its historic whaling district holds a certain fascination, both in its early form (square, rather than rectangular blocks) and in its classic nineteenth-century toponymy: numerically named north-south streets (some evidently renamed), as well as cross-streets such as Pleasant, Walnut, and School. The truth is that New Bedford proves to be extraordinary in its ordinariness: while New Bedford the town was not set off from Dartmouth and incorporated until 1787, Bedford village as it was called began to spring up on the Russell and Kempton farms even before the Revolution. Joseph Russell III, "the Father of New Bedford," in 1764 "devised a plan whereby a prosperous village should be built upon the lower portion of his farm. This plan was not only profitable to him, but was comprehensive and far reaching in its results, and laid the foundation for our whole street system."[22] The plan called for four streets (today Union, Spring, School, and Walnut) perpendicular to the waterfront intersected by eight numerically named cross-streets (First through Eighth) to form a grid. (Russell's First Street was called Water Street when accepted by the town in 1769.) Thus, far from being the hallmark of a nineteenth-century boomtown, the New Bedford street-grid is much older. Russell was engaged in precisely the same sort of land speculation as underlay Bulfinch's Bangor ME grid, but almost forty years earlier. If every artifact has a template, then where did the New Bedford street-grid come from? What may be the prototype was laid out in 1680 at Bristol RI, then a part of Plymouth

Colony. A grid of four roughly north-south streets and nine cross-streets enclosed blocks of eight acres (four lots) each on the landward side; the shoreline compromised the rectilinearity of the waterfront street, and lots there were smaller (and more valuable). This grid plan is unique in Rhode Island, and given its very early date, perhaps the first grid port in New England.[23]

Another intriguing possibility is that this combination of both grid and numbered streets derives at least in part from the well-known and widely influential "Philadelphia plan" adopted by William Penn in 1682. (The New York plan of numbered streets and avenues was not finalized until 1811.)[24] Quakers were not respecters of persons, and numbered streets accorded well with their practice of denominating both days and months by ordinals (First Day, First Month, etc.). Probably in a more abstract way the simple, tidy, orderly grid appealed to them on a spiritual level. The plot thickens when we realize that, like Nantucket, New Bedford was dominated until the early nineteenth century by prominent Quaker families, the Russells among them. In a nutshell, then: Was the New Bedford street-grid laid out in 1764 about forty years ahead of its time by a Quaker sea-merchant and land-speculator in distant emulation of Philadelphia, the chief planned Quaker city on the continent? Surely, New Bedford was well positioned in time (founded a little before the post-Revolutionary maritime boom) and in place (seaports are natural points of diffusion) to serve as a prototype for other nascent ports in southeastern New England and beyond. More research is needed to establish the grid port as a definite settlement type. A word of warning: many old ports mushroomed as manufacturing cities due to their amassed investment capital, a strong workforce swelled by immigration and maritime downturns, cheap sea access to coal, and even, in the case of textiles, damp sea air that kept fibers from breaking in high-speed machinery.[25] Furthermore, the rise in seaside resorts saw an explosion of gridded cottage cities on postage-stamp lots. In short: not all coastal towns with street-grids are grid ports by any means.

If the rise of the seaport gave the grid its first real foothold in New England, the birth of the manufacturing city and dormitory suburb assured its dominance until 1930. Successive mappings bracket the appearance of street-grids in inland factory towns like Fitchburg (ca. 1831–1854), Athol (ca. 1840–1880), and North Easton MA (ca. 1851–1886). Some were composed of piecemeal blocks not strictly rectilinear, closure often being made by an elbow street that kept its name as it turned the corner. Early blocks were often nearly square and less than five hundred feet on a side; later were longer and narrower. Square blocks might be bisected by cross-streets and blind courts that ran east-west here and north-south there.

But to take up the neglected question: exactly why the grid? The gridiron had latent elements in the cadastre in its favor. Lot-layers had always prized rectilinearity, and colonial pastures broke the ground for early grids in Greater Boston (as on Beacon Hill, Boston and Spring Hill, Somerville),[26] a trend that the piecemeal development of old farm fields in the nineteenth century would only reinforce. Early range-lotted townships that boomed in the nineteenth and early twentieth centuries provided a ready-made squarish framework to infill with grid streets. The boot-and-shoe towns of Brockton and Marlborough MA benefited from such a quadrangular bias to their cadastre. True range towns that grew into grid cities are rarer because historically less populated: Pittsfield MA and Burlington VT perhaps qualify. In general, the grid, whether in the layout of eighteenth-century farms or nineteenth-century cities, appealed especially to land speculators.

Sam Bass Warner saw the grid street in Boston as a fashionable import from Georgian London justified by the dominance of wealthy row houses (connected townhouse blocks) from 1793 to 1873 in Beacon Hill, the South End, and Back Bay. This trend carried over into the platting of cheap tracts of wood-framed houses in East and South Boston. The demise of the row house after the Panic of 1873 and advent of the detached house in the outer districts of Dorchester and Roxbury did nothing to shake its hold. The costly provision of sidewalks, curbed forty-foot-wide streets, and an increasingly modern panoply of utilities and amenities (water, sewer, gas, and latterly telephones, electricity, even shade-trees), coupled with a popular insistence on frontage lots and abhorrence of anything that smacked of a rear tenement, embedded the street-grid all the more deeply in the economics of real estate development. To reduce his outlay, the developer kept cross-streets to a minimum, which increased buildable acreage and decreased the running feet of utilities. This accounts for the shift from the square block to the rectangular. Streetcars took housing out of the dense urban core and into outer districts of cheap land, allowing small developers to run up houses on small tracts, creating piecemeal street-blocks quite unlike the large, orderly grids imposed by earlier nineteenth-century land syndicates. The width of a typical 90-foot deep lot could be tailored from 50 feet for a substantial middle-class house to

27 feet for a three-decker, allowing a speculator to keep his feet in a change-able market.[27]

But beyond Warner's streetcar suburbs there were the more affluent steam-car suburbs joined by rail to the commercial downtown. The first significant residential street-grid in Arlington, described at the time as "the new village at Arlington Heights," was laid out and improved by a syndicate "composed mostly of gentlemen doing business in Boston" in the years 1872–1878. It had two very typical attributes: airy location on a rocky hilltop above a plain still widely farmed, and the requisite passenger depot at the foot of the new, sweeping 80-foot-wide Park Avenue. It had another characteristic not altogether uncommon in the earlier, experimen-tal tracts, the fact that the developers' intention was "to build up a village as a place of residence for themselves and others similarly situated." This last may account for the scheme's ability to weather the Panic of 1873, which sent real estate values plummeting.[28] Arlington Heights is note-worthy because Park Avenue and the circular summit park to which it led wrapped the grid behind a dramatic curvilinear facade, a contoured design that spoke of things to come.

Pretzel

The successor to the grid, what I call the pretzel, was deliberately curvi-linear and at least at first, wedded to the terrain. In time it manifested itself even where relief was minimal, both from blind fashion, and with the advent of the automobile, to thwart fast through traffic in residential areas. Curvilinear layouts had long been dear to English landscape gar-deners like Capability Brown and Humphry Repton, but the influential New England prototype was clearly the road plan for Mount Auburn Cemetery in Cambridge MA laid out in 1831 by Henry A. S. Dearborn. The intricate morainal topography was a rough diamond ready to be brought out by a sensitive hand. (Glacial sand and till were also easily dug for graves.) Such glaciated terrain appealed strongly to nineteenth-century romantic tastes and in Concord MA inspired names such as Fairyland, Sleepy Hollow, and Weird Dell. Also for topographical reasons, curvilinear design was well suited to many railroad suburbs, which were built on rocky uplands, for the sake of the air, commanding views, and the cheapness of the land: the levels around Boston was still intensively farmed as a market-garden belt.

The subdivision of once-rural estates engulfed by the urban tide often brought together taste, talent, and a fastidious concern for the fate of the land not usual in ordinary speculative development. For example, the subdivision of Norton's Woods in Cambridge would occupy the talents of both F. L. Olmsted (1868) and his pupil Charles Eliot (1887–1888). In some early Greater Boston examples as Fisher Hill, Brookline (1884), Hubbard Park, Cambridge (1889), and Rangeley, Winchester (1876) such subdivisions were tightly covenanted estate enclaves in which the gentleman-developer not only controlled what might be built, but also chose his neighbors. At Rangeley, David N. Skillings's control was even greater: he rented to hand-picked friends and relatives the houses he built within the walls of his private park. (Rental, not purchase, was not an unusual arrangement, even for the wealthy: those of the Wedgemere syndicate as well as the Firth development, both also in Winchester, were likewise rented.) By 1915 when the Larchwood tract was platted in Cambridge MA, the pretzel (and Larchwood is the quintessential pretzel) was the plan of choice for an upper middle-class neighborhood of substantial neo-Georgian houses, even on essentially flat land.[29]

The taste and money requisite to the early classic pretzel was rarely found outside of metropolitan areas.[30] Brookline as an estate-suburb probably has the largest number of pre-1930 pretzel tracts in New England; by contrast none are found in Brockton, and only one, probably fairly recent, in Worcester. Brookline was also home to the influential F. L. Olmsted and his firm, and their philosophy, which wedded roads to terrain, was of immense importance in the triumph of curvilinear design. Hopedale MA, described as "a model company town," provides an unusual instance of an extra-metropolitan pretzel and also an Olmsted connection.

The Draper family of textile machinery magnates took a deep if pragmatic interest in the worker-housing issue: good houses attracted good workers. Over the period 1896–1903, they built the Bancroft Park neighborhood of some thirty handsome shingled double-houses on curved streets in part to demonstrate what might be feasibly accomplished in this direction. Five different house plans were employed, many with the then-fashionable cross-gable and gambrel roofs. Bancroft Park wound around a knoll in an inner and outer loop and was screened from the machine works by a berm (an old Olmstedian trick). This site plan was the creation of Warren Henry Manning, whose first acquaintance with the Drapers had come a decade earlier while employed by Olmsted, probably in work on the Eben S. Draper estate and the village cemetery. Manning sold the

Draper Company on the merits of contour planning and over the next three decades undertook several independent commissions for them. Another thirty company-owned double-houses on a picturesque thumb of land thrust into Hopedale Pond (a millpond) called Lake Point (1904) was also contour-planned, this time by Arthur A. Shurcliff, another Olmsted disciple. Both of these were a far cry from earlier Draper company housing, such as the "Seven Sisters" (1874 and after): boxy, nondescript if solid double-houses on a grid street that were much more typical of a mill-village.[31]

The prewar pretzel evolved into postwar curvilinear suburbia. Large lots loosened the often intricately tied road-knot, sidewalks disappeared, topographic contours in many cases ceased to justify and inspire the layout because the tracts were on level, abandoned farmland. Cartographically (and admittedly, such a perspective is to a degree irrelevant) their lines seem to sprawl artlessly like stray spaghetti, with many loose-ended strands and little of the tight formality left to justify the sobriquet "pretzel."

{ *Five* }

HOUSES

BUILDINGS are the most visible and enduring artifacts in the cultural landscape, and together make up the richest physical record of New England social history. The appraisal of this record is both micro- and macrogeographic: at one moment, it can ask us to count the panes in a window sash, and at the next, to contemplate sixty thousand square miles of architectural geography as though it were but an intricate pattern in the mind's eye.

Houses are artifacts. To apply our standard clay-pot analogy, houses are squarish vessels, fashioned of cheap or choice materials of discernible provenance, displaying datable techniques, local or regional schools of workmanship, styles, and motifs, suited to particular people for particular purposes, whether as elaborate showpieces or for simple daily use. The styles of houses can reveal the cultural influences and aspirations of the users, while an analysis of their form tells something of how those users lived. And, as with excavated pots, context and locality, and not status as objets d'art, are critical to their interpretation. Some are virtual icons, so readily recognized and clear in their associations that they become index types for a certain era, social class, or economic order. A firm grasp of architectural chronology enhances the study of settlement geography: houses become critical datemarks with which to delimit cultural hearths and spheres of influence, to chart the tides of settlement, urban growth, and suburban sprawl over the face of the land. Without this awareness of time-depth, the builtscape can seem very flat indeed.

Buildings, like all physical entities, exist both in space and time, and their architecture can often be categorized by which of the two—space or time—has exerted the deeper influence. What might be called time styles are eminently datable and reflect the fashions of the moment; place styles are "placeable," grounded in customary building practice, and more enduring. Traditional societies show diversity over space balanced by stability over time; for example, the ancient provinces of France, each with its own patois and traditions of costume, cookery, and building, persisted for centuries within a country smaller than Texas. Modern societies show quite the reverse: uniformity over space but rapid change through time. People in Boston today dress more like their contemporaries in New York than they do like the Bostonians of even a few years ago. Traditional European landscapes are marked by regional or place styles, while modern American landscapes are marked by period or time styles.

For Americans, Old World regional architecture is enviably varied and picturesque, that of England being perhaps the most familiar. No survey of English architectural geography could omit the thatched cob cottages of Devon, the half-timbered houses of Worcestershire, the honey-gold stone of the Cotswolds, the harsh red brick of Lancashire, the rush thatch of Norfolk, or the tile-hung, weatherboarded Weald. Were the landscapes of New England so dramatically diverse, we should certainly have noticed —or would we? Nothing truly comparable to this county-scaled region-alism exists in New England, but not because New England is devoid of place style. Rather, it is because New England, while two-thirds the size of all Great Britain with its numerous cultural regions, itself constitutes but a single American cultural region, one with its own place style that comes as the legacy of the preindustrial era. This vernacular architecture, what Henry James called the New England Homogeneous, serves as a unifying background to the entire area.[1]

The founding of New England knew nothing of the cultural diversity of the Middle Colonies where New Jersey alone saw early settlement by the Dutch, Germans, Swedes, English Quakers, and even emigrant New Englanders. In result, while New Jersey has a rich architectural legacy of patterned brick Quaker houses, bank-end German houses, flared-eaved Dutch houses, and New England saltboxes, it also lacks a visually unified vernacular landscape. By contrast, the early settlers of New England were largely from southern England and culturally homogeneous. Many were from East Anglia, where wood was still abundant in the early seventeenth century and the timber-framed house was the norm: quite unlike much of Britain at that time.

The East Anglian roots of New England house-building are easily over-looked because hidden beneath clapboards, and because the prevalence of wooden construction on a well-forested continent is misjudged as inevi-table. Felling timber and raising houses runs deep in the saga of North American settlement. Yet the notable rarity of stone or brick building outside of fire-swept cities in pre-1850 New England is atypical of the eastern seaboard. A well-traveled Virginian such as George Washington could not help remarking upon the lack of masonry in Portsmouth NH in 1789 "as the country is full of stone and good clay for bricks." In the prosperous Connecticut Valley, with its abundance of brownstone, even the eighteenth-century high-style architecture is of wood, despite the countervailing example of the Hudson Valley to the west, where stone and brick farmhouses were built by the Dutch and Germans according to their own cultural dictates. In New Jersey, a line drawn from Princeton to Wilmington DE demarcates a region of stone buildings to the north and brick to the south. In Pennsylvania, a "stone triangle," with corners at Philadelphia, Allentown, and Harrisburg encloses a concentration of stone-built barns and houses. Even today, in much poorer parts of the country such as western North Carolina, brick enjoys a cultural ascendancy over wood unparalleled here.[2]

The cultural geographer's interest in vernacular architecture focuses on house types: on the evolution of floor plans, chimney placement, and room usage, and not the thin stylish veneer that marks a house as early or late Georgian, Greek Revival or Italianate. At the time of the Great Migration (1630–1642) English standards of housing were in revolution: people still lived in chimneyless houses with central hearths vented through smoke holes in the roof. Second floors were mere lofts or nonexistent. The primitiveness of early New England housing, which we might glibly ascribe to pioneer conditions, was actually the lot of many in Old England as well. Unfortunately, knowledge of pre-1725 building practices in Massachusetts rests primarily on the study of roughly 150 extant or documented houses in the Bay Colony and half that number in the poorer and less populous Plymouth Colony.[3] And this is undoubtedly the richest trove of any New England state. One might legitimately wonder whether the study of some tenscore seventeenth-century houses is truly an overture to the New England builtscape, or merely a curious footnote to the British postmedieval timber-framed house. However, there can be no doubt that the eighteenth-century vernacular house types that emerged from them are critical to an appreciation of the living landscapes of New England and beyond.

Geographers recognize three westward cultural streams that emanated from three eastern seaboard "hearths": New England (Boston), the Middle Atlantic (Philadelphia), and the South (Tidewater Virginia). Much has been written about the dominance of the Middle Atlantic hearth in the formation of general American culture, and its pivotal role in the development and diffusion of the log cabin. Thus it occasions considerable surprise to hear the cultural geographer Peirce Lewis characterize the architecture of the Middle Atlantic stream as one of "marked conservatism," and further assert that the house types of the Southern stream showed "a poverty of imagination," and conclude, in fact, that "only in New England and in New England's mighty extension into the Mohawk Valley and the upper Midwest, can one find evidence of large-scale innovation in vernacular housing." Certainly this is a novel slant on what we have come to regard as the staid, white-clapboard New England Homogeneous. Lewis hones his argument further: "the transplanted New Englanders of New York State invented new house types with enthusiastic abandon and planted them across the land"; and again more pointedly, "between 1800 and 1850, the strip of territory between Albany and Buffalo sprouted more architectural ideas than any other region of the country, and the innovations took all kinds of forms." (It was also a hotbed of religious, social, and even toponymic change: the Classical name-fad first took hold here.)[4]

Joseph W. Glass has stated that "a cultural hearth is the area in which a culture was born or where many of its components were originally assembled, but not necessarily where it matured." In his view, the "core" of the Middle Atlantic cultural stream was not in Philadelphia or Baltimore, but in southeastern Pennsylvania where the two influences came together.[5] Similarly, the Mohawk Valley may have lain in the cultural confluence of Boston and New York: perhaps accounting for its ferment. (New York is conventionally mapped as part of the New England cultural stream.) Standing at the headwaters of a great settlement stream that flowed westward across the northern tier of states to Michigan and beyond, the house types of New England have strong extraregional implications.

Place styles, as has been said, are stable (although not static) over time, in part because they rest on traditional practices of building. In New England, an expanding but thinly settled region, where houses were raised by builders and often volunteer laborers of varied origins, society greatly benefited from simplicity and standardization. For reasons not unlike those for which the Army once built Quonset huts and Bailey bridges, the vernacular builtscape was characterized by a limited number of highly adaptable house types that evolved slowly and enjoyed long-running popularity. Before we look at these house types more closely, three characteristics that contributed to their long-term success—modularity, mobility, and conformity—will be reviewed.

Modularity

The customary builder's measure was the *bay*, the interval between two frame posts or crossbeams, often taken to be 16 (15–20) feet. The one-bay house was something like 16 feet square and was the standard module of vernacular construction, one variously explained as room enough to house a span of four oxen or be heated by one fireplace. In the English tradition, the one-bay house was square or nearly so; in the Scotch-Irish and German, it was rectangular, usually at least five feet wider than it was deep. Chimneys were reckoned separately, so while the one-room Dean Winthrop House (1638–1650) might measure 24' × 18' overall, it was all but square minus the chimney bay. Even in seventeenth-century Massachusetts Bay anything less than a 15-foot-square bay seems to have been substandard. As we shall see later, the one-bay or Hall house could be expanded to the side, the rear, and overhead by whole or half modules, to yield, as in quasi-genetic sequence, other vernacular house types, such as the Hall-and-Parlor and the Saltbox.

More generally, modularity refers to the capacity to enlarge a house repeatedly, whether by customary annex types such as the jut-by of the Saltbox, or more amorphously, by ramshackle ells. Characteristic programs for the extension of houses are a part of the regional vernacular. The Connected Farmstead was a common pattern in New England. By contrast, the "telescope" house of the eastern shore of Maryland was expanded laterally, with each successive unit stepped up a half-story in height, a formula not seen here. The repertoire of annex types, while governed by vernacular tradition, was also limited by the practicalities of post-and-beam construction. Only once balloon frames (invented in Chicago in 1832) had revolutionized carpentry could wooden houses sprout the wealth of polygonal towers and bays that were in such stark contrast to the boxy and purposeful warts, jut-bys, and ells of the vernacular.[6]

Connected architecture was very much a thing of its time, and of its place. While many modern architects have been clearly inspired by it, the fact that so few have convincingly reproduced it suggests two things: that the vernacular housewright worked on certain principles that the modern architect honors in the breach, and furthermore that the architect breaches them for technical and socioeconomic reasons every bit as cogent as those that compelled the housewright to observe them in the first place. Beyond the study of the methods of modularity lies an appreciation of its motives, which are sociocultural rather than architectural. The freedom and repeatedness with which additions were made to old houses suggests that family life functioned differently then; that families were more sizable and extended, that the line between homeplace and workplace was differently drawn, that other social customs prevailed for dealing with widows, in-laws, hired help, destitute relations, paying boarders, and the town's poor. (Ironically though, families seem to have grown statistically smaller as dwellings grew larger.)[7] Unfortunately, the man-in-the-street awareness of the groundplan of old houses is dismally shallow. Often the most we carry away is the approved image of their facade: so memorably strong is its simple symmetry and so difficult the task of geometrically reconciling with it the three other complex elevations hidden from the street. This two-dimensional view is unfortunate; for while the facade expresses the aesthetic aspirations of the time, the rear is more telling of how life was actually lived.

Mobility

A critical factor in early house-building was the actual cost or labor-invested value of the materials, and economic necessity enjoined their wise reuse. Salvage was so much the custom that beams, joists, nails, windows, and the like when individually dated by technical criteria, cannot be taken without corroboration to indicate the age of the building as a whole. On a larger scale, economics also dictated that it was cheaper to move a build-ing than to build one; in fact, as long as the custom of "changing work" (mutual aid) persisted in rural communities, moving was essentially free. Because the timber-framed house was in effect a solid box that merely sat on its foundation, it could be jacked up, put on skids, and drawn to a new site by oxen or horses. Winter, when the ground was frozen, and farmwork at a standstill, was a favorite season to move. Farmers who spent the cold months "swamping" in the woods had developed techniques, tools, and the needed experience to transport heavy loads of timber; thus it was noth-ing strange to assemble twenty teams of oxen to move a building as much as five miles.

Motivation varied from the purely personal, whether altered family cir-cumstance or perhaps even whimsy, to what verged on a collective social phenomenon. Sometime before his death in 1852, Lemuel Burnham of Bolton MA moved his house across a frozen pond and two roads for no evident reason other than, as his great-grandson speculated, "because he didn't like it where it was." Mary Vorse in Provincetown MA recalled a neighbor who over the years moved down a two-room cottage from his back lot to accommodate his aging mother, then detached and moved it back on her death "because . . . it made him feel lonely," only to reannex it afterwards to house his widowed sister and her children. Thomas Hubka, on the other hand, regarded the question "why Tobias Walker moved his shed," as symptomatic of a broader phenomenon, central to the evolution of the New England Connected Farmstead. The appeal of this scheme rested on its not entailing a complete reconstruction of farms, but only a reshuffling of existing buildings. Sheds, stables, and barns skidded from scattered locations and joined in an interconnected whole enabled farmers to capitalize on surplus buildings occasioned by obsolescence and rural abandonment.[8]

Timber-framed houses were not always moved in one piece, but might be disassembled into numbered, framed components, a procedure known on Cape Cod as "flaking." (Balloon-framed houses might be flaked, but this required sawing and splicing.) Knocked-down houses were a tradi- tional New England export to the Caribbean, California, and perhaps Nova Scotia. Along the coast, whole houses were often moved by sea: indeed, Cape Codders have been reputed to regard their houses, like their boats, as movable property. Tradition has it that numbers of Cape Cod houses were floated over the sound from Nantucket in the mid-nineteenth century, and forty-eight condemned houses were floated off Long Point on wrecking barrels when the land was taken to protect Provincetown Harbor. If this seems an unlikely or fluky enterprise, recent history suggests otherwise: Newfoundlanders displaced in the 1960s by the provincial pol- icy of resettlement (that is, the discontinuance of mail and medical services to remote outports) responded by skidding their houses to the water's edge and barging them to more populous locations. Only the cellar holes and cemeteries were left behind. Local historians have traced as many as sixty- seven house removals in Williamstown MA and eighty-nine in Bristol RI; Henry Forman offhandedly estimated that more than one-half of early Nantucket houses were "peripatetic." Thomas Hubka concluded that "when the history of building movement in a particular New England town is accurately recorded, as in the towns of Fryeburg and Cornish, Maine, it appears as if the entire town was constantly being moved about." Because houses are not deeded, and can be bought and sold like personal property without record of title change, there is no legal paper trail by which to establish the full magnitude of the practice.

House-moving was by no means confined to skinflint farmers. Public buildings also were moved. The old Lincoln town hall (1848) was moved twice, once ca. 1891 and again some years later, which explains why this Greek Revival templefront stands so far from the town center. It would have moved considerably further in 1928 had the chief selectman not flatly declined Henry Ford's offer to buy it for Dearborn Village. The old Lexington town hall was shorn of its Greek portico, then severed into two houses now on Vine Street. In Cambridge, in 1867, the North Avenue Congregational Church was moved from Harvard to Porter Square. In 1922 the old Concord Manual Training School was sold to make room for the new armory, then moved over Mill Brook with oxen and turnbuckles. The work took three days and cost three hundred dollars.[9]

While in themselves these events may seem trivial, collectively they portray an afterlife of buildings starkly different from anything known today. Of the thirteen documented pre-1835 houses in northwest Cambridge, only one remains on its original site. Such a high percentage of house-moving among the survivors should come as no surprise, as house-moving was a critical survival mechanism amid the upward spiral of urban land values: twelve-thirteenths of pre-1835 houses did not move; twelve-thirteenths of pre-1835 houses that *survived* moved, and survived because they moved. Likewise, the extant building stock of colonial New England is disproportionately representative of more substantial and stylish dwellings, because precisely such qualities assured their greater rate of survival. Antique buildings should not be seen as miraculously spared or forgotten by time. Survival is more fruitfully regarded as something concretely explicable by some special adaptability or circumstance. The traditional practice of house-moving largely died out in rural areas in the 1940s. Heavy traffic, the hazard of utility wires, the disappearance of cross-lots routes, of oxen, of rural cooperation, and any number of factors have made house-moving a rarer and more newsworthy event. A house floated across Pleasant Bay from Orleans to Chatham MA in 1994 commanded a front-page color photo in the *Boston Globe*: it was not so rare a thing.[10]

Conformity

The domestic vernacular is often admired for the austere harmony of its external details, which by the mid-nineteenth century had become codified to a vocabulary of clapboards, cornerboards, and modest classical details rendered vanilla-plain by white paint. Particularly in smaller and remoter places, architectural display was tempered by vestigial Puritanism, poverty, frontier egalitarianism, and old-fashioned neighborly jealousy to make for a communally enforced stricture against ostentation. Early Nantucket is rich in anecdotes that illustrate a Quaker hostility to architectural "extravagance." One man was prevailed upon to cut down the rear posts of his two-story house frame to make it a lowly saltbox like everyone else's; another defied the opposition and built his, but his own father fulfilled a sworn vow never to cross the threshold. In the village of 'Sconset, Henry Forman observed that "nonconformity was evidently frowned upon, so that nearly every home developed somewhat around the same plan." In fact, one villager (sometime about 1800) was ostracized by his neighbors for plastering the inside of his house, an unheard-of luxury.[11]

J. Ritchie Garrison suggests that the dynamics of this austerity were more complex than commonly supposed. In Deerfield, exterior architectural conformity arose when the gulf between rich and poor became more acute, and was intended to preserve social cohesion:

> Beginning in the 1730s, as class lines sharpened, politics became more stratified, and the community became wealthier, architectural *forms* on the old village street became, paradoxically, more alike. In a village setting, where neighbors' judgments were close at hand and where competition to maintain standards of material respectability was most visible, the reduction of sharp differences in building forms encouraged families to organize their lives in similar, more predictable ways.[12]

The precise form this conformity took is hard to assess at present, as the "New England Homogeneous" was further homogenized in the early twentieth century by the Colonial Revival, which forever remolded the New England landscape into its own semimythical image and likeness.

House Types

The term house type will be employed here rather freely, but the concept requires some qualification. The development of related house types should not be blindly construed as in genetic sequence, the Hall-and-Parlor house growing into the Saltbox, and the Saltbox into the Central Chimney Large, and so forth. Nor, in my estimation, should house types be viewed as prototypes, so much as the refined end-products of trial and error. The many were winnowed to the few. The most perfect are chronologically quite often the latest, while the earliest refuse to conform to type. The sightseeker is handicapped by the fact that he can rarely set foot in the houses he is ostensibly classifying by floor plan. Indeed, in the course of scrutinizing the builtscape, he will soon suspect that the whole concept of house types is a cruel hoax: artificial, inadequate, and without theoretical rigor. Many, if not most houses can only be classified as hybrids, which is certainly unencouraging. While the types here discussed do not begin to classify the true variety of the pre-1900 builtscape, as index types they do reflect the complex spatiotemporal dynamics of New England settlement history as a whole. However incomplete, to my knowledge it is the most comprehensive typology in print. Were it larger, it would be less workable as a *videnda* of things to see, of artifacts to actively seek in the landscape. Without the rose-tinted glasses this typological framework provides, the sightseeker, like the common lot of travelers, would often see little to note in the builtscape. Roadsides scanned with devouring curiosity through these lenses would be otherwise barren, and while they may distort, we would be blind without them. Unfortunately, these house types have long gone unnoticed and unnamed, and efforts to christen them in retrospect are often awkward. A few have memorable names, but for most we must resort to descriptive phrases like "twin rearwall chimney house," phrases so prosaic as to become quickly lost in the text. To emphasize that these are indeed proper terms not casual combinations of words, the names of principal house types have been capitalized.

In the pages that follow, we shall define the form and, as far as possible, trace the architectural origins, geographical diffusion, and social history of some two dozen New England house types. Telling contrasts between the long-and the short-lived will readily emerge, and through these, a sense of what qualities made for success. In this light, subregionalisms often look suspiciously like regionalisms that failed to meet the mark.

Hall House

The one-room or Hall house was by physical and economic necessity a characteristic element of the settlement landscape from 1620 until the early nineteenth century. Abbott Lowell Cummings found that somewhat more than half of the seventeenth-century Massachusetts Bay houses documented from probate records were of one room, and that their owners were generally, though not always, of the poorer class. As late as 1798, a strong case can be made that one-seventh of the Worcester County MA housing-stock consisted of one-room dwellings.[13] Nor was the one-room house strictly a frontier or rural product: in 1814 a one-room (albeit two-story) house was built in Portsmouth NH, the tail end of a seventeenth-century urban building tradition. Few if any survive unaltered, and most have long since been enlarged out of recognition, annexed to more substantial dwellings, or razed. Many seventeenth-century houses nonetheless can be shown from physical evidence to have passed through an initial one-room phase that might consist of an eighteen-square-foot Great Room or Hall, with a chimney and garret stair in a six-foot end bay. Enlargement was typically longitudinal, with a second, symmetrical room added on the opposite side of the end chimney. This addition created the Hall-and-Parlor house, a one-pile (that is, one-room deep) house in which Hall and Parlor flanked the central chimney with sleeping lofts or chambers above. The Hall house is often called the Half house implying such transitionality, but other modes of enlargement were possible.

Conjectural replicas abound of the earliest one-room houses, which we must stress from the outset to have been timber-framed, "fair" English houses, and most emphatically not log cabins. George Francis Dow undertook the first such replicas at Salem Pioneer Village in 1930 for the Massachusetts Tercentenary, "small framed cottages, each provided with a brick or 'catted' chimney, and roofed with thatch." The end chimneys can be seen to rise from behind the ridge, perhaps to bypass the ridgepole (if any) or allow for a larger entryway; whatever its rationale, the arrangement is typical of many such replicas (see fig. 12). Research and excavation undertaken by Charles R. and Sydney T. Strickland for Plimoth Plantation in 1947 resulted in houses that confirmed (or conformed to) Dow's interpretation, and that seem, externally at least, identical to those at Salem. The Plimoth houses possess a half loft over one end of the hall, as derived from likely English prototypes: a feature for which Henry Forman subsequently found physical evidence in the late seventeenth-century 'Sconset Whale houses discussed below. Likewise, the one-room West Hoosac "regulation house" reconstructed in 1953 for the Williamstown Bicentennial strikingly resembles those at Plimoth Plantation, although the roof was shingled, not thatched. While without doubt incestuous in design, these reconstructions argue for a long and now largely invisible tradition of one-room, initially one-story, end chimney houses in New England. (Replicas at Jamestown VA are also remarkably similar.) Strands of this tradition can be recognized in the Rhode Island Stone-Ender, 'Sconset Whale house, and Charter houses of the eighteenth-century frontier now to be examined. In general, the one-room house is too small for modern revival, although "historic" house plans for a so-called New England hearth room house (with an integral addition) have been marketed.[14]

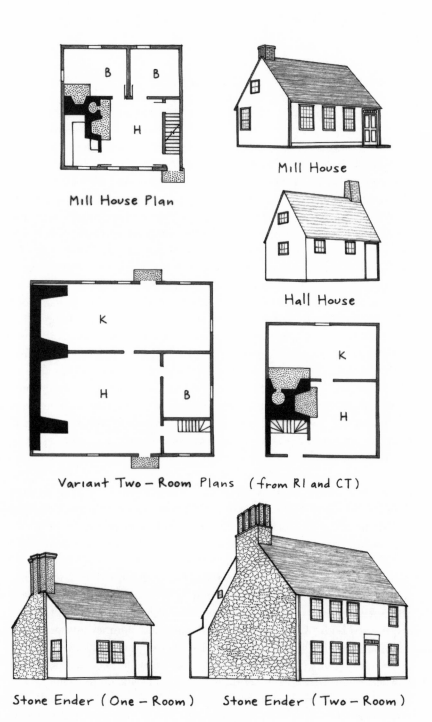

Mill House Plan

Mill House

Hall House

Variant Two-Room Plans (from RI and CT)

Stone Ender (One-Room)   Stone Ender (Two-Room)

FIGURE 12

House Types I

Stone-Ender plans redrawn from Isham 1967, 7, fig. 3. Mill house plan
redrawn from Garner 1984, 95.

Stone-Ender

The Stone-Ender was a short-lived seventeenth-century house type peculiar to northern Rhode Island and adjacent areas of Connecticut and Massachusetts (see fig. 12). Most of the earliest were burned in King Philip's War (1675–1676); now fewer than a dozen, mostly built between 1678 and 1720, survive in Rhode Island. It is notable for its early use of stone and mortar and its novel variation on the one-room house plan. While a Welsh origin has been suggested, Downing and Scully implicate certain half-timbered story-and-a-half Sussex cottages which, while not of stone, otherwise strongly resemble the Stone-Ender. (Sussex was the native county of some Providence and Newport settlers.) The state preservation report for Lincoln RI more circumspectly ascribes it to English rural tradition. The type originated as a one-room, one-and-a-half story timber-framed house with a massive stone chimney that made up all or most of one end wall, typically on the north. The stone was local schist or granite fieldstone (both good firestones) and the mortar made of shell lime or limestone quarried and burned locally at Dexter's Ledge, in the village of Lime Rock RI. The outer end wall was left exposed and not clapboarded over as were stone chimneys in southern Rhode Island and Connecticut. As was the custom in northern Rhode Island generally, and probably derived from Plymouth Colony, the body of the house was plank- and not stud-framed.

Enlargements of the Stone-Ender were idiosyncratic, suggesting the house type never evolved systematic solutions essential to its long-term viablity. The stone end wall might be widened to accommodate two side-by-side fireplaces, or opened at the rear for two back-to-back ones; sheds and lean-tos might be annexed to gable ends or rear wall, a second story added, or a whole new house built abutting it. Isham and Brown regarded side-by-side hearths as most characteristic, thereby strongly differentiating the (northern) Rhode Island two-room plan from the Hall-and-Parlor plan prevalent in both Massachusetts and Connecticut. (Newport, in the south, used both.) One-room, exposed stone end chimney houses, as the Thomas Lee House (1664) in East Lyme, were not unknown in early Connecticut, but when they were enlarged, as this one was ca. 1690, it was typically lengthwise to the central chimney Hall-and-Parlor plan. Likewise, the one Massachusetts example, now demolished, the Waite-Potter house (1677) in South Westport, was also enlarged laterally in 1760 when a brick chimney was built back-to-back with the old stone end.

It should be stressed that, apart from the choice of stone as the chimney material, the original Rhode Island nucleus differs little from the one-room end chimney plans of the earliest houses believed to have been built at Salem, Plymouth, and elsewhere. As Roger Williams himself was expelled from Salem, and as many of the first settlers of Providence derived from Plymouth, there is little surprise in both the prevalence of this plan and in the use, likewise common at Plymouth, of planked walls. While after about 1700 the end chimney was more often built entirely of brick, or of brick on a stone base (perhaps the chimney stump of a dwelling burned in King Philip's War), the one-room nucleus remained an authentic stage in the development of many houses until 1730; after that, at least in Isham's and Brown's view, the five-room plan rapidly revolutionized house-building in Rhode Island. Allen Noble sees a more evolutionary transformation; as few examples survive, however, there is clearly room for interpretation. The Stone-Ender is a defunct house type, much admired as a picturesque relic, but architecturally unsuited to the sort of twentieth-century revival seen by the Cape Cod house. Rare exceptions are a private residence in Scituate RI and an out-of-the-way replica in Bozeman MT built by the Field School on Early American Building and Crafts.[15]

Although the Stone-Ender as such did not survive the mid-eighteenth century, a prolongation of the tradition can be discerned in the floor plan of certain loom-weavers' cottages built in 1812 by Caleb Fiske in his cotton-mill village at Fiskeville RI. An end chimney served front and rear rooms, with the door located at the opposite end, a plan in accord with elements of both the Rhode Island and Connecticut Stone-Ender as recorded by Norman Isham (see fig. 12). While mainly wood-framed and clad in clapboard or shingle, some were built of stone. Mill houses of this plan are to be found also at Harris, Hope, and elsewhere, so much alike as to prompt speculation that they were "probably constructed by the same builders, if not for the same company."[16] Such harmony of design might also stem from acquiescence to a common tradition. Because these cottages were built within as little as ten miles of Providence, in an era when the legacy of the Stone-Ender was likely still much in evidence, and given the tendency of obsolete house types to persist at the lower end of the economic scale, an influence is quite plausible. After all, even in its prime, the floor plan of the Stone-Ender was never rigidly codified, and such mill-houses may have been its ultimate variation.

Whale House

The 'Sconset Whale house is a curious variation on the one-room, end-chimney Hall house, about forty of which survive on the east end of Nantucket Island. Siasconset was established as a whaling station ca. 1676, one of four such on Nantucket about that time, and the only one still standing. The fortunes of these stations rose and fell with the alongshore whale fishery (1670–1760), that throve only so long as whales were plentiful enough to pursue in small boats launched from land. Many of the Whale houses now at 'Sconset were dismantled and brought there by boat, largely from Sesachacha. In its day, the foremost whaling station, by 1888 this village had dwindled to three or four dwellings. In time, 'Sconset became the last gathering ground for these seasonal fishing-shanties once found more generally on the island, not only at Sesachacha, but also Cisco and Miacomet.

The nucleus of the Whale house is a one-story, roughly 12' × 18' dwelling, two-thirds of which was given over to the open-raftered great or fire room with north end chimney (see fig. 13). The door aligned with this chimney and looked east onto the sea, customarily opening outwards, it is said, to keep the house from blowing down. The remaining third of the house was partitioned into two low, tiny bedchambers or "staterooms," overhung by a half loft reached by ladder. These would have originally provided tight sleeping quarters for the six-man whale-boat crews. Roofs were pitched at 35°, and from sills to ridgepole the Whale house might be as low as 13½', contrasted with the more generous 44° pitch and 21 height typical of the Cape Cod house—itself regarded as snug! Initial enlargement of the Whale house in the eighteenth century was by offshoots at the south end called "warts," tiny lean-to wings on the sides of the staterooms that carried the roof down to within four feet of the ground and gave the house a T-plan. Warts were later integrated into the basic design. In the nineteenth century, the house might be lengthened by a bay, or enlarged by a so-called porch or kitchen-shed added at the chimney end, and these in turn might sprout warts. Later, the overall dimensions might be increased by raising the roof, widening the warts, or extending the gables.

Henry Chandlee Forman, on the basis of eleven years of investigation, exhaustively charted this typical evolution in his study *Early Nantucket and its Whale Houses*. He regards the type as firmly within the British vernacular tradition of the but-and-ben or hall-and-bower house. We might also note that something very like warts, alias bed-outshots, are found in north-

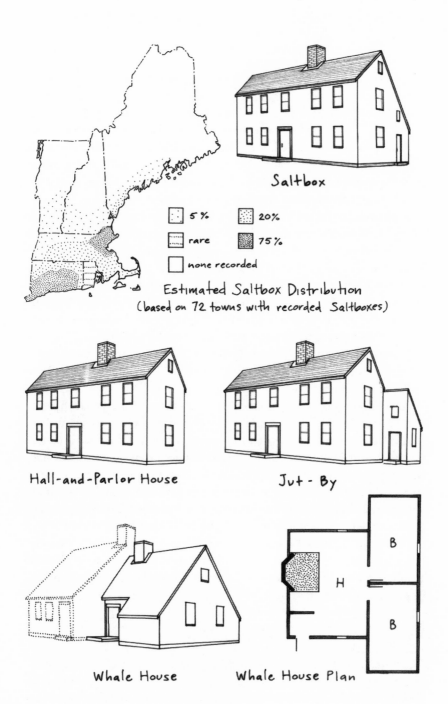

Saltbox

5 % 20%

rare 75%

none recorded

Estimated Saltbox Distribution
(based on 72 towns with recorded Saltboxes)

Hall-and-Parlor House

Jut - By

Whale House

Whale House Plan

FIGURE 13
House Types II
Whale House plan redrawn from Forman 1966, 116.

west Ireland, northeast Scotland, and in a coastwise band from Norway to Belgium. However, as the origins of the Nantucket whale fishery lay off-island, on eastern Long Island where it was pursued by the English and Dutch as early as the 1640s, the possibility exists that the Whale house originated off-island as well. Furthermore, in both places, the fishery was partly or largely manned by Indians, and the early Whale houses of Long Island have been described as "small thatched huts or wigwams." On Nantucket, the growth of the whale fishery coincided with the decline and dispossession of the local Indian population, culminating in the sickness of 1763–1764 when a number of wigwams were burned to forestall contagion. Many of the materials of the thirty-eight to sixty wooden Indian houses that stood on the sheep commons in the eighteenth century undoubtedly found their way to 'Sconset. Rose Cottage legendarily originated as a wigwam; Nickanoose may have as well. It is perhaps not irrelevant that the houses were often framed with saplings and that some originally had earthen floors, open hearths, and smoke holes in the roof. Thus it is with some point that Elizabeth Little asks: "Could Indian architectural concepts applied to English building materials have produced the Siasconset architectural style?"

In the present context, however, the importance of the Whale house is as a minor New England house type—only fifty-six are known to have existed in 1835—but one that all the same evolved as a unique variant on the one-room Hall house. It was expanded according to its own equally unique program of outshuts or "warts," which became integral in time, then freely adapted the early nineteenth-century kitchen ell. Use was made of relocated structures and salvaged building materials. The type even underwent a local revival of sorts with the resort development of Pochick Street in the 1880s by Edward F. Underhill, and deliberately " 'Sconsety" summer cottages crop up occasionally elsewhere on the island. Thus the 'Sconset Whale house reflects in miniature many key facets of the New England vernacular at large.[17]

Charter House

The term *Charter house* will be used here for a dwelling of specified minimum dimensions that each settler was required to build within a stated time to secure his title under the terms of a grant or town charter. It has also been called a regulation house, claim house, or possibly little house. Its construction was usually but one of several "settling duties," such as the obligation to set aside lots for the meetinghouse, the first minister, and the support of a school. By an act of 1733, settlers of Petersham MA were required within three years to build a house eighteen feet square with a seven-foot stud and to clear three acres of tillage and three of grass; in 1737 the settlers of Cornwall CT had two years to accomplish virtually the same thing. Elsewhere, similar duties were imposed. In Williamstown MA (1750) five acres had to be cleared and fenced, and the regulation house of 18' × 15' × 7' constructed. Settlers of Richmond MA (1762) were required within five years to have "cleared, fenced and brought to English grass or ploughed seven acres" and to have built a dwelling 24' × 18' × 7'. In Fryeburg ME (1762) these dimensions were slightly more modest, 20' × 18' × 7', while in Jamaica VT (1780) the charter provision was for the old standard, 18' × 18'.

Mere dimensions do not make a house type, but the replica regulation house erected on Williamstown Common in 1953 suggests that they were typically the initial Hall half of the pervasive Hall-and-Parlor plan. As reconstructed, the Williamstown "1753" House was a one-room, end-chimney house with the door aligned with the chimney. The chimney rose behind the ridge, and was of stone, laid up in mud until mortar was available; sides were clapboarded and roof shingled with ash shakes. The house might be framed over the sawpit, so work could proceed under cover; the pit served as the cellar afterward. The Seth Hudson house (1752) in whose one room the Williamstown proprietors first met in 1753, shows just how much such a Charter house might be transformed: moved twice, it grew from one room to two, then to two-over-two under a gable roof, and to four-over-four under a hipped roof, to which was subsequently adjoined an ell and finally a garage. Only the telltale timbers, exposed in the dining nook, now betray the size of the original house.

Not all Charter houses resume so methodically in their history the overidealized evolution of the New England dwelling. Perhaps more commonly, once he was established, a settler built an entirely new house, with the old house annexed as a shed or kitchen ell. Thomas Hubka asserts that many such ells in Fryeburg ME conform to the 1762 charter-imposed 20' × 18' standard, and that these in fact constitute the original dwellings that were later incorporated into the "big house–little house" scheme of the Connected Farmstead. In Jamaica VT, the ell of the Aaron Butler homestead is built of square-hewn logs and represents the original ca. 1780 claim house.

The specified dimensions of Charter houses appear simply to codify good common practice, differing little from those of one-room houses built before dwelling size came to be regulated in newly granted townships. They are directly descended from the one-room houses of the seventeenth century, but represent a later stage in that tradition, one in which through endless experiment and repetition, an initial diversity of English practices was becoming simplified and standardized into a New England vernacular. Settlement involved fewer uncertainties and relied more on the confident replication of proven models. However much the "1753" House in Williamstown and the tools with which it was built may call to mind the reconstructions and methods demonstrated at Plimoth Plantation, the Williamstown settlers had behind them the benefit of five generations' hard-won knowledge: they could set to work more expeditiously, and proceed to a higher stage of material comfort more assuredly, than ever had been possible for their counterparts of a century and a quarter before.[18]

Other One-Room Houses

In the 'Sconset Whale house we can see the nexus of well-documented architectural traditions and the lesser known ones of seasonal and makeshift housing, which all too rarely survive to be studied. Yet the one-room house is concerned with minimal shelter, a question vital to new people in a new land, be it in 1635, 1844, or 1900, and to the poor of any age. It has been asserted that a twelve-foot (rather than a sixteen-foot) module, typical of shotgun houses in the South and in Haiti, can be discerned in freed slaves' dwellings excavated at Parting Ways, Plymouth MA, and at Black Lucy's Garden, Andover MA. This twelve-foot module "may represent a distinctive Afro-American architectural tradition." (Mud wall-and-post construction, reminiscent of West African methods, is also present at the former site.)[19]

An accident of history immortalized the one-room shanty (ca. 1845) of Irish railroad laborer James Collins that Thoreau purchased to provide boards for his Walden house: "It was of small dimensions, with a peaked cottage roof, and not much else to be seen, the dirt being raised five feet all around as if it were a compost heap." The inside was dark, with but one high, small window, a stove, a few sticks of furniture, a largely earthen floor, no doorsill, and was home to Collins, his wife, infant child, and cat. It took Thoreau but a morning to dismantle; ironically, the boards were used to enclose quite another version of the minimal dwelling, which despite its modest 10' × 15' size had a traditional New England hewn frame with posts six inches square. Other writers in search of solitude took note. On Cape Cod, what began perhaps in the late nineteenth century as half-way houses for coastguardsmen on beach patrol had developed by the 1920s into the dune shack, made famous by Eugene O'Neill and others, with the best-known colony at Peaked Hill in Provincetown. These were generally of one room and might run 11' × 8½', although the Fo'castle (1927), the custom-built two-room dune shack on Eastham bar celebrated by the naturalist Henry Beston in *The Outermost House*, was 20' × 16'. Dune shacks were often built of salvaged materials and could be moved about on skids; some are of traditional appearance, and like Thoreau's Walden house, serve to focus attention on otherwise unsung marginal housing within the vernacular tradition. (Possibly related are fish houses or "stores," which consisted of a single room for mending nets and a loft overhead for storing gear.)[20]

Of quite different origins are the ruins of the "the mud hut colony" inhabited by Italian dam-workers ca. 1907 to be found alongside the Newton Reservoir in Athol MA. Since mud houses are unknown in twentieth-century New England, it is perhaps fair to conclude that this sort of temporary housing was something with which the workmen themselves were familiar in Italy, and it may be that such makeshift shelters were a typical adjunct of the "padrone" system whereby an Italian labor-broker assembled and delivered gangs of unemployed men to the job site. A photograph of another such Italian work camp for construction of the Fitchburg and Ayer Street Railway ca. 1900 reveals tent- and teepee-like hovels of turf, bark, and poles erected in ways that suggest woodcraft and not random desperation. These represent a very different tradition from the Irish shanty. Charcoal burners, who supplied fuel to iron furnaces before the ready availability of coke, lived in the woods in burrows of sticks and turf while they tended their kilns in late spring. As late as 1978 the *Appalachian Trail Guide* asserted that "the ruins of charcoal burners' shacks" were to be seen on the slopes of Coltsfoot Mountain, near Cornwall Bridge CT. Such workman's shanties, work camps, and temporary hutments, often housing immigrant labor, were commonplace in the expansive years of the nineteenth century. The need was dramatic, as exampled by the come-and-go of railroad laborers on the Boston and Albany, which reads as a blip in the decennial census of Middlefield MA from 1830 to 1850: 720, *1717*, 737. Clearly such transients lived somewhere, but their dwellings largely survive only in descriptions and photographs, and are rarely studied. John Coolidge devotes but half a page of *Mill and Mansion*, his architectural study of Lowell, to the shanties of the Irish laborers who built the city. Our treatment of the architecturally respectable one-room house fails to do justice to the little-known traditions of minimal shelter in New England, traditions that clearly did not end when the first planters of Concord emerged from their hillside burrows in 1636 and began to build themselves a town.[21]

Hall-and-Parlor House

The one-room, or Hall house, as we have seen was typically expanded sideways to create the central-chimney Hall-and-Parlor house. This formula was already well established in East Anglia, and some very early New England houses (like the Fairbanks house in Dedham MA [1637]) were built on this two-room plan from the first. In the Bay hearth core, the Hall-and-Parlor as such rarely survives and is treated in the literature largely as transitional to the Saltbox, or as the discernible nucleus of a much enlarged seventeenth- or early eighteenth-century house. However, further upcountry, in southern Maine, its true longevity becomes apparent in the one-story 18' × 35' Eliphalet Walker house (ca. 1780) in Kennebunk, and in the two-story Nutting house (ca. 1825) in Otisfield (probably built with an integral one-story kitchen ell). Thus, the Hall-and-Parlor plan could serve equally as the basis for a one-story "starter" house (comparable to the Cape Cod, also widely used in Maine), or a fashionable two-story Federal. Whether this persistence of the Hall-and-Parlor was peculiar to Maine or in fact quite general is unclear. The presentation of house types as if in an evolutionary modular series may have clouded perceptions; in this scheme, as we shall see, the Hall-and-Parlor had already started to "evolve" into the Saltbox in the mid-seventeenth century (see fig. 13).[22]

Saltbox

By about 1650 the two-story Hall-and-Parlor plan began to be enlarged by the addition of a one-story rear lean-to. This enlargement conferred the characteristic Saltbox profile: two stories in front, one story behind, with a long rear roof slope (sometimes called a catslide) that descended to a point seven feet (or less) above the ground. The roof still peaked as on a one-pile house, so the ridgeline lay nearer the front than the rear. Early Saltboxes often show a telltale break in the rear roof slope where the new timbers were married in at a slightly lower angle. By 1700 or 1725, when lean-tos had come generally to be built, not as afterthoughts, but integrally with new construction, the roofline was continuous. In England, the lean-to, or outshut as it was called, first appeared ca. 1525–50, and was still innovative in East Anglia at the time of the Great Migration, emphasizing that the same fundamental problems in the early modern vernacular house were being worked out simultaneously on both sides of the Atlantic. Emigrant carpenters were young, recently apprenticed, and imbued in the new directions of their trade.[23]

The orientation and layout of the Saltbox in New England owed much to the severe climate. The two-story housefront faced south to take fullest advantage of the sun; the low rear lean-to, at times as low as five feet at the eaves, was snugged down against the north wind. (Examples on eastern Long Island face west, perhaps the lee side there.) The floor plan was quite commodious; some were traditionally said to be "two rods square." Such spaciousness gave rise to a new standard room plan by the late seventeenth century. In Massachusetts Bay the rear lean-to was customarily partitioned into three: a large middle kitchen with fireplace and bakeoven focused on the central chimney stack, and two smaller end chambers, a cooler northwest buttery for food storage, and a warmer northeast bedchamber near at hand for the care of the sick, the elderly, and women in childbed (hence the term "borning room"). Hall, parlor, and rear range of three rooms made for five altogether, and this influential plan, common to all central chimney houses, is known as the Massachusetts Bay or five-room plan (see fig. 14). This layout was by no means the only disposition of space; in England, for example, outshuts often had a staircase in place of the kitchen hearth. Rarely documented in Massachusetts Bay (Hapgood House, Stow MA, ca. 1726), but standard throughout northwestern Rhode Island was the absence of a kitchen in the rear range. In as many as 95% of five-room plan houses (not necessarily Saltboxes) in Foster RI, the kitchen is found in one of the front rooms, a likely holdover from the earlier Hall-and-Parlor plan. A bedroom occupied the middle rear room.[24]

A not untypical adjunct to the Saltbox (and some other house types) was the jut-by or Beverly jog: a lateral one-bay extension of the lean-to that jutted beyond the side of the main house (see fig. 13). It might have windows or door; some houses have them on both ends. By no means unique to Beverly MA, the jut-by occurs in Salem, Marblehead, Ipswich, Concord, and Harvard, as well as elsewhere in Essex, Middlesex, and at least parts of Worcester County MA. Several are known in Temple and elsewhere in southern New Hampshire. An unusual brick jut-by on a ca. 1830 store is

Cape Cod House
(Full, Three-Quarter, and Half)

Cape Ann Cottage

Three-Quarter Large House

Nantucket House

Central Chimney Large House

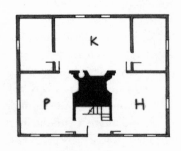

Five-Room Plan

FIGURE 14
House Types III
Five-Room plan redrawn and adapted from Rhode Island Historical
Preservation Commission 1982a, 13, fig. 17.

found in Gloucester, and one dubious example on a much altered house is known from Nantucket Town, which may be related to what Henry Forman has called the half-gable, split gable-end, or "flounder wing" in 'Sconset (this last apparently a southern term). However, from the available evidence, the jut-by appears to be peculiar to the Massachusetts Bay hearth: examples elsewhere are not documented in standard sources, even in Connecticut where the Saltbox is rife. Lean-tos might also be added to smaller houses, and story-and-a-half Saltboxes appear commonest in Connecticut and Connecticut-settled areas of Vermont and perhaps Ohio.[25]

The Saltbox originated between 1650 and 1675, became standardized about 1700–1725, and obsolete in the Boston core by 1750, being supplanted by a full two-story, two-pile house. Saltboxes continued to be built on the peripheries into the early nineteenth century, as was an 1803 example in Warwick MA. On Nantucket, later versions were downsized to one and three-quarters stories, a tendency observable in obsolescent house types. Documented case histories defy overneat ideas of their evolution. The Job Lane house in Bedford MA was erected ca. 1715–1720 as a two-story Hall or half-house with integral lean-to, literally a half Saltbox. To this, on the northwest corner, was appended a shed (possibly a back kitchen) later in the century. Only about 1825 was the corresponding "parlor" half of the main house added, belatedly converting it into what we regard as the archetypal full Saltbox a century after it was first built, and considerably after this house type had fallen from fashion. This sequence is neither unique nor novel. The Cooper-Frost-Austin house (ca. 1689) in Cambridge MA was originally such a half Saltbox (that is, hall–and–lean-to) that was not enlarged laterally to its full size until after 1718. On Nantucket, such a line of evolution was normal.[26] So this "unprogrammatic" scenario is both common and perhaps even as old as the Saltbox type itself. Ironically, once the "ideal" form is belatedly achieved—as much as a century belatedly —it beguilingly invites its wholesale retroprojection into a mythical past. (Such retroprojection is a frequent source of error in reading the landscape, as when the stereotypical prim white clapboard New England village of the mid-to-late nineteenth century is taken to be a Revolutionary relic.)

The actual distribution of Saltboxes is not precisely known. Because the Saltbox was early established in both the Connecticut and Massachusetts hearths, it was well placed to diffuse inland along both main settlement streams and no marked east-west divide is to be expected. Two noteworthy early concentrations are coastal ones, at Guilford CT and Ipswich MA, but the type must potentially extend inland to anywhere within the 1800 settlement frontier. No Saltboxes are found below Brewster on Cape Cod. Nor do they occur on Martha's Vineyard: a lack that Jonathan Scott accounts for by the fact that the Hall-and-Parlor plan (which led to the Saltbox) never existed on the island. Instead, early Vineyard houses of moderate size were one room wide and two rooms deep; these would be extended sideward, not rearward, and the lean-to was thus obviated.[27] (Ironically, "revived" Saltboxes now dot the island, among them a bank and the Chilmark firehouse.) Recorded examples are also slim in Rhode Island, where the Massachusetts five-room plan was not adopted until late (after 1720), and then largely in its full two-story form. A possible pattern emerges in which Saltboxes are deficient wherever the Hall-and-Parlor house was not a basic "starter" home. The map in figure 13 is meant to give a rough idea of this distribution. It is based on my handlist of seventy-two towns known to have at least one Saltbox: percentages thus refer to the relative density of "Saltbox towns," not Saltboxes themselves.

In terms of dissemination outside of New England, the Saltbox was not well positioned for export. Diffusion favors the right house in the right place at the right time, and the Saltbox was never quite this. During the century 1675–1775 when the Saltbox was in its prime, New Englanders were preoccupied by their own internal development and did not overspill their boundaries. Three minor out-migrations did occur, from Connecticut across the Sound to eastern Long Island ca. 1648, to East Jersey after 1664, and finally to Nova Scotia in a pre-Loyalist efflux from 1760 to 1775. To some degree, extant Saltboxes can still be found in all three areas. Saltboxes as early as 1720 also occur in Westchester County NY and probably overspill the New England–New York border elsewhere. However, by the time of the general New England westward exodus of 1790–1830, the house type was virtually obsolete. It had always been something of a half-measure, a stepping-stone to a two-pile, two-story house, and it was in fact this later development, in the form of the Center Hall and Central Chimney Large house, that spread across upstate New York in the years after the Revolution. One Ohio Saltbox, the Hale farmhouse (ca. 1830) identified by Stanley Schuler, is little more than a One-and-a-Half with lean-to; if this is the best the Connecticut-settled Western Reserve has to offer, it is clear proof that by the end of the Revolution the Saltbox was too moribund to travel west. Revival of the type has been limited.[28]

The origin of the term *saltbox*, first recorded in 1876, is obvious yet obscure. It possibly refers to a slant-lidded bin that once held coarse salt for curing meats; a smaller article, however, is sometimes depicted, and even Wallace Nutting confessed ignorance on the point. Pattern-book plans for the Saltbox house itself appeared as early as 1887, and derived, asymmetric rooflines (often reversed) might sweep picturesquely down to shelter a porch in Queen Anne houses. Something nearer the true Saltbox shape (but still reversed) was used for carriage houses. A greater historicism asserted itself in 1923 when the builder's association Better Homes in America opened a show house in Washington DC, published plans, and proclaimed the Saltbox the "most talked of house in America." Still it remained the province of architect-designed, upper-middle-class houses; widespread imitation came only much later, largely as a novelty after the postwar Cape and garrison waves had crested. Often, the historic rationale for the rear lean-to has utterly vanished, and it may instead conceal a two-car garage, enfold a rear porch or patio, or serve merely as a curious historical motif.[29]

Central Chimney Large House

This two-story, two-pile, central chimney house was built on the five-room plan and can be regarded as the end result of the dubiously "typical" modular sequence of Hall, Hall-and-Parlor, and Saltbox. Since the layout upstairs and down was often five-over-five, this may represent the proverbial "ten-room house" (see fig. 14). It generally appeared after 1740, although examples as early as 1699 are known.[30] Externally, the house emulated the five-bay, bilateral symmetry of the Center Hall (discussed later). Early examples of Large houses that simply "grew" to size from an original seventeenth- or eighteenth-century core, often exhibit less than perfect balance: the chimney may be off-center, the newer half larger-scaled, with later windows, ornate window caps, and other refinements. Seams in the clapboards and contrasts in the foundation may demarcate the original build.

The remorseless five-bay standard brings all departures from it under sharp scrutiny. Some three- and four-bay variants were locally fashionable, while others were simply poor-man's five-bays. A center-entry, three-bay facade was well established in early Connecticut Saltboxes, as at Guilford (1670) and Farmington (1660)—perhaps originally to conserve window glass, as the houses appear to be no smaller. The practice continued into the eighteenth century as an acceptable alternative on full two-story houses, and spread up the Connecticut Valley at least as far north as Greenfield MA. (Three-bay facades are also found in the Bay hearth, so no strict monopoly of this subtype exists; indeed, they often occur on stylish houses in the British Isles.) More often, however, the three- and four-bay versions are physically smaller and represent modular fractions of the full Large house, much as in the Cape Cod house. Particularly in thickly built coastal towns, examples of what might be termed the half (or three-bay) and three-quarter (four-bay) Large house appeared in the eighteenth and early nineteenth centuries, perhaps in response to tighter houselots. In Rockport MA, the Hannah Gott Giles house (1814) is one of at least thirteen such half Large houses in the village, some of which were enlarged by jut-bys. On Cape Cod, the Central Chimney Large house debuted with the Moody house (ca. 1699) in Sandwich, but was fairly rare and largely confined to the upper Cape until two-story houses became more general in the prosperous 1820s. As it is, the more modest three-and four-bay variants occur chiefly on the poorer lower Cape from Yarmouth to Provincetown. On Nantucket, all variants are known, but a four-bay subtype, on a high brick basement, with a dead-center chimney and an off-center doorway in the second or third bay, became sufficiently established in the early nineteenth century to earn the name of the Nantucket (or Quaker) House today (see fig. 14).[31]

The stylishness of the Central Chimney Large declined through the eighteenth century in favor of the Center Hall. In Portsmouth NH, it opened the century as an elite house and closed it as the abode of the artisan class. Provincial examples of some pretensions continued to be built fairly late: in Concord (1784), in Swampscott (1789), and most exceptionally, in Duxbury MA (1826). In the hinterlands, the Central Chimney Large persisted into the 1820s, but even in rural southwest Maine it was essentially a "pre-1830" house type. Its goodly size and the cachet of its five-bay facade (which was but the reflected glory of the Center Hall house) favored its early westward spread through the Mohawk Valley of upstate New York. However, beyond their common clapboard vernacular traditions, the architectural affinity of the Old Northwest with New England rests more solidly on later, more innovative Greek Revival house types, such as the Upright-and-Wing and variants on the Templefront.[32]

Cape Cod House

The Cape Cod house, despite its name, is neither provably original to the Cape, nor in any way unique to it, except that in its cheapness and simplicity it persisted longest and survived best in impoverished areas, such as southeastern Massachusetts. Ironically, the earliest known example on Cape Cod, the Saconesset homestead in West Falmouth, postdates by some dozen years two in Old Lyme CT, believed to have been built in 1666. The geographic specificity of the name should not blind one to the great ubiquity and longevity (ca. 1660–1860) of this house type throughout New England, not simply along the coast, but equally in the hardscrabble hinterlands of Vermont, New Hampshire, and Maine. This presumed early universality of the Cape Cod house renders perplexing the fact that the type was first authoritatively recorded in 1800 by Timothy Dwight, a Northampton-born, Yale-bred descriptive traveler otherwise generally content to qualify houses as "neatly built" or "good farmers' dwellings." Why, after five years and thousands of miles of travel through New England and New York, suddenly sit up and take notice, describe in detail and name what might "be called with propriety Cape Cod houses," if they were the salt of the earth?[33] Yet so they seem to have been.

Socioeconomically, the Cape Cod is what Anthony Garvan would term a "husbandman house," as opposed to a "yeoman house," and its admirable ability to fill precisely this niche assured its longevity. For such reasons, the Cape seems to have undergone a post-Revolutionary surge; in the late 1700s and early 1800s, "Cape Cod type houses were among the earliest and most common dwelling houses built in Vermont." In Washington NH, the most common early farmhouse was the "small one-story center-chimney 'Cape Cod,'" about a half-dozen of which dating to the 1780s still stand. As late as the 1850s, the half-Cape was still a viable type in inland Maine, particularly for a young farmer starting out. Thomas Hubka estimated that 25% of all pre-1900 houses in West Bridgton ME were Cape Cods and felt the same would hold true for much of rural New England. Strange, then, is the fact that there are only one or two historic Capes on Nantucket, and those but half-houses at that. The modern revival of the Cape Cod house in the Depression era, and its dramatic rise as a postwar "starter home" with a G.I. mortgage, is entirely in keeping with its social history.[34]

The Cape Cod was a story-and-attic house built in three modular sizes, now generally called half, three-quarter, and full Capes, as gauged by whether the facade is composed of a door and two, three, or four windows (see fig. 14). These sizes typically measured 20, 28, and 34 feet across the front and 23–28 feet along the gable. The plan was thus squarish but not rigorously square, and under its low broad roof the Cape is conspicuously snug. In all three sizes, from the end-chimney half Cape to the center-chimney full Cape, the front door always aligns with the chimney-stack and gives onto a small stair lobby. Usually the stairs were steep and straight-run; dogleg stairs were more typical of eighteenth-century two-story houses. Full Capes used the five-room plan, and half and three-quarter Capes followed suit to the extent that their reduced dimensions allowed. Only the full Cape achieved the classically balanced five-bay facade, and many Capes exhibit a pre-Georgian indifference to the new Anglo-Palladian aesthetic slowly codified in the colonies in the eighteenth century. (Full Capes in Connecticut, like Saltboxes, often had only three bays.) As a house type, it was immunized by custom, poverty, and its own modular logic to such dictates. The half-to-full Cape sequence does not represent literal growth-stages through which every house passed; rather it is proof of the house type's ready adaptability to a range of owners' needs (and purses) over time. Indeed, growth did not necessarily stop at the canonical five bays. Examples of up to eight or nine bays of what has been termed the lengthened house are known, constructed in stages, or by splicing together earlier dwellings. Two half houses could create a six-bay house with a chimney at each end.[35]

On Cape Cod, cellars were no more than small, circular pits dug in the sand, nine to twelve feet in diameter and brick-lined. Access might be by an interior trapdoor, or more characteristically by an exterior cellarway that took either the form of a "cellar house" a small, peaked, almost doghouse affair, or that of a slant-doored "bulkhead." The nautical origin of this latter term, and the fact that its usage in this sense, as mapped by *DARE*,[36] is chiefly restricted to New England, suggests that both the word and perhaps the thing itself are of coastal and possibly Cape Cod origin. Indifference to modern aesthetics is apparent in old photographs, which show bulkheads in plain view on the front facade of the house. Exterior finish and detail is modest. On Cape Cod, shingling is traditional, and might be varied with clapboards on the front (and perhaps rear) facade. Inland, clapboarding is the norm. Ornament is generally austere Federal or Greek Revival, with simple transoms and sidelights. (Studs were too low for elaborate overdoor treatments, such as fanlights.)

A house type that throve for two centuries and spread throughout New England (and beyond) not surprisingly evidenced variations on its classic formula. Many sought to remedy the dark, cramped garrets. On Cape Cod proper, one local hallmark is seen in the attic gables, where the usual complement of windows is eked out by two or three small, fixed, occasionally triangular lights. Such eccentric fenestration is also found well inland in western Franklin and Hampshire Counties MA, where early settlers had roots in southeastern Massachusetts. Another less common, but well-known (and exaggeratedly imitated) feature largely peculiar to Cape Cod is the bowed roof, although the actual camber may be scarcely visible. Customarily, dormers on Cape Cods existed only as additions, but what appear to be original five-bay shed dormers are found in Plymouth County MA on the fronts of eighteenth-century Capes. Dormered Capes with steep gambrel roofs are widespread in Connecticut (as at Old Lyme); the triple Gothic dormers often seen in coastal Essex County MA in all likelihood date to the nineteenth century. In Connecticut, end attic-story overhangs are fairly common on Cape Cod (and other) houses; this peculiarity was carried northward to Vermont by settlement migration. (While virtually absent on Cape Cod, such overhangs can be found not far inland, as at West Bridgewater MA.)[37]

A gambrel-roofed half-Cape—specifically termed a Cape Ann cottage—flourished in Gloucester and Rockport MA about 1690–1760 (see fig. 14). Gambrels increased headroom and allowed shorter timbers to be used for rafters; both "flat" and "steep" gambrels (52°–59° in the lower pitch) were employed on Cape Ann; the flat being the more usual. Thoreau noted both types on his 1858 Cape Ann walking tour: "The narrow road—where we followed it—wound about big boulders, past small, often bevel-roofed cottages where sometimes was a flag flying for a vane. The number and variety of bevelled roofs on the Cape is surprising. Some are so nearly flat that they reminded me of the low brows of monkeys." In its 1933 photographic survey of Cape Ann, the *White Pine Monograph Series* illustrated eleven gambrel-roofed Capes, at least six of which were specifically half Capes; to believe the writer, for every one included at least a dozen were "passed by" as "irretrievably spoiled." Whatever its present numbers, the Cape Ann cottage no longer asserts itself on the landscape as it did in Thoreau's time, but the term is still used in the building trades to describe a gambrel-roofed house of a revival design probably loosely derived from the pages of the White Pine series. However, any claims as to its uniqueness to Cape Ann must be taken with a grain of salt: gambrels were generally fashionable in the eighteenth century, and even one-story gambrel-roofed houses were not uncommon elsewhere in Essex County (as at Rowley), or further inland, as for example the Pliny Freeman farmhouse (ca. 1810–15), removed to Old Sturbridge Village from elsewhere in the town. Some were even enlarged by a rear lean-to to form a story-and-a-half gambrel-saltbox, such as the ca. 1760 "Quaker" house in Petersham MA and the ca. 1798 David Sexton house in Deerfield MA.[38]

In Vermont, the Cape Cod house derives more directly from Connecticut than Massachusetts, and constitutes, at least in local eyes, "a parallel development," sufficiently distinct to be called the Vermont story-and-a-half. In 1835, such houses made up 75% of the housing stock of a town such as Royalton. Notable features may include gable-end overhangs, roofs with pinned pairs of rafters instead of ridgepoles, and three small square lights in the corners of the attic gable. The cramped entry-hall stair was often replaced by an enclosed straight-run stairway in a corner of the rear kitchen. The Vermont story-and-a-half might do duty as the main house of the Connected Farmstead, with outbuildings running rearwards or sidewards to the barn. (Calvin Coolidge described his Plymouth VT birthplace as a "five room, story and a half cottage attached to the post office and general store.")

Further north still, an efflux of New Englanders in the 1760s brought the Cape Cod house to Nova Scotia, where it became particularly well established along the coast from Yarmouth to Halifax. (Nova Scotian telephone directories are said to be replete with "Cape Cod" names: Nickersons, Bournes, and the like.) The type was adopted or adapted by succeeding settlement groups down to the 1840s, sometimes altered with Irish or Scottish floor plans. By 1860 a distinct Maritime cottage emerged, a story-and-a-half house with a four-room central hall plan, three-bay facade, and a trademark central dormer. This might be a peaked Gothic dormer, a five-sided Scottish dormer, or the even more elaborate bracketed Lunenburg dormer, which overhung the entry porch (see fig. 15).[39]

Demise of the historic Cape Cod house has been variously dated to the mid-1860s in interior Maine and slightly later on Cape Cod itself. Obsolescence of the central chimney, competition from more modern styles and framing methods and postbellum economic stagnation in the peripheries to which the type was ultimately confined were all factors. Furthermore, the adaptability of the type meant that it evolved to transcend its own inherent limitations of design. As early as 1800 the Captain Solomon Howse house in Chatham MA was built with twin ridge chimneys. A common means to increase the headroom of a Cape was to "raise the roof" by carrying the walls up about four feet, and steepening the rafters. (These raised walls, technically called kneewalls, give rise to the term "kneewall cottage" sometimes used to distinguish later Greek Revival cottages from the true Cape.) Gablefront versions with more elaborate Greek detailing, such as embellished corner pilasters or even columns, and locally termed "captain's houses" on Cape Cod, began to appear in the 1830s. Curiously,

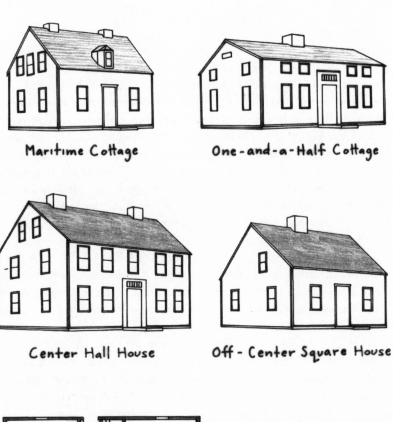

Maritime Cottage   One-and-a-Half Cottage

Center Hall House   Off-Center Square House

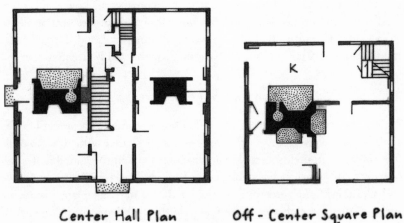

Center Hall Plan   Off-Center Square Plan

FIGURE 15

House Types IV

Off-Center Square plan redrawn and adapted from Stanchiw and Small
1989, 137, fig. 3; Center Hall plan redrawn from Garrison 1991, 159, fig 7.8.

in the 1870s and 1880s, as the Cape Cod house itself died out, a notably smaller (25' × 10' or less), one-story, two-room, center-entrance, central chimney cottage began to appear on the back roads of Plymouth County MA. Whether year-round dwellings or seasonal (read cranberry?) workers' housing is unclear, but perhaps they can be seen as the ultimate devolution of the Cape Cod tradition. However, even as the Cape Cod house adapted to the nineteenth century it also ceased to be a Cape. Very small houses were frequent casualties of progress; they might be enlarged, altered, obliterated, or adjoined to the rear quarters of a later house, often as a kitchen ell. The late nineteenth century advent of summer colonies on both Cape Cod and Cape Ann enhanced the economic niche of old small houses that might elsewhere have been razed or left to rot.[40]

Revival of the Cape Cod house in the 1930s and after is strongly associated with the work of Boston architect Royal Barry Wills. Specimen plans of his design published in *Life* magazine in the early 1950s extolled the Cape as a sort of everyman's dreamhouse. In private commissions, Wills's trademarks were oversized chimneys, gambrel roofs, hand-riven roof shingles, graduated clapboards, and pine-paneled and rough-plastered interiors with exposed beams.[41] Generally speaking, the revived Cape Cod house features unhistorical dormers, picture windows, breezeways, and a disregard for door-chimney alignment. The cheapest are perhaps not surprisingly the most faithful. While it is the iconic small house, the Cape Cod is but the best-known member of a whole misunderstood family of story-and-attic houses that are rarely carefully distinguished. Two of these, the One-and-a-Half and the Off-Center Square, will be treated next.

One-and-a-Half Cottage

This is a story-and-a-half, gable-roofed house, some 32 to 34 feet long by 22 to 26 feet wide (see fig. 15). Early examples shared with the Cape the five-room, central chimney plan and differed only in the added half-story (4–5 feet) that gave it its name. Later One-and-a-Halfs had one or two end chimneys and something closer to the center-hall plan, distancing them further from the textbook Cape, but not perhaps from how actual Capes came to be modified to meet the needs of the nineteenth century world. Like the Cape Cod, it ran from three to five bays; unlike it, the three-bay version had a center door, and the bays themselves seem not to have been modular. The added height allowed for more elaborate door-heads and cornices on the facade and ampler attic rooms within, and seems merely to institutionalize the nineteenth-century practice of raising the roof on existing Capes.

Let into the typically wide Greek Revival entablatures were two to five so-called half, knee, or lie-on-your-stomach windows, which furnished welcome light and air to the attic rooms within. While such rectangular under-eave windows had appeared by the 1790s, they proliferated under the Greek Revival and became a hallmark of the classic One-and-a-Half. (They are also seen on other cottages and farmhouse ells.) In New York and the Midwest, knee windows often took the form of ornately grilled vents, while others in southern Maine were glazed, but fancifully shaped: wreathed ovals, flattened octagons, and scrolled edges are found. Knee windows were readily converted to through-the-wall dormers, and houses may display an eccentric combination of both. A less common feature was the funeral door, said to permit the decorous removal of the dead in the days of home burial. (Exit through the cramped front entry hall required upending the coffin.) Such doors were little used and lacked steps. (A similar rationale is given for the enclosed vestibules seen on many Cape Cod houses.) The One-and-a-Half might be enlarged by ells, wings, or even lean-tos, making effectively a small, knee-windowed Saltbox. Curious story-and-a-half Saltboxes crop up far and wide: the John Benedict house (1750) in Ridgefield CT; an undated house in Newfane VT; as well as the Hale farmhouse (ca. 1830) in Richmond OH.[42]

Paradoxically, what the literature terms the One-and-a-Half *New England* cottage has gone little remarked upon in New England itself, and the elucidation of its origin and diffusion has fallen to two midwestern scholars, Allen G. Noble and Nancy J. Bracz. The specific designation "New England" was first employed in Noble's *Wood, Brick, and Stone* (significantly, a book researched at the University of Akron) and the name seems to be given from an Ohioan viewpoint; like the "Loyalist" architecture of Canada, the attribution may be only half accurate. All examples that Noble specifically locates are in the Akron area, which lies within the Connecticut-settled Western Reserve. In his view the fullest development of the type was in Ohio, although present in New England settled areas further west, and further east in a presumably earlier concentration in the "upper Hudson valley and its major tributaries." Bracz asserts that while the type first appeared in the Ohio Western Reserve ca. 1833, earlier examples from the mid-to-late 1820s are to be found in central New York along the Syracuse-Utica axis. She traces the popularity, but not the origin of the house to plate 63 of *The Young Builder's General Instructor*, a pattern-book published by New York architect-builder Minard Lafever in 1829, where a more commodious version of the house, with twin end chimneys and a rear kitchen ell, appears. Lafever described it as a farmer's cottage, one characterized by "frugality, convenience, and neatness in a plain style." Since Lafever trained in upstate New York, his published plans may have well elaborated and codified a successful vernacular house type already familiar to him.[43] But what of its ultimate origins?

In one line of argument, the One-and-a-Half is an evolved subtype of the Cape Cod house, and indeed, Fred Kniffen, who seems to have coined the term One-and-a-Half, employed it more broadly, lumping both together without a qualm. A ca. 1750 Cape in Haddam CT is known to have had its roof raised in the mid-nineteenth century and knee windows incorporated. Overall, twelve-foot studs succeeded to the traditional eight-foot stud around 1830, a development that would have favored the One-and-a-Half. Les Walker, in *American Shelter*, suggests an alternate origin about 1690 in a three-bay, story-and-a-half, Hall-and-Parlor house with knee windows, which he regards as the first stage in the forty-year evolution of the Central Chimney Large house. Regrettably, Walker indicates no concrete basis for his pencil sketches of houses dated 1690 and 1720. However, as a Yale-trained architect based in New York City and Woodstock NY, one might surmise that his area of observation was western Connecticut and eastern New York.[44]

Similarly, one of the best descriptions of the type comes from Henry and Ottalie Williams, known for their western Connecticut bias. "Somewhere between 1800 and 1840 there were built a great many small one and a half story houses on the Early American, central-chimney plan, with modified classical features." These had "a broad, plain frieze into which was let a series of "eyebrow" windows, innocent of any grating."[45] Perhaps closely related is the fact that true Cape Cod houses in southwestern New England seem to be rarely documented in published photographs. This absence, if not a mere sampling fluke, argues that the rival One-and-a-Half occupied its accustomed niche. Overall, methodical thumbing of illustrated local histories and other works turned up twenty-nine presumed examples of the type, with the densest concentrations within seventy-five miles of New York City (including areas of NY, NJ, and CT). The oldest with available dates are found in Huntington, Long Island (1710, 1732) and Ridgefield CT (ca. 1730–1750).

If the One-and-a-Half with knee windows did evolve from the Cape

Cod house, it is paradoxically rare on Cape Cod itself. Observation on Martha's Vineyard, however, reveals many gablefront, Greek Revival cottages with such windows unobtrusively on the sides: had these preserved their traditional flank fronts, they would be One-and-a-Halfs. While seldom seen in eastern Massachusetts, bicycle surveys in central New England recorded examples in Athol, Ashburnham, Bolton, and Winchendon MA; Winchester NH; and several in Newfane VT. Bibber illustrates one in Maine.[46] This evidence suggests that the type emerged in the eighteenth century in southwestern Connecticut and eastern New York, as either a variant of the Hall-and-Parlor or Cape Cod house. It achieved visual distinction with knee windows, and later, Greek Revival details, and was carried by settlers up the Hudson and the Housatonic and west along the Mohawk. The slow spread of this folk house took wing when it was promulgated in a pattern-book by Minard Lafever in 1829. Much more fieldwork on the One-and-a-Half is required before its true place within the New England settlement landscape can be assessed.

Off-Center Square House

What in the literature has been called the Hemenway, or New England square plan, I choose to call the Off-Center Square house (see fig. 15).[47] It was first formally identified in 1985–1986 by researchers investigating the Emerson Bixby house (ca. 1802–1809) in Barre Four Corners MA, prior to its removal to Old Sturbridge Village. Five such houses, at least two the known work of local housewrights named Hemenway, turned up in a survey of the vicinity and were dated 1786–1809. These houses are characterized by a squarish plan, an off-center chimney, two unequal front (or south) rooms and a long rear (or north) kitchen. A narrow chimneyside passage (sometimes taken up by a bakeoven or closet) joins sitting-room and kitchen. Notable is the lack of the stair lobby of central chimney plans; front doors opened directly into the intimacy of sitting room or kitchen with premodern disregard for formality or privacy. Stairways in the Off-Center Square were off the kitchen, generally in the back of the house. In the examples surveyed by Old Sturbridge Village, the kitchen always lay on the north side, and the roof ridge always ran parallel to the road. Thus the kitchen was found on the streetfront of the Bixby house, where best room and sitting room might customarily be expected. This layout is not everywhere the rule and perhaps was a quirk of the Hemenways themselves. Compromises between the organic (north-side kitchen) and the formal (road-aligned roof) were common among country builders who strove to balance old customs and new trends.

A paucity of recorded examples makes it difficult to dogmatize on how this house type externally differs from the readily confused Cape Cod house. Published plans for the Bixby and Holland houses indicate them to be roughly 24' × 24' making them mathematically squarer than roughly commensurate half- or three-quarter Capes. In the Cape, the chimney always aligns with the front door; in the Off-Center Square almost never. As the Off-Center Square is preeminently a floor plan, the overall house shape, the number and placement of doors and windows, even height in stories, all so codified in the Cape, are highly variable. Furthermore, it is a small house, soon outgrown, with no standard program of enlargement as had the Cape; thus surviving examples are often so highly altered and ramshackle as to defy anything but expert identification. The Bixby house was enlarged laterally in two phases between 1815 and 1845. Range and origin of the house type have yet to be worked out. At present two clusters are known, five in Barre-Petersham, dated 1786–1809, and four in Deerfield (of which only two are standing) where J. Ritchie Garrison has correlated the plan with that of the earlier David Saxton (ca. 1761) and the two-story Nims (ca. 1739) houses. The nineteenth-century Deerfield historian George Sheldon ranked it alongside the central chimney plan as one of the two commonest early house types in the town. Two two-story examples that do not survive, the Arms house (ca. 1720), and the Smead house (before 1747), are known from photos, floor plans, and verbal descriptions.[48]

Outside these two clusters, photographs and floor plans of a heretofore neglected example were published as early as 1957 by Henry and Ottalie Williams, scholarly if practical-minded house restorers. The Williamses unfortunately give no location for this house (which they restored, and perhaps lived in), but it seems likely to be in western Connecticut, perhaps in or near Sherman or Bloomfield, where they resided about this time. The house, which they describe as a "1790 wreck," was still in 1951 a splendid, textbook Off-Center Square if ever there was one: "a really unspoiled house that had nothing—no water, plumbing, heat or electricity." The original core of the house appears to have been periodically enlarged in a manner very different from the Bixby house in Barre: a rear lean-to, a side porch, and a smaller "house" and shed adjoined to the back corner. In a chapter ironically entitled "Remodelling vs Remuddling," the Williamses describe how they relocated the stairs to a new front entry hall: "This little entry eliminated the inconvenience of having the front door open directly into the dining room. All of these operations, it will be noted, resulted in a floor plan that was much more typical of an old-time house than the original one was." In short, they sought to transform an Off-Center Square plan into a central chimney plan, of course, long before scholars appreciated the distinction. To their credit, the Williamses methodically documented their find, and were clearly cognizant of its anomaly, but regarded it simply as a "wreck" to be "restored," not an artifact to be typologically analyzed. If in fact this house is in western Connecticut, then it would complement the theory that this was an early eighteenth-century Connecticut Valley type, originally proper for even substantial houses, which by century's end had slid down the social scale and spread as a one- or one-and-a-half-story house into the poorer, later-settled uplands. But the type may be more general: James Garvin has documented one as far afield as Pittsfield NH.[49]

Awareness of the Off-Center Square is important because it compels scrutiny of many small houses hitherto dismissed as "imperfect" or altered Cape Cods, or one-of-a-kind anomalies. Photographs make amply clear that before it was moved to Old Sturbridge Village no one would have looked twice at the Bixby house. While story-and-attic houses sprout everywhere in poorer uplands, these small, plain, and mean houses of small farmers, now often ill-kept, make little impression. Yet they are every bit as much a part of the New England landscape heritage as the more substantial farmhouse, which had perhaps five times the survival rate. Tax-census data indicate that in Worcester County MA, in 1798, 66%, of the houses were one-story or "low" houses; in some towns this percentage ran as high as 80%. Indeed, it has been quipped that Shays' Rebellion (1786–1787) in this region was waged between two social classes: the one-story small farmers against the two-story gentry-merchants. The economic cleavage between the core and the periphery could be particularly glaring: while the housing stock of prosperous, old-settled Essex County towns was as much as 80 % two-storied, comparable figures for Berkshire towns ran as low as 10%.[50]

Center Hall House

This house, sometimes known as the four-over-four, was built on a four-room plan with two equal-sized rooms on either side of a central entry hall (see fig. 15). This hall contained the stairway and might serve as a through passage to a rear door, or be curtailed to allow for an additional room in the back range of the house. The house had two chimneys symmetrically paired. By 1730, interior chimneys that rose from the roof ridge on either side of the center hall were the norm; these served back-to-back fireplaces built in the walls between front and rear rooms. End wall chimneys (sometimes doubled to make a total of four) became stylish in the Federal period after 1780. Thus was eliminated the five-room plan, with its cramped stair-lobby and bulky central chimney. The Center Hall was the beau ideal of the New England two-pile, two-story "box house"; it promulgated both the balanced five-bay facade and the classical vocabulary of ornamentation. The Center Hall plan is first documented in Boston ca. 1690 in houses of the elite; by 1730 it graced country seats in surrounding towns such as Cambridge, Milton, Roxbury, and Medford. After 1750 it spread to notable rural buildings such as parsonages and taverns, and within seventy years came to supplant the central chimney plan for two-story houses in many rural areas; while "nearly universal" in the uplands of west-central Massachusetts, it had scarcely touched the ordinary farm country of southwest Maine.[51]

The Center Hall savvily maintained its stylistic prestige for almost 150 years, no small feat as timber frames lacked the polymorphous possibilities of the later balloon frame. Instead, fashionability expressed itself through roof shape, chimney placement, and most plastically of all, applied ornamental detail. The sheer volume of enclosed space often required more ambitious roofing schemes than the simple gable. The gambrel enjoyed considerable favor for all classes of dwellings from early in the eighteenth century until the Revolution; on the fully developed Center Hall house, however, the hipped roof in its various forms held sway. The simple hipped, the M- or parallel hipped, the decked hip, the hip-on-hip and the gable-on-hip were all employed. The simple gable was of course also found. (Subregional variation in roof type will be discussed under Localism.) In the Federal era the Center Hall house assumed a shallower hipped or gable roof and four end wall chimneys, and was called a brick-ender when these end walls were built entirely of brick. Brick-enders were a considerable status mark in the country towns about Boston. In the Concord of the 1820s and 1830s they were within the means of only a handful: Squire Samuel Hoar, Captain John Adams, a few prosperous farmers, as well as (in a stunted one and a half-story version) a mill-owner in the new-sprung Factory Village, an avatar of the emergent industrialist class amidst the old agrarian order.[52]

Without doubt, the readiest way in which the Center Hall could stylistically renew itself was through facade details, initially of Georgian, and later of Federal inspiration. While 1780 might be taken to mark the divide between these two periods, the style change was more transitional, and as the Federal refined, rather than broke with the classicism of the Georgian, the distinctions are relatively subtle. Only a thumbnail summary can be attempted here. An early Georgian house might have a dentiled cornice, corner quoins, twelve-over-twelve windows with pedimented crowns, a transom-lit doorway surmounted by a triangular or broken pediment. Later, after about 1750, the High Georgian might express itself in two-story corner pilasters, roof balustrades, pedimented dormers, perhaps a Palladian stair hall window above the front door. The Federal style encouraged a more restrained facade, modillioned cornices, flat-linteled six-over-six windows with thin muntins, fanlighted doorways with elaborate classical porches or porticos, as well as a taste for circular or oval shapes, and carved motifs such as swags, garlands, and urns.[53]

To a great extent, all contemporary house types, and in particular the Central Chimney Large, sought to emulate the Center Hall in style and details. Some tried too hard: the central chimney Thomas Hubbard house (ca. 1789) in Concord MA, with its hipped roof, quoined corners, and pedimented door porch looks strangely overdressed, like an oaf in a tailcoat. The prestige of the Center Hall inspired many modified and derived forms. In towns, where space was at a premium, a narrow, urban, three-storied subtype emerged, often of brick, often with end chimneys. Its two-room plan was in essence the front half of the conventional four-room Center Hall, and this half-sized scheme proved widely influential, especially as embodied in the Twin Rearwall Chimney house (to be treated next). Early in the industrial era, the Center Hall devolved further to serve as the basis for a working-class double-house. The ascendancy of the asymmetrical three-bay side-hall plan effectively brought the long reign of the Center Hall to a close, but square-plan houses, built in the Greek Revival, Italianate, and Mansard styles (these latter often in the cupola-on-cube formula) kept the spirit of the type alive until it was bodily resurrected by the Colonial Revival at the end of the nineteenth century. As the epitome of sophisticated Colonial architecture, with its potential for a grand foyer and imposing stairway, the Center Hall was the chief inspiration for this movement, which fancifully borrowed and blended both Georgian and Federal elements with little regard for strict historicism.[54]

Twin Rearwall Chimney House

Distinguished by its paired rear chimneys, and narrow, single-pile profile, the Twin Rearwall Chimney house is a neglected icon of the New England landscape (see fig. 16). In general, the diagnostic chimneys have survived remarkably well, although often lowered. Where the stacks no longer rise loftily above the ridgeline, the houses may appear chimneyless from the street, and identification requires neck-craning and sallies up driveways. Where one or both chimneys have been removed, an oil furnace flue or windowless blank may yet tell the tale. On later examples, the chimneys, which in the eighteenth century rose flush from the rearwalls, even at times being built integrally into rearwalls of brick, began to creep inward two or so feet, betokening the advent of the stove. Initial recognition often comes from the slim one-room deep profile, particularly arresting when the fenestration from top to bottom is 1:1:1. Also common is 1:2:2 (the extra windows may have been a status mark), as well as 0:1:1, where the attic gable was left blank or merely vented. This profile serves quickly to distinguish it from the two-pile Center Hall and other plans, with which it shares the balanced five-bay Georgian facade.

Floor plans are variations on the Hall-and-Parlor. Some have a central passage and straight-run staircase, while others exhibit the same cramped lobby and winding stair as central chimney houses, but with a rear chamber where the chimney would stand. (Indeed, some are documented to be remodeled central chimney houses.) The occasional lean-to or jut-by annexed to the Twin Rearwall further entangles it with the Saltbox–Central Chimney tradition. More common was the kitchen ell, often with a massive chimney that served back-to-back fireplaces. This subtype (which can meld the Twin Rearwall with other types) came to be integrally built in the nineteenth century and was known as the ell house, from its very popular L-shaped floor plan. In Portsmouth NH, the ell house found favor at all social levels, and two- and even three-storied versions were built. When capped by a quarter-hipped roof, it conjured up, from most viewpoints, the substantial presence of a hip-roofed, Center Hall house. In the countryside, the quarter-hipped ell house might do duty as a Federal-era tavern, perhaps at a turnpike crossroads.[55]

The Twin Rearwall is an I-house, a loosely defined family of two-story,

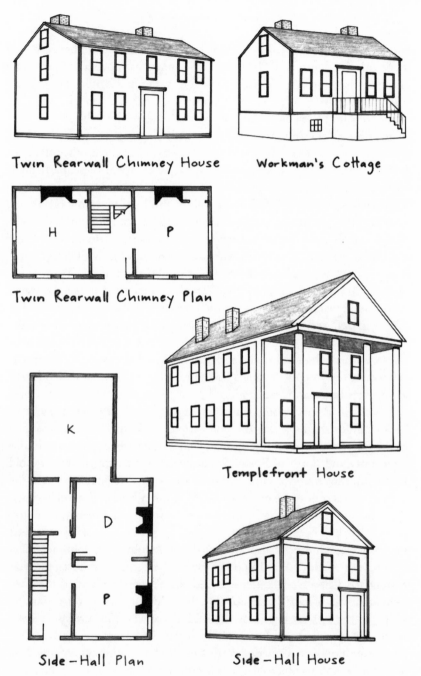

Twin Rearwall Chimney House

Workman's Cottage

Twin Rearwall Chimney Plan

Templefront House

Side-Hall Plan

Side-Hall House

FIGURE 16

House Types V

Twin Rearwall Chimney plan redrawn from Cambridge Historical Commission 1965–1977, 5:64. Side-hall plan redrawn from Cambridge Historical Commission 1965–1977, 3:39.

single-pile houses carried over from Britain and more commonly associated with the South and Midwest. The I-house with paired, exterior, end chimneys is a virtual trademark of the Tidewater and Midland South and was built into the early twentieth century. Central chimney I-houses are more typical of the colder Midwest. (It is actually from the states of Indiana, Illinois, and Iowa that the I-house took its name, not its silhouette.) The central chimney I-houses in New England is none other than the Hall-and-Parlor House, and the structural similarity of these, particularly with 1:1:1 fenestration, to the Twin Rearwall raises a fundamental question. To what degree did the Twin Rearwall in the Bay core grow out of and supplant the seldom-surviving Hall-and-Parlor, and indeed, retrospectively transform it through chimney renovations? Dated early-to-mid eighteenth-century Twin Rearwalls are to be found in Middleton (1714), Marblehead (1715, 1721), Lexington (ca. 1730), Woburn (1733), Westford (ca. 1735), Lynn (1757, 1750?), and Lincoln (mid- eighteenth century). However, renovation means that the date of the house is not always that of the plan. The Massachusetts Historical Commission classified it as an "innovative Georgian design," "essentially a half-size version" of the Center Hall house, and concluded that the earliest "were probably built in the 1740s." (However, the Joseph Reynolds house, ca. 1698, in Boston-settled Bristol RI suggests this plan is older still.)[56]

The Twin Rearwall only truly flourished after the Revolution, when it became arguably the most common early nineteenth-century house plan in the Boston area, built throughout, "but particularly popular in Middlesex county and north of Boston." The secret of this appeal was twofold. Internally, the advent of improved fireplaces and stove heat favored its flue arrangements over those of the Central Chimney house. Externally, it was the most modest house still to possess the prestigious five-bay, two-story facade that had become de rigueur in prosperous eastern Massachusetts. Discreet rear chimneys also conferred the "unobtrusive roofline" much prized in the Federal era. (In poorer southeastern Massachusetts, the well-entrenched Cape Cod house occupied the economic niche of the Twin Rearwall, which most likely accounts for why the type never penetrated south of the border towns of the Old Colony.)[57] The pre-nineteenth-century examples showed an evolutionary variety, from the six-bay Thomas Dane house in Concord, with mismatched chimneys, enlarged organically on a 1656 core; to the simple farmhouses in Lincoln; to high-style Federal townhouses set endwise to the street, and graced by hipped roofs, or three stories, or rear-brick or all-brick construction as in Marblehead or Cambridge.

From about 1810 to the effective demise of the type around 1850, elite examples are fewer and the form became standardized into two subtypes. The first, a yeoman's dwelling, common to farms and villages, was a two-story clapboard house with 1:1:1 or 1:2:2 fenestration in the gable-ends and simple Federal or Greek Revival details, although late incarnations might have Italianate brackets and strapwork, or Gothic bargeboards. The second subtype was an urban workman's cottage, with only one and a half stories (though often mounted on a high brick basement) that sometimes stood in narrow lots with gable end to street (see fig. 16). These characterized the neighborhoods of early nineteenth-century East and North Cambridge, and would influence the layout of both the double-house and rowhouse. A similar degeneration of the type to tenant and artisan housing is observable in Portsmouth NH; by 1834 Twin Rearwalls were built there specifically as rentals.[58]

The southerly and westerly limits of the Twin Rearwall have been fairly well charted. Fieldworkers for the Massachusetts Historical Commission (M.H.C.) reported it throughout Greater Boston, most frequently to the west and north in Middlesex and Essex Counties. My own counts amply corroborate this, showing both the earliest, and the densest concentrations in Middlesex-Essex towns, with the type being notably numerous in Lexington (31), Bedford (14), Marblehead (14), and Rowley (13). Further M.H.C. fieldwork has also established that the type is "almost unknown in southeast Massachusetts. The only examples observed are clustered in the Easton/Norton area." To the west, the type flourished from 1800 to 1830, mainly in the northern tier of Worcester County towns and extending as far south as Fitchburg, Rutland, and Hubbardston. The date of one anomalous pre-Revolutionary example in Southborough has been questioned, but adjacency to Middlesex County makes the attribution certainly plausible. Only four one-story examples are known. Further westward, in the Connecticut Valley, a few Colonial examples were noted, principally in Deerfield, where some were built before 1750 (although some may possibly be central chimney conversions). In the Federal period, it occurred "with some frequency," specifically in western Franklin County, Hatfield, and Northampton.[59]

But what of its northward and northwestward extent? In Portsmouth NH, the Twin Rearwall was "the most common house-type during the first half of the 19th Century"; often sited endwise to the street, it might have a kitchen in a rear lean-to, ell, or basement. Some were made-over Central Chimney houses, as one remodeled about 1810 in conjunction with the addition of a (kitchen?) ell: both date and motive are perhaps typical. Thomas Hubka, whose focus is southwest Maine, classifies the Twin Rear-wall rather vaguely, as a "two-chimneys-behind-the-ridge, kitchen-ell house," of either one or two stories, with an innovative and original kitchen ell (an interesting point). In his view it arose in New England seaports sometime about 1800; however, as his remarks are limited and only one schematic drawing and none of his painstakingly analyzed houses are of this type, presumably it did not much penetrate the ordinary nineteenth-century Maine farmscapes under study. I-houses have been catalogued as a late eighteenth–early nineteenth-century Vermont house type, but chimney placement was not emphasized, and those illustrated were not Twin Rearwalls.[60]

My own vigilant but geographically limited bicycle census turned up over 220 examples, many in places far afield of the Bay core. Fryeburg ME, Jaffrey NH, Deerfield MA, and Townshend VT lie at the known outer limits of its range. However, as this is the area of New England with which I am most architecturally familiar, the well-known caveat must be called to mind: maps of findspots more faithfully record the distribution of over-zealous archeologists than they do of the artifacts themselves. All the same, the adumbrated distribution is entirely consistent with dispersal from the Massachusetts Bay hearth, but in the absence of definite information from Connecticut (where it should be insignificant in number) it is impossible to be absolutely certain. (Connecticut architectural literature is silent on the subject, but silence can speak volumes.) Kurath's maps of New England settlement history indicate that Bay colonists had settled in all the indicated territory by 1790. However, because much of this area was commercially integrated, and exemplary Twin Rearwall Chimney houses lined the great roads and turnpikes north and west of Boston, it is not necessary to posit that the house type reached the limits of its domain with the first outswarms of settlers. With its early nineteenth-century resurgence, the type could easily have spread much as any other fashionable innovation, by imitation, and the influence of prominent local builders. The effect of a style center is not necessarily circularly concentric; it is perhaps more commonly exerted as an axis of influence, elongated along well-established transport corridors.

Local observation suggests ways in which development, diffusion, and decline of the form may have occurred. A tight row of thirteen examples in East Lexington, and another thirteen along U.S. 1 in Rowley Center argue for the work of either a single or close-knit group of housewrights. Several in Bedford were the known work of Joshua Page, but unlike most in East Lexington (which bear an almost company-town sameness), the Twin Rearwalls in Bedford Center vary greatly in style, status, and age. Among them figure an enlarged half-house, a one and a half-story cottage, as well as plain and fancy two-stories with hipped roof and jut-bys, or recessed Federal doorway, or Italianate doorhood and double doors. This ferment in Bedford may record the early nineteenth-century upsurge of the type in its birth and perhaps death throes; the homogeneity in East Lexington its calm apogee as a modest but repectable yeoman dwelling ca. 1830. Two Twin Rearwalls are now local museums: the relocated Cutter House (1833) in Arlington (a so-called Federal Saltbox, possessing a rear lean-to and jut-bys), and the Amos Blanchard house in Andover (1819). Neither, however, is interpreted to visitors as an exemplar of the type.

The Twin Rearwall affords a splendid introduction both to the pleasures of house types and to architectural geography in general, precisely because the mere existence of the term makes a nondescript house suddenly descript. While there are but two Templefronts in East Lexington there are thirteen Twin Rearwalls, yet for the want of a word thousands pass them daily, stone-blind, registering only the high-style exceptions and ignoring the vernacular rule. Consequently, I find it a delicate tribute to the research lavished by the Pitman Studios on their dioramas of Old Cambridge displayed in Widener Library at Harvard that they meticulously replicated many tiny Twin Rearwall Chimney houses (little bigger than Monopoly pieces) at a time when the type was nameless and no architectural historian had given it a second look. This is the same craftsmanly fidelity that often makes the "crude" eighteenth- and nineteenth-century townscapes rendered by self-taught artists such peerless documents, precisely because they noted shapes, counted chimneys and windows, and recorded exactly what they saw.

Gablefront Houses

In the early nineteenth century, a new generation of house types appeared in which the ridgepole was turned perpendicular to the street. This gablefront, rather than flankfront, orientation was a radical but pervasive change; foreshadowed in churches, it spread to houses and ultimately to barns. Even older buildings might be rotated into fashion. The shift was given impetus by the Greek Revival style, which likened the peaked gable end to the triangular pediment of classical architecture, a conceit that found its noblest expression in the Templefront house, with its formal, two-story colonnade surmounted by an attic pediment. In the modest vernacular, the pediment was merely suggested by heavy cornice returns, and columns by cornerboards or at most wooden pilasters. The most enduringly popular gablefront was the Side-hall house, a three-bay plan with off-center entry well suited to narrow intown lots (see fig. 16). With superficial changes in ornament (such as brackets and doorhoods; stickwork and spindles), it rode the stylistic waves into the early twentieth century. Rear kitchen ells were common, but most characteristically this annex took the form of a lower one-and-a-half story wing, often setback and fronted by a porch in the crook of the L: a configuration known as the Upright-and-Wing (see fig. 17). In Maine, the formula shifted, with the wing more commonly set forward, while the porch spanned the setback upright. In time, the gablefront door might disappear altogether, with the only entry in the kitchen wing, thereby sparing the front parlor from mud, and acknowledging the kitchen as the focus of farm family life.[61]

The Upright-and-Wing was at its height of popularity ca. 1840 to 1860 and flourished in New England Extended, that Yankee-settled northern tier of states which ranges westward to the upper Midwest. Such farmhouses embody the last phase of the "white clapboard" vernacular and are key to landscapes that at least in Midwestern eyes, evoke New England. Like other such "New England" features, its easternmost flowering may have occurred in upstate New York. The Upright-and-Wing is particularly dear to cultural geographer Peirce Lewis, who grew familiar with the type as a child in Michigan, and who has asserted that in his professional career it has taught him "more than any other material artifact." On the basis of a field survey conducted in 1968–1969, he concluded that the type originated in Massachusetts, "somewhere between Boston and the Berkshires" and became "thoroughly institutionalized in central New York, where it flowered in classical glory." Lewis further believes that the prototypes began with the "wing," usually a New England one-and-a-half (that is, Cape Cod–type) house, to which was later appended a fashionable two-story Greek Revival "upright" when the owner's fortunes allowed. Anecdotal evidence of precisely this scenario can be read from the text and photo of a recent historic house real estate ad: "South Berkshire County, MA. A 2½-story Greek Revival, ca. 1820, attached to 1½-story Cape, ca. 1782."[62]

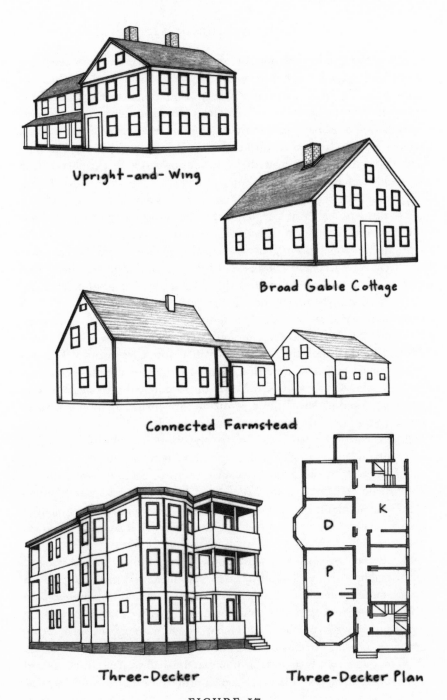

Upright-and-Wing

Broad Gable Cottage

Connected Farmstead

Three-Decker

Three-Decker Plan

FIGURE 17
House Types VI
Three-Decker plan redrawn from Cambridge Historical
Commission 1965–1977, 5:76.

Allen Noble takes the opposite view, that the wing was an extension that added a new kitchen, bedrooms, and pantry to an existing gablefront farmhouse. This contradiction may result from observations made at opposite ends of the house's range: Noble in Ohio and Lewis in New York–New England. A perhaps naively literal interpretation of Fred Kniffen's tabular bar-graphs ("based on field counts") indicates that the Upright-and-Wing makes up 9% of the folk housing stock in peripheral New England, 20% in upstate New York, 50% in Ohio, and 73% in Michigan. The house type peters out in Minnesota, but has a westernmost outlier in the Willamette Valley of Oregon, where it was carried by emigrants from the northern tier of states. Kniffen's statistics leave no doubt why the Upright-and-Wing is not generally regarded by New Englanders as a distinctively New England house, even if it may have originated here. The type emerged too late to impress the older-settled core, but conceivably struck deeper in the peripheries (as Vermont).[63]

A more elaborate version, the upright-and-double-wing, also exists. The most stylish, such as those promulgated in 1832 by New York architect Minard Lefever, took the form of a pavilion with ells: a central Temple-front flanked by two symmetrical wings. While these are more common in the Mohawk Valley of New York, some infiltrated into western New England. Landmarks of the type are the (demolished) Joseph Bowers house (1825) in Northampton MA, designed by Ithiel Town, and also the Joel and Josiah Hayden houses (1830–1834) in nearby Williamsburg MA; less monumental examples with single-story porticos, are seen in Wells and Pawlet VT, while modest versions, shorn of columns, are found in Townshend and Grafton VT. Federal and Italianate versions are not uncommon. As might be expected, the form is exceedingly rare in eastern New England; in Maine, but one high-style example at Augusta is known, and that only from old photographs. Yet another, derived type, the tri-gable ell—an L-shaped two-story house with both upright and wing of equal height—may be typical of later Upright-and-Wings, built not in stages, but as a unit. This subtype was the most common late nineteenth-century house in western Vermont towns such as Fair Haven.[64]

One last house type, the Broad Gable cottage, a strange hybrid of Georgian five-bay and Greek Revival gablefront dictates, merits note (see fig. 17). This gablefront is nominally of one-and-a-half stories, with a ground story of five bays, and a large triangular attic story that is comically disproportionate in size. The fenestration, from top to bottom, may be 1:3:5, but often varies; all windows are full-sized, and the upper two rows of windows lie within the gable triangle. Entry is generally through the center bay of the ground story. Broad Gable cottages may be as shallow as two bays in depth, looking as if ready to keel over. Most are frame, but two stone examples are found in Chester VT, and at least four brick ones in Fitchburg MA, double mill-houses with double rather than single center entries. This type is known to be common in central New England, in the counties of Franklin and Worcester MA, Cheshire NH, Windham VT (and probably further north). Very few are to be found in Maine, and Joyce Bibber deems the form "more common to New Hampshire and Vermont."[65]

Connected Farmstead

This is not so much a house type as a building plan that joins the main house to the barn by a string of ells (see fig. 17). The scheme is nicely caught by the old children's jingle "Big house, little house, back house, barn," which Thomas Hubka took as the title of his landmark study of the type. It was chiefly popular in northern New England from 1830 to 1880, with densities of greater than 20% found in an area roughly circumscribed to the west by the Green Mountains, to the north by the limit of agriculture, and to the south by the present-day Massachusetts Turnpike, although this southerly limit may need to be readjusted in light of numbers of U-shaped Connected Farmsteads reported in Plymouth County, a locality—and possibly a configuration—unfamiliar to Hubka. (Hubka, Noble, and Zelinsky all demarcate this southern line differently.) Its densest concentration occurs in southwestern Maine, whence it may have in some sense diffused, or where conditions were ripest for its adoption. Its distribution is very much spatiotemporally controlled: the type developed too late to be carried west with the main surge of New England emigration (1790–1840), and in an era when New England agriculture was in low national esteem and little likely to be imitated by outsiders.[66]

Even on its home ground, it had to overcome fears of quick-spreading fires and an old New England prejudice against anything that even smacked of mixing people and animals under one roof. It took hold most strongly in later-settled areas where the countryside was still in a state of becoming: still rural, homogeneously Yankee, and imbued in the vernacular tradition, prosperous after a fashion but alert to innovations that might stave off impending decline. The earlier-settled towns of Franklin County MA failed to adopt it, while the later did. The Connected Farmstead is a belated agricultural adaptation of the high-style Federal manor house of ca. 1800 with its attached, often ornamented, outbuildings, such as the familiar carriage (chaise) house with elliptically arched bays. Connectedness was sometimes achieved only for show, with a short, high board fence between buildings. The manor house prototypes often ranged parallel with the road; by 1830, however, the rearward progression of Greek Revival Side-hall house, kitchen ell and attached carriage house had become popular and remained a small-town fixture until as late as 1910.[67]

The Connected Farmstead, where it gained a foothold, revolutionarily reordered the eighteenth-century landscape of detached house and barn. The earlier detached farmstead, still observable in southern New England, was characterized by a south-facing house oblique to the road, and an unattached, 30' × 20' (extendable) "English" barn with broadside doorway that was often set tight along the road across from the main house. All about was a random scatter of rarely surviving outbuildings such as corncrib, wood house, and privy. By contrast, the Connected Farmstead lay parallel or perpendicular to the road, with house joined to barn by a straight or staggered run of ells and outbuildings, among them the kitchen (often the original dwelling), wood house, and privy (set on a cold northwest corner to minimize odors). The barn was of the gablefront, largely post-1830 "New England" type, with slightly off-center door, cow tie-up on the sheltered, southerly side, and hay storage on the colder side opposite. This shallow L- or U-shaped plan enclosed a warm southeasterly dooryard and barnyard, the workaday core of the farmstead, handy to the ell-kitchen, its doorway often fitted with a side porch or "piazza" when these became fashionable in the 1850s. Clapboards and white paint in time extended the architectural formality of the housefront as far as (and often no farther than) the front of the barn.[68]

The Connected Farmstead has come to be regarded as the epitome of New England vernacular architecture, but not simply because of its spare, white-clapboard classicism, traditional timber-framed carpentry, or any of its indisputable agrarian scenic and symbolic power. It is the great New England vernacularism because it exhibits to perfection the three great tenets of vernacular architecture: modularity, mobility, and conformity. It appealed to the parsimony of the rarely prosperous New England farmer because it allowed him to modernize and increase efficiency by redeploying existing structures in a new connected pattern. In an age of agricultural abandonment and consolidation of farms, many old buildings stood empty that could be given new life for the trouble of the moving. Otherwise ramshackle runs of buildings with a little white paint might be unified into impressive wholes that serenely harmonized with the builtscape at large.

Three-Decker

With this house type we clearly pass from the vernacular into the popular era (see fig. 17). Like other popular artifacts, the interest it holds for the sightseeker is enhanced by the light it sheds on a fundamental question: how effective are techniques formulated to analyze rural landscapes when applied to the urban ones that followed? The Three-Decker flourished between 1880 and 1930 and was clearly not the work of rural housewrights but of industrial-era speculative builders. In the early and middle years there was considerable stylistic and economic variety—high-style examples are known—but by 1915 it had lost its architectural pretensions, and became increasingly plain, standardized, and stigmatized as an urban blight. In Greater Boston, some towns banned them by bylaw, and in places the flood of Three-Deckers seems arrested near townlines as if by Aaron's rod. Where not outlawed, it continued to be built until the dead halt of the Depression, although the more respectable two-family house overtook it in many neighborhoods. Some revival after 1970 has occurred as sympathetic infill or condominium townhouses.

The Three-Decker is a three-story house with one six- or seven-room flat to each floor and the characteristic triple tier of porches or decks in the rear. Facades vary, but often a full-height window-bay is flanked by a stair-entry bay with a two- or three-tiered stack of porches that, unlike the matchstick rear decks, might approach the monumental. With Colonial Revival millwork, these porches conferred status on an otherwise simple design, and imposed an impressive rhythm on the streetscape. Room arrangement has been described as a "double-barrel shotgun" plan, but no strict formula obtains. A bay-fronted parlor *en suite* with a bay-sided dining room was typical, with a rear kitchen, and back porch for hanging laundry. While often thought of as flat-roofed (actually slightly sloping), a variety of roofs are found, often geographically identified with certain cities but perhaps better explained temporally in terms of the prevalent fashion when the Three-Decker boom hit town. (Snow may have been a factor.) Gabled roofs have been associated with Roxbury, flat with Dorchester, hipped with Worcester, and gambrels with Springfield-Chicopee. Many cities lack sufficient visual cohesion to claim a Three-

Decker type: Brockton has a blend of flat, hipped, and gable roofs, while New Bedford and Fall River have both hipped and gable roofs, although with a greater harmony of overall design. Three-Deckers in Rhode Island mill towns were often singularly drab boxes with three tiers of matchstick porches across the front, as in heavily French-Canadian Manville.[69]

Like almost any house type, the Three-Decker is the well-named, readily recognized member of a whole family of dwellings—flat-topped two-deckers, four-deckers, brick-built Three-Deckers, mansard-roofed Three-Deckers, double Three-Deckers—that defy simple typology. Its origins are not clearly known. In whatever place (or places) the Three-Decker may have originated one should expect to find early, varied, evolutionary forms that typically signal a center of innovation, forms that are easily ignored because they do not look like a "classic" Three-Decker. Many places such as Brockton that only display the fully developed type with no evidence of its birth-throes can be discounted on these grounds. Boston-Cambridge has the best-documented mix of prestandard varieties (such as an 1875 double Three-Decker in Cambridge),[70] but it also has the best-documented architecture generally.

Arthur Krim mapped seventy-one largely southern New England towns where Three-Deckers are found and estimated their importance in the local housing stock. In only twenty-one did he deem their numbers to be "moderate" or "substantial," and in only four were they "predominant": namely, Boston-Cambridge, Worcester, Fall River MA; and Waterbury CT. While Boston as the metropolis of New England may be admirably placed as a center of innovation, it is difficult to read here a record of concentric diffusion. The Three-Decker seems to have cropped up, if not simultaneously, then within a decade or two, wherever conditions were ripe, such as: rapid rise in population due to immigration and industrialization, inauguration of streetcar lines, increased urban land values. Krim regarded the conflicting claims of origin advanced by Boston, Fall River, and Worcester as evidence that the type developed "simultaneously in a number of cities in eastern New England."[71]

Three broad hypotheses for the origin of the Three-Decker exist: that it emerged from the bottom up, from the top down, or from outside; and all three have been advocated. In the first, the "bottom up" view, the Three-Decker developed naturally from the wooden tenement block and the urban townhouse or rowhouse. Like the earliest multifamily housing of the 1820s and 1830s, it emerged from ideas implicit in the actual housing stock, often pragmatically derived from carving up older houses into flats.

As one investigator concluded, "Even the triple-decker, designed from the start as a three-family house, was related to the old house types in both plan and evolution. The standard triple-decker is nothing but an evolved form of the side-hall plan house; it was even built in the same combinations, singly, doubly, and occasionally also in rows." Early examples from the late 1870s display modest door-hooded or porched entries like tenements as well as the mansard characteristic of rowhouses of the time. One apocryphal story attributes the idea to a Worcester builder who set out to build a three-story mansard but was forced to economize with a flat roof.[72]

The second, "top down" view, as argued by Douglas Shand Tucci, regards it not so much as a popular building type as a popularization of the fashionable "French flat" apartment house, an American adaptation of the Parisian residential hotel, of which the first appeared in Boston in the Hotel Pelham (1857). A French flat had all rooms arranged horizontally on the same floor, rather than vertically as in a Beacon Hill townhouse. The concept hung fire until about 1870, then gained steadily; tall apartment blocks of as many as ten stories were built between Copley and Symphony. Indeed, by the 1890s flat-dwelling had penetrated fashionable single-family residential districts, sometimes as high-style Three-Deckers or double Three-Deckers, but more ingeniously as opulent "two-suite" houses disguised as complexly massed, picturesque mansions that blended into the streetscape. After 1900 rows of double Three-Deckers began to line fashionable trolley-served boulevards such as Commonwealth Avenue. Yet seldom do these elite houses occur singly as frank, detached three-flat dwellings; they seem, at most, patrician parallels and not prototypes of the classic Three-Decker.[73]

A third logical possibility remains, that of an "outside" origin. Strange as it seems, a virtually identical house type (externally at least) called the *maison étagée* has been identified by Richard Milot in the eastern townships of Quebec. A typical Sherbrooke example has the familiar bay front balanced by a triple front porch, and differs only in the external dogleg front stairway leading to the upper flats (a common Quebec feature, and a lucrative source of revenue to injury lawyers in the winter months). In New England upstairs access is inside; external stairs, where they exist at all, are as a rule confined to the rear decks. (However, in North Cambridge, some anomalous Three-Deckers were built in the 1890s with outside front stairways by French Canadians and perhaps exist elsewhere.) Milot is of the opinion that the type developed equally in Canada and in New England. While the appearance of the Three-Decker coincided with heavy

French Canadian immigration to New England (1860–1920), often precisely to those mill towns where Three-Decker housing is so characteristic, there is nothing specifically to link French Canadians with its presumed origins, photographically corroborated, in Roxbury–South Boston in the late 1870s. Gerard Brault makes no mention of the question in his *French-Canadian Heritage in New England*; yet he clearly personally regarded the Three-Decker as emblematic of "the Little Canadas of New England." It is curious, however, that many of the cities such as Fitchburg, Southbridge, and Woonsocket with "substantial" Three-Decker stock also have a large French Canadian element in their population, and that, overall, 85% of Massachusetts Three-Decker towns supported at the very least a French Canadian Catholic parish. (In Connecticut the figure is only 38%.)[74]

A thumbnail history of one such Little Canada is instructive. Cleghorn, alias Petit Québec, the French Canadian section of Fitchburg, is characterized by French-surnamed streets (Blais, Martel, Legros, Delisle, Dumais) and by three- and four-deckers (both single and double). Cleghorn Mill was established in 1885 to weave cotton dress goods with a workforce of two hundred, 80% of whom were French Canadians specifically recruited for Andrew Cleghorn by agents in Canada. A neighborhood sprang up to house them, more than half of it the work of New Brunswicker Magloire Sawyer, an Acadian French builder-contractor newly arrived in Fitchburg in 1886, who "despite his lack of education . . . could quickly estimate the cost of a three-decker without paper or pencil or setting down a figure." The economics of the time were quite simple: a 50' × 100' houselot cost $200–$500, a Three-Decker cost $2000–$3000 to build, and a five-room flat cost $10 a month to rent.[75]

In all, we have established something more than a casual correlation between Three-Deckers and French Canadians: they were built by them and lived in by them; the two turn up together with surprising regularity in the mill towns of both New England and Quebec. Yet the facts might be otherwise explained by supposing that both were merely key elements of personnel and matériel frequently brought together by postbellum New England industry—particularly textiles. As did Andrew Cleghorn: start up cotton mill, import docile workers, spawn boom in cheap, minimally acceptable housing for workers who were often only railroad immigrant "guest workers" with no firm plans to stay. In this view the noted affinity of French Canadians for Three-Deckers is no different from that of New Hampshire farm girls for Lowell boardinghouses in the 1830s: no one would claim the farm girls invented the boardinghouses. The affinity might be further explained if various immigrant groups could be shown to have had their own standards for acceptable housing. Does the presumed Frenchness of the "French flat" reflect housing customs so ancient or influential as to have affected French Canada? In Taunton MA of the 1920s–1930s, Irish and Polish neighborhoods are said to have been marked by two-story houses and cottages, while Three-Deckers were only characteristic of French Canadian neighborhoods.[76]

But how best to explain the Quebec Three-Deckers of Sherbrooke? Because as many as half of all French Canadian immigrants returned to Canada, might not the influence quite as likely have run the other way: to Quebec and not from Quebec? Sherbrooke lies within 250 miles by rail of both Portland ME and Manchester NH, and its industrial base was not unlike many New England cities: textile and knitting mills, foundries and railroad shops. As industrialization was new to Quebec, what could be more natural than to look southward for successful housing solutions? While Milot only vaguely dates the *maison étagée* as an industrial-era house (post–1880), his illustration exhibits details that in New England would be interpreted as Colonial Revival, perhaps ca. 1905. My own limited observations made largely at Beauport (outside Quebec City) suggest that two-, three-, and occasionally four-decker housing with external stairs is indeed widespread (even in rural areas). Yet no examples closely resembled the New England Three-Decker, nor had this architectural tendency, however prevalent, crystallized into a single dominant type. Virtually no two were alike. While none appeared to predate the New England type, many postdated its demise. In Montreal, clearly related three-storied, flat-roofed, brick rowhouses with external stairs and wooden porches were built in the Hochelaga-Maisonneuve section, ca. 1890–1930. To close if not conclude, as if this tangle of claims were not enough, Peter Ennals wonders aloud whether "Maritimers imposed their own special stamp on the triple decker tenements and suburban housing of urban Massachusetts." Indeed, it is true that many late nineteenth-century carpenters and contractors were Nova Scotian (among them, Edward Dix, who built Three-Deckers in Cambridge MA). Everyone, it would seem, wants their measure of credit for the origin of the humble Three-Decker.[77]

Localism

While architectural localisms hold undoubted allure, their actual role in the variety of the New England landscape is only minor. The great builtscape pattern-maker is the economic localization of period styles: High Georgian confined largely to late Colonial seaports and country seats, the mansard style thickest near railroad depots and wherever business boomed ca. 1860–1880, Tudor Revival in elite suburbs of the 1920s, and so forth. Places are marked by the time styles of their boom years, overlaid on the residual regional vernacular, not by the salient hallmarks of local building traditions. Yet these in fact do exist.

Localisms are frequently inconspicuous variants on little-recognized standard forms: clearly it is hard to spot the exception when one has failed to grasp the rule. To discern them requires a critical vocabulary and a visual familiarity with thousands of square miles of New England, a familiarity rarely acquired for more than the few restricted places one has called home. While the wealth of architectural photographs in print, from coffee-table books to calendars, can enhance one's awareness, these lack the systematic focus that only fieldwork can supply. Without this, however much the sightseeker travels, looks, and grows in wisdom, when suddenly brought face to face with the quirky fenestration of one house or the odd chimney placement of another, it is quite impossible to be certain whether this is in fact the first time he has seen it, or merely the first time he has noticed it. Without a program of observation—a veritable *videnda* or checklist of things to see—and a dutiful effort to record everywhere a feature does and does not occur, no reliable distribution of even the most blatant feature can be plotted.

What are localisms? The hallmarks of an influential local architect or school of housewrights, the pattern-book preferences of a widely employed carpenter, even the stock millwork of a jerrybuilder or the blueprints of a contractor, can all create pockets wherein a higher than average incidence of certain features is found. Details derived from local landmarks can suffuse a town or cluster of towns. A stone quarry or brickyard can mark the masonry within its market radius. Rural backwaters can preserve once-general features obliterated elsewhere by prosperity or urbanization. Scattered handfuls of high-style houses or isolated exotics (as an octagon house) often assert themselves inordinately in the landscape. Any and all of these might be construed as localisms: something characteristic of a particular locality. Still other features, such as the plank-framed house, crop up in widely disjunct clusters all over New England and strain the powers of definition. Unfortunately, distinctions between words such as regionalism, subregionalism, and localism are only vague and contextual; indeed, the whole vocabulary of cartographic distributions in general is surprisingly ill defined. Geologists, botanists, strategists, linguists—all struggle with their own terms: outcrop, disjunct colony, pocket, relict area, etc. Localism, with its vernacular overtones and its suggestion of a small, neatly circumscribed area, is inadequate to describe and historically explain the incredible range of map-patterns that a nonubiquitous feature might take.

Here I can only offer the reader a sample of features broadly comprehended by localism or subregionalism: some well documented, others only incidentally and perhaps erroneously alluded to by architectural historians. Even the most sober scholar delights to record their existence, albeit too frequently in curt and ambivalent asides. Conversely, many salient peculiarities (as bowed roofs and jut-bys) have been so trivialized as revivalist motifs that experts seem to shun discussion of them. I offer them as I found them, as valid illustrations of what localisms are, or are claimed (or wished) to be, of the forms they likely may take, and as testimony to their high appeal. Localisms will be conveniently, if arbitrarily, cataloged here as variations on a theme (as roof types), or grouped by subregion (as Connecticut Valley features), or otherwise causally classified (as the hallmarks of local builders). Many localisms have been more profitably discussed in other contexts, and no effort will be made to repeat them all here.

Roof Types

Roofs, as one of the most expressive elements of the timber-framed house, were susceptible to local influences. Roof shape varied from time to time and place to place. Steep Colonial roofs with strong profiles yielded to lower-pitched Federal ones hidden discreetly behind balustrades. High-hipped roofs, long useful in spanning large spaces, became fossilized as emblems of civic prestige. The Old Ship meetinghouse in Hingham MA (1681) is among the earliest extant examples, but the Concord courthouse (1794, demolished) and Westford Academy (ca. 1792) of more than a century later show how long its cachet endured. Several early nineteenth-century midwestern capitols (notably in Ohio) attest how hardy a transplant it could be.[78]

Both north and south of Boston, as in Swampscott, Marblehead, and Duxbury, the pyramidal hip roof continued to dignify important Central Chimney houses at a time when both high roofs and central chimneys had fallen from fashion. At Duxbury in particular, maritime prosperity, outmoded provincial taste, and (one assumes) a close-knit, rivalrous world both of owners and housewrights conferred upon this shallow-draft port a remarkable cluster of some dozen central chimney hip houses ca. 1798–1826. (The 1798 Hall tavern was moved to Cambridge in 1930.) This cluster well illustrates the ambiguities in "localism": here is an undoubtedly unique concentration of an otherwise widespread, if scattered feature. Central chimney hip houses are found upcountry in Hollis NH (1771), Temple NH (1797–1798), Wilton NH (1810) and Manchester VT.[79]

A curious group of early nineteenth-century hip-roofed houses believed to derive from West Indian sources exists in Roxbury and Brookline MA. These Federal farmhouses had a low-pitched, widely overhanging hipped roof supported by square full-height pillars that formed a two-story veranda on three sides. These "Jamaican planter's houses" are well exemplified by the Samuel Gardner Perkins house (1803) on Cottage Street (sometime home of H. H. Richardson). The specific attribution to Jamaica is all the more intriguing in that these houses stood only a stone's throw from Jamaica Pond and Jamaica Plain; one oft-heard explanation of these place-names is that the early residents made their fortunes in the Jamaica trade. This former section of Roxbury was known as Jamaica as early as 1676, but the earliest house in this vein I find recorded was a West Indian–style cottage built on the estate of Benjamin Fanueil sometime before 1760.[80] Perhaps the placename called forth its architectural complement in a sort of self-fulfilling prophecy. (It should be noted that the John Holmes house, 1806, Alfred ME, externally resembles these houses.)

In Portsmouth NH, a high-hipped roof adorned with dormers that alternated triangular and broken scrolled pediments, was in vogue from about 1750 to 1820. On less pretentious gambrel-roofed houses, one roof slope might give place to a clerestory of sash windows or "running dormer." While none of these clerestories survive locally, two downcoast at Portland and Machias ME probably derive from them. A notable pre-1750 cluster of quarter-hipped houses is found in York ME. In Rhode Island, the gable-on-hip was popular in Newport, Greenwich, and the south of the state in the early to mid-eighteenth century. In Providence, the monitored hip roof appeared in the 1770s; given sanction by architects such as John Holden Greene, it would hold strong for more than a half century, being used as late as the 1830s on monumental Greek Revival work. The fashion spilled northward into Massachusetts, as can be seen in the Governor Martin house, Seekonk (1810) and the Reuben Fish house, Fairhaven (1831); it possibly extended westward as well: witness the Salem Towne house, Charlton (1796) (moved to Old Sturbridge Village), the demolished Jennings house, Brookfield, and the monitored hip-on-hip William Norcross house, Monson (ca. 1785).[81] As many localisms are in effect provincialisms, they not unexpectedly took root remote from the stylish seaboard and flowered most lushly where agrarian wealth was amassed. Most remarkable in this regard is the Connecticut Valley, to which we shall next turn.

Connecticut Valley

The Connecticut Valley, because of its mid-seventeenth-century settlement and early wealth, its circumscribed, inland location at the cultural confluence of the two main colonial hearths, as well as its all-important role as the main natural corridor into the northern interior, is predictably fertile ground for localisms. And to be sure, the most celebrated New England localism is without doubt the Connecticut Valley Doorway, with its monumental scrolled pediment mounted over a typically double-leaf door (see fig. 18). It was the mid-eighteenth-century totem of the River Gods, the wealthy of this fertile valley, long the granary of New England. It achieved its apotheosis in 1916 with the installation of the Daniel Fowler House doorway (1760) from Westfield MA in the Metropolitan Museum of Art in New York.[82] From a provincial motif it was elevated to the status of a Colonial Revival icon and widely signaled that New England had a plunderable past. This doorway is rich, showy, exhaustively documented, and exemplifies all that architectural localisms are wished to be, but rarely are. Despite this, the surviving numbers involved reveal that as a present-day landscape pattern-maker it might be dismissed as a drop in the ocean.

The scroll-pedimented form flourished only briefly, from about 1750 to 1770. Amelia F. Miller (whose researches I here encapsulate) has documented and mapped sixty-six of these fully developed Connecticut Valley doorways, extant or destroyed, forty-two of which stood within ten miles of the Connecticut River from Old Lyme CT north to Deerfield MA. Of these, twenty-three survive, thirteen *in situ* and the remainder in museums. Against these as a backdrop, are one hundred fifty related triangular and flat-topped doorways, but despite their numbers, the scroll-pedimented doorway is universally regarded as the true and quintessential form. Such elaborate joinery represented a month's work for one or two itinerant craftsmen. Like gravestones (as we shall see), the doorways were a sort of

Balcony Distribution
(17 examples)

Recessed Balcony

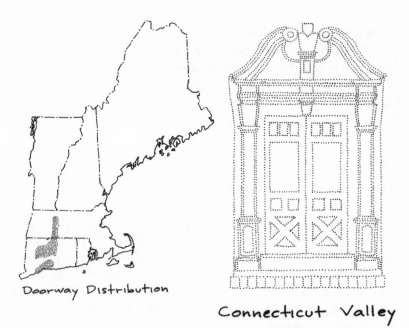

Doorway Distribution

Connecticut Valley
Doorway

FIGURE 18
Recessed Balcony and Connecticut Valley Doorway
Connecticut Valley Doorway distribution derived from
Miller 1983, 20, fig. 3.

finely wrought out-of-door furniture that betray to a richer degree than most landscape artifacts traceable hallmarks, elements of craftsmanship and style, ascribable to a particular master's hand or school.

One joiner, Thomas Salmon, "ingenious architect" per his gravestone, emigrated about 1720 from Wiltshire, England to Stratford CT, where he is credited with the design of Christ Church, and on it, the first known scroll-pedimented doorway, executed ca. 1744–1749. Little more than his name and occupation can be yet adduced in favor of the tantalizing hypothesis that he was a kinsman-apprentice of William Salmon, English carpenter and author of *Palladio Londonensis* (1734), a widely circulated pattern-book in which an engraved plate of a prototypical scroll-pedimented door appears. It is tempting to see him and his Christ Church doorway as (to borrow Abbott Lowell Cummings's term) the "fountain-head" from which the motif diffused.[83]

The work of another joiner, Samuel Partridge, has been traced by Miller in part through the evidence of a scroll-topped clock case known to be by his hand. Its peculiar rosettes match those that adorn the doorway of the Elijah Williams house (1760) in Deerfield; wonderfully, Partridge is credited in Williams's account books for work on this house. This attribution unlocks a deeper understanding of Partridge's preferred motifs, proportions, and moldings, and on this basis thirteen other doorways, mostly in the Deerfield-Hatfield area, have been assigned to him. While the classic scroll-topped doorway died out before the Revolution, a local taste for wide entrances persisted long after in both Federal and Greek Revival architecture. The door itself was also a locally favored motif of the Colonial Revival.[84]

A customary feature of less pretentious Connecticut Valley houses was a side entry called the corner, end, funeral, or coffin door. A Massachusetts traveler, Samuel Davis, remarked in 1789 upon Connecticut houses with "a door on the end near the front corner, which looks awkward." Henry and Ottalie Williams, known for their Connecticut bias, show it as a typical feature in floor plans of the Central Chimney house. (It should be cautioned that the John Buttrick house, ca. 1715, in Concord MA also has such a door.) Unlike the conventional side door, the coffin door did not enter the back range of the house, but rather the front, and (to judge from pictures) typically the righthand or east parlor. The traditional rationale for such doors we have already encountered: the need to carry a coffin straight in and out of the front room in which the body was waked. The fact that two Connecticut Valley features happen both to be doors—the humble coffin and the grandiose scroll top—may not be pure coincidence. There may be some homology of function between the two, and perhaps folk usage of the humble coffin door helped to usher in the grand double-leaf doors. They are not known to co-occur. Coffin doors are conceptually identical to the funeral door characteristic of the One-and-a-Half cottage of the Ohio Western Reserve, alias New Connecticut. Does this argue for a Connecticut origin for the One-and-a-Half house type, rather than the Hudson Valley one implied by Allen Noble? Perhaps, but such doors also exist in northeastern New York, and the waters are further muddied by the fact that "funeral" door in folk usage seems to describe almost any superfluous entry.[85]

The framed overhang, adorned with brackets and pendant "drops," was derived from the Elizabethan jetty and in New England enjoyed a belated vogue from 1670 to 1700 as a status symbol on the houses of the newly wealthy of both the Connecticut Valley and Massachusetts Bay hearths. Obsolete in England, to a largely New England–born generation of carpenters it conjured up the architectural splendors of the half-known homeland. Poor cousin to the framed overhang was the hewn overhang, a structurally simpler affair, not actually cantilevered, but formed instead by a slight outward jog cut into the wall posts. Hewn overhangs flourished in eastern Massachusetts from about 1670 to 1725, but persisted as an outmoded provincialism in the Connecticut Valley until after the Revolution. An overhang "revival" occurred in Farmington CT in the 1780s, and its use did not die out in Connecticut until the early nineteenth century. Further up the valley in Massachusetts, overhangs are commonest in western Hampden and Hampshire counties, where they are met with through the Federal period. Late eighteenth-century houses further east at Athol, and west at Williamstown MA, testify that the style, and perhaps a Connecticut Valley influence, was more pervasive. Not simply front, but end (often attic) and even rear overhangs became part of the ornamental repertoire of the eighteenth-century central chimney house. These became shallower as the century wore on; never projecting more than half a foot, they customarily measured three or four inches, and ultimately, only about an inch. At times, two indigenous status symbols ran afoul, as on the Captain Charles Churchill house in Newbury CT (1754 or 1763), where the scroll-pedimented doorhead had to be fitted over a six-inch overhang.[86]

Recent research in Deerfield by Susan McGowan and Amelia F. Miller (prompted by the redoubtable A. L. Cummings) has shed light not only on several other Connecticut Valley features, but also on the mechanism by which they were spread. These are gambrel-roofed ells; simple, round-columned front-entrance porticos; side porches on gabled ells with roofs married into the pitch of the main rafters; and wide, overhanging second-story cornices. The Deerfield flowering of these features (1798–1815) has been correlated with the arrival in town of the Wethersfield CT house-wright and joiner William Russell in 1797. Perhaps six or seven gambrel-roofed story-and-a-half rear ells made their appearance during this time, and at least six porticos. These features are typical of Connecticut and some may ultimately derive from the Long Island Dutch. Here we must have a principal mechanism for axial diffusion: generations of William Russells who took up their toolboxes to find work further inland, elongating the influence of subregional style centers along main-traveled roads. Another, at least nominally, Connecticut Valley feature on which the usual authorities are ominously silent is the Deerfield blind, a single, double-wide shutter that swung over the entire window, documented to have been hung at the Bishop Huntington house in nearby Hadley in 1799. Old photos reveal only one Deerfield building with them, the Williams-Ware store, operated since before the Revolution, so Deerfield blind may simply be a misnomer inspired by the memory of the Indian raids. (They are also known from old photos of the Friends Meeting House in Adams and a store in Duxbury MA.)[87]

One final nineteenth-century flowering of this Connecticut Valley genius is found further upriver in Vermont and New Hampshire. Here are seen arched balconies recessed into the attic gables of Greek Revival houses, areas normally fitted with windows, louvers, or left as a flush-boarded blank (see fig. 18). A prominent prototype overlooks the Connecticut River in Bradford VT, where a round-arched balcony is gracefully recessed into the pediment of the templefronted Low mansion (ca. 1797–1807). Later, lesser versions might be square-headed, as that of the Messenger house (ca. 1845) in Meriden NH, or presumably pointed on the reported Gothic Revival examples. The Fifield millinery store (1840) in Orford NH was a particularly small and spare vernacular adaptation. I have identified some nineteen such buildings standing or razed, five of which have served as hotels or inns: the balconied upper floors made ideal ball-rooms (in private homes they saw use as music or Masonic lodge rooms). A curious string of Vermont hotels illustrates both how such a concept might spread, and how the geography and chronology of diffusion rarely neatly dovetail. From the (razed) Saxton's River hotel (1817–1823) in Rockingham it is seven miles uproad and upriver to the Eagle hotel (1839) in

Grafton, and from here another seven miles to Rowell's Inn in Simonsville (Chester), which nonetheless bears the earlier date of 1820. (West of the Green Mountains, recessed balconies are considered to be "very rare.")[88] These architectural localisms, amplified by other artifacts such as -field town names, street villages, and gravestones (discussed later), together imprint the Connecticut Valley with a strong vernacular identity.

Hallmarks

The distinctive handiwork of housewrights, rural carpenters, and builder-contractors frequently endures as a recognizable local legacy in the towns where they plied their trade. The study of such hallmarks, whether original, copied from pattern-books, or bought as ready-cut millwork, can give voice to the artisans, local customs, and influences that lie behind the otherwise mute fact of builtscape. For every known hallmark, ten can be surmised. Housewright Anthony Raymond of Brunswick and Bath ME is believed to have "signed" his buildings with distinctive cut-out Gothic arches at the tops of the cornerboard pilaster panels, and the practice was probably not uncommon. In Georgetown MA, a finialed, ornamental fence at the White Horse tavern was the likely harbinger of a townwide trend: "in isolated communities such architectural details seem to have spread from house to house, thus establishing a local style."[89] Sidelights and transoms on Greek Revival doorways in Gray ME followed very local fashions. Sea-serpent porch brackets in Bristol RI, or the jigsawed gableboards in a jerrybuilt neighborhood of Franklin MA bear the imprint of some forgotten builder. Dog-toothed cornices on brick mills and mill-houses in the Springfield-Holyoke area were the hallmark of Charles McClallan.

Some echoic motifs (such as a prominent trio of bull's-eye attic windows that look down on the square in Concord MA) are unlikely to be the work of the same hand, but rather represent conscious emulation. In an age before architects, sagacious copying was the rule. The influence of pattern-books, such as Asher Benjamin's *Country Builder's Assistant* (1797), is well known, but their effects were not localized (although certain titles might enjoy a regional vogue). Details referable to specific plates and pages in such books might crop up anywhere a copy fell into willing hands. Direct imitation of actual buildings was more likely to be local, although a landmark such as Boston's Old North Church (Christ Church) was quoted as far as Providence RI and Wethersfield CT. Design by imitation abetted the work of town building committees: as late as the 1870s, Harvard MA wavered between Acton and Westford as the model for its new town hall.[90] Meetinghouses make an ideal study of design by imitation, one in which stylistic diffusion was leavened by parochial rivalry. Happily, town records often specify what models were to be followed, models that were usually local, and often still documentable even if no longer standing.

In a classic example, Elias Carter's masterful First Parish Church (1811) in Templeton MA spurred a series of imitations that spread northward through six towns to Newport NH (1823), one hundred miles away. Each freely copied the one before, "in a 'nearest neighbor' manner of dispersal" that slowly transmuted the original design, notably in its porch and Wren-Gibbs steeple. The Templeton Run, as Peter Benes has called it, illustrates how stylistic influences are often suffused not concentrically, but linearly along the "wellworn settlement paths," that were the great pre-rail corridors of communication. (It is no accident that the Fitzwilliam NH townhouse stands on the Templeton Turnpike.) The Templeton Run also sheds light on the unseen workings of other, vaster trends, such as the sweeping adoption of the "church plan" across rural New England after the Revolution, in which the old plain-style meetinghouse (The Lord's Barn) was reoriented gable-end to the road and graced with a steeple. There is in Benes's findings grounds to believe that while high-style innovations diffused linearly and involved permanent marks of prestige (steeples, pulpits, overall plan), vernacular ones created "well-defined, locally-oriented pockets of usage," and were confined to mundane or oft-renewed details (stairwells, pews, paint colors). This has profound implications for where and what we seek when we study localisms.[91]

The most blatant localisms involve bastardizations of well-known architectural standards. Square-headed Palladian windows began to appear about 1772 on houses in and around Guilford VT; the earliest was the work of a builder who once lived in Massachusetts and Rhode Island, whence he perhaps imported the notion. The design was more readily executed than the classic arched version and gained local favor. Further simplified with all three parts of equal height, it could be squeezed into low-studded houses. Even a pointed Gothic version is known from a church in Highgate Falls VT.[92]

The introduction of the Greek Revival was particularly volatile because it brought into play design principles in direct collision with the slowly but surely established Georgian style. Georgian had long inculcated five-bay, flank-front, center-entry design, while Greek Revival now expounded the three-bay gable-front with off-center entry. Building was still largely

in the hands of housewrights, country carpenters, and self-taught architects, and their aesthetic trials and errors stand out all the more starkly
against the uncompromising integrity of the classical prototypes. (Later in
the nineteenth century the situation was much altered once the "battle of
the styles" heated up and confusion, hybridization, and rapid changes of
taste ruled the day.) The offspring of these contradictory dictates might
combine Greek gable-ends and Georgian flanks, or broad five-bay gable-
fronts with narrow sides, or be graced by stylish porticoes that were mys-
teriously self-effacing, either "temple-sided" by design, with their columns
on the flank, or else textbook templefronts turned sideways to the street.
Classical temples had no windows in their pediments, so the handling of
these, as well as stove chimneys or steeples on churches, was a matter of
inspired innovation.[93] Attic story windows vary and may reflect local tastes:
many-paned rectangles perhaps in western New England.

Amidst this considerable variety, divers localizations have been nebulously alluded to by architectural historians. A string of high and narrow
templefront houses is said to stretch from Cambridge to Newton and
Natick; and while not specifically cited, the prominent Allen Curtis house
(1828) in Newton (the so-called Pillar House) is probably included. One
scholar makes extravagant claims for central Massachusetts as the "Georgia
of the North," due to the numbers of "hip-roofed, cubical, columned
houses" designed by Elias Carter to be found in Worcester, Barre, and
elsewhere. These represent "nearly all the peripterally colonnaded houses
found in the North," although none he illustrates are in fact "peripterally
colonnaded"; that is, with columns on all four sides. Since Carter practiced
for a time in Georgia, the idea that his Greek Revival manner originally
derived from the South is surely attractive, but chronologically untenable
in the view of a more sober-minded historian.[94] North of Boston, templefront houses with full-height porticos and cast-iron or wooden second-
story balconies were locally fashionable. Templefronts with five columns,
rather than the customary four or six, have been singled out as the hallmark
of Thomas Royal Dake in Vermont, but are seen also in Vassalboro ME.
Also in this corner of Maine, a variant "end-hall" plan for both templefront
and more modest Greek Revival houses is encountered "with some frequency on the lower Kennebec." Plymouth County MA is known for Side-
hall houses with recessed entry porches supported by one or two columns.[95]
Such built-in corner porches are also common in Maine, as at Topsham,
which lies, presumably only coincidentally, within the Kennebec grant
noted earlier for its Plymouth placenames and gravestones.

Materials

In colonial times, local stone was early worked for rough masonry: Cambridge slate has upheld the Somerville Powderhouse (1702; originally a windmill) for three centuries, while Connecticut Valley sandstone underpinned houses long before it was properly quarried. By the mid-nineteenth century brick and finished granite often banished these rude materials from sight, if not from use: Cambridge slate and Medford diorite were employed overtly or covertly into the early twentieth century within a few miles of the quarries. Stone mill buildings, whether of (stuccoed) granite ashlar or cheaper random rubble (quite as able to withstand machine vibration) demarcate an early to mid-nineteenth-century Rhode Island sphere of influence. By contrast, brick was used at Waltham and Lowell MA and widely imitated to the north and west: adherence to the mill-village or manufacturing city model often translated into masonry preferences. Stone mills and workshops are (or were) common in cities like Providence RI, Fall River, and New Bedford MA, but are also found in a host of towns and obscure mill-villages like Quidnick and Georgiaville RI that lend credence to Melville's dictum that true places are never down in any map. Complexes such as the Ames shovel shops in North Easton and the Chilson iron works in Mansfield MA are particularly impressive. As late as 1905 a large stone mill was rebuilt at Sprague (alias Baltic) CT.[96]

Vermont quarrying towns prodigally flaunted their wares: marble for curbs, sidewalks, and foundations in Proctor; slate similarly in Fair Haven, Poultney, and Castleton. Immigrant Welsh slaters locally made their mark with bold polychrome roofing, and built unusual "stacked slate" warehouses and double houses that may betray traditional Welsh methods and plans. A fashion for the picturesque and for rustic informality in the latter nineteenth century brought local stone once more into the open, even stone never before used for building, such as the rounded beach cobbles (locally, "popples") that veneer cottages and gazebos at Rockport MA. The grayish-pink Roxbury puddingstone and the tawnier Brookline variety enriched silk-stocking Protestant churches in Greater Boston from 1840 to 1880. Longmeadow brownstone and pink Milford granite were hallmarks of H. H. Richardson. Native schist distinguishes the Williams College chapel (1905) and the low, windswept Bascomb Lodge atop Mount Greylock (ca. 1935) nearby. The post–World War I collapse of the granite industry echoed high and low in the builtscape. In Rockport MA, the post office (1938), planned as brick, was built of local stone only after popular outcry,[97] while the startling numbers of granite garages found everywhere on Cape Ann must surely have been another sort of makework for unemployed quarrymen.

The fashionable use of native stone in architected landmarks is easily documented; subtler preferences for less striking materials over a nebulously large area are harder to establish and assess because they involve latent or, let us say, "subvisible" patterns (although "supervisible" might be closer to my meaning; they are in fact too vast, rather than too small, to be seen). Exactly what is meant by subvisibility is well illustrated by paint color. Paint is the most conspicuous and universally made choice of building material. It is also impermanent, and its usage has been historically shaped by factors such as availability and cost of pigments, prevalent fashion, and community attitudes towards outward display. Could the colors of each of the hundreds of thousands of wooden structures that form the New England builtscape be mapped, they would make a vast mosaic marked by slow, steady, individual changes over time. Unfortunately, a lack of color in the historic photographic record means that we know far less of the colors of past builtscapes than of their forms. It is a lamentable void, and curiosity is only piqued the more by placenames such as Peachblow House ca. 1828 (Stoneham MA), Canary Cottage (North Easton MA), White Village (Lower Waterford VT), and Red Village (Lyndon VT).[98]

Peter Benes has made a study of mostly written records of eighteenth- and early nineteenth-century church colors and found evidence of significant localizations. Yellow was the most popular and broadly distributed of these colors, yet the better half of yellow churches were distributed axially along a line from Lexington MA to Walpole NH, a well-known migration and trade corridor. If one graphs the firm dates for these yellow churches decade by decade, they exhibit a battleship-shaped curve from 1750 to 1820, with its peak at 1790. While the fifteen examples strung along this axis are by no means chronological from east to west, the earlier (1750–1780) do cluster downcountry and the later (1789–1822) upcountry, with Keene NH (ca. 1789) placed pivotally between. It is intriguing that the paint for the Bedford MA meetinghouse was mixed locally from sour milk and a native yellow ocher to make a color known as Bedford yellow; true commercial exploitation of this deposit is not recorded until 1812. If this ocher was "mined" more widely, or was traded along this axis in the latter eighteenth century, it might have given impetus to the fad. Benes also noted a tight cluster of five orange meetinghouses within a sixteen-mile radius of Pomfret CT, its presumed origin and center point, which arose between 1762 and 1769. A remote outlier in Gilsum NH (1791) is ascribed to out-migration.[99] While not precisely mirrored, these yellow and orange color distributions are linguistically reflected in Kurath's isogloss maps for *spoonhunt* (mountain laurel) and *apple slump* (see fig. 19).[100]

If our knowledge of the colors of past builtscapes suffers from too little data, our present awareness is hampered by too much. For example, it has been stated that (with exceptions) in southern Vermont white clapboards and green shutters predominate while in the north, possibly under a French Canadian influence, colors are brighter and more varied. It might seem that only a team of fieldworkers could validate such an assertion. However, statistical techniques to survey paint colors over equally vast areas have been employed to excellent result.

Joseph W. Glass, in *The Pennsylvania Culture Region*, methodically sampled barn colors throughout the southeast of that state and the culturally dependent Shenandoah Valley in Virginia. Colors were recorded for sample clusters of ten barns each, located nearest to fifty-three regularly determined grid corners laid off at twenty-mile intervals. This data was analyzed and plotted on range maps to show where the principal colors— white, red, and unpainted—had "site frequencies that exceeded the mean."[101] Note that this is an extremely subtle measure. An average of 39.4% of barns were found to be red; thus a sample cluster of ten with four red barns falls within the red-barn zone as mapped, while a sample cluster with three falls outside it. Clearly, such a distinction is purely statistical and is visually imperceptible. Yet, despite some anomalies, this method delineates three coherent color and style zones: a fashionable white zone nearest Philadelphia, a middle ground of red, and an unfashionable hinterland of unpainted barns. I suspect that, while statistically clear-cut, this zonation would be elusive in the field; it is real, but subvisible. These outward style ripples, the demarcation of which would be largely economically determined by richness of soils and proximity to markets, were subvisibly fossilized in the landscape, like the faded pattern in a rug that can only be brought out by minute examination of the yarns. Might such faded patterns likewise lie hidden in New England builtscape? Four "figures in the carpet" that haunt me, whether real or imagined, weave their spells in shingles, clapboards, bricks, and tin.

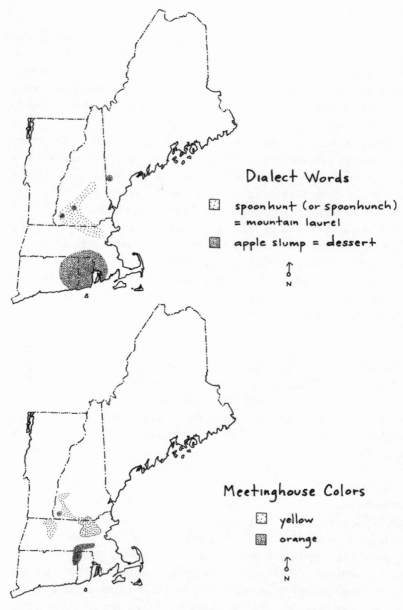

FIGURE 19

Words and Things

Cultural artifacts as diverse as dialect words and paint colors can reflect
the same settlement migrations and spheres of influence.

Meetinghouse distribution from Benes, Peter 1979a, 60, fig. 5–6; dialect
isoglosses derived from Kurath 1939–1943.

Figures in the Carpet

The first, a blind article of faith among Cape Codders, is the *shingled coast*, a thesis for which can be conjecturally sketched: a salty, damp atmosphere inimical to clapboards; a dearth of paint and an immunity to fashion due to poverty; a local scarcity of oak timber offset by an abundance of white cedar swamps; the potential for white pine shingles shipped coastwise from Maine; a renewal in construction in the late nineteenth century when the Shingle Style reigned in summer colonies, and when shingles proliferated on the cheap, small outbuildings of poulterers, cranberry growers, and fishermen. All might have contributed to visibly higher ratios of shingles to clapboards in Barnstable, Plymouth, and Bristol Counties MA. (Here and there they blended, with clapboards giving a stylish facade to otherwise shingled houses.) In New England, white cedar swamps occur most commonly in southeastern Massachusetts, Rhode Island, and eastern Connecticut, and were a cheap source of wall-sheathing: shingle mills in Plympton MA custom-cut local cedar for farmers until the early 1920s. Certainly the postwar explosion of residential building and restoration, anxious to achieve an "authentic" Cape Cod look, has overwhelmed by dilution and alteration any historical evidence to the contrary.

Timothy Dwight in 1800 unequivocally stated that Cape Cod houses were covered roof and sides with "pine shingles, eighteen inches in length." However, the full breadth and depth of this tradition is uncertain: reputedly introduced from Dutch Long Island, wall shingles were rare in seventeenth-century Massachusetts Bay and unusual in early colonial Connecticut outside Milford and Stratford; while considered an "old Rhode Island practice," no early examples survive. On barns, shingling dates to 1830 and perhaps not commonly until after 1860 (which may implicate mechanization). By contrast, the case for our second conjectural subregion, the *clapboarded inland* simply takes many of these same arguments and stands them on their head. There is little more to say. Curiously, however, Alan Gowans contends—incorrectly, it would seem—that the Cape Cod house was commonly clapboarded in southern New England and shingled in the north.[102]

The *brick midland* is little more than a conjecture, prompted by an old photo of my own great-great-grandmother's brick story-and-a-half farmhouse in Lempster NH and by the sighting of its dead ringer in Limerick ME. Until then, my feelings had been those of Peirce Lewis, when he asserted that "a proper upcountry New England farmer would as soon have sold his daughter into slavery as he would have built a brick house."[103] However, closer study reveals that a vague area of inland New England harbors disquieting numbers of penny-plain Federal brick farmhouses, stores, and churches. Brick anywhere in New England outside of cities sticks out like a sore thumb, and brick in the preindustrial landscape most of all. After about 1850, cheap hydraulic-pressed brick carried far and wide by railroads, would obscure the traditional patterns. In eastern Massachusetts, the Federal brick-ender, modest in its use of brick and often rendered more so by whitewash (which blends with the clapboards), is the most masonry one can decently hope for outside of a mill village. The idea of a brick-built house came slowly to older rural districts; brick end walls (and rearwalls) popularly recommended themselves only when integral with chimneys. The downcountry New England landscape was consolidated a generation or two earlier than the later-settled north, and less affected by the Federal taste in brick and the rise in brick-making. Small, unpretentious brick farmhouses, such as a brick Cape in Fryeburg ME, and another that anchors an otherwise white-clapboard Connected Farmstead in Lovell ME, are rare as hen's teeth in downcountry Middlesex County MA. The four brick houses (two with tin roofs) in tiny Lovell village, one dated ca. 1847, are more than a downcountry village such as Lexington Center could boast.

Several factors were at work. Obviously, clay for brick, lime for mortar, and men with brick-making and bricklaying skills had to come together. This critical mass could be achieved surprisingly early on the northern frontier, where brick farmhouses were not infrequently built from clay dug and burned on the farm itself. In Sudbury VT, the Noah Merritt farmhouse (ca. 1790) was laid in Flemish bond with gable-ends embellished with diaperwork of burnt brick; the farm became known as Spunkhole from the fire pit in which the brick was made. Such artistry was rare, though, as the "special idioms of brick" never flourished in Vermont, where masonry was handled "carpenter-fashion." Brick burned on the spot might prove more expeditious than hauling timber to a sawmill miles away— which stands the perceived economics on its head. By 1793 two brickyards in Bolton MA produced 200,000 bricks annually; lime for mortar was also quarried locally. No fewer than seven houses in town were built of Bolton brick, most of them between 1810 and 1825. Robert Brooks pointedly notes that "in the 1820's and early 1830's there was a little flurry of brick domestic dwellings in Williamstown [MA] as well as elsewhere in New England." Williamstown lies in the Hoosic Valley, and an abundance of clay in river valleys may also explain why brick houses are commoner in the Winooski Valley than elsewhere in Vermont.[104] Indeed, explanations for the geography of brick may lie in the Pleistocene geology of clay deposits.

A fourth conjectural subregion is the *tin-roofed backland*. More so even than the other three this domain is an amorphous territory of my imagination. To start with the obvious, tin roofs are all but anathema within forty miles of Boston. On the other hand, they are common upcountry and often stand out in idyllic picture postcards if one looks closely: the much-photographed crossroads hamlet of Waits River vT is at its center entirely tin-roofed, including the church. Around Conway NH, the lowly white clapboard farm cottage with its green tin roof holds the foreground of many classic White Mountain vistas. The boundaries of any such *tin-roofed backland* must fall somewhere between these upcountry and downcountry extremes, and will have had a century and a half to sort themselves out. Zinc, lead, slate, and machine-made wooden shingles were all available in Maine by the 1830s. Sheet metal roofing is cheap, durable (if properly painted), fireproof, and (in the words of the 1908 Sears, Roebuck catalogue) "any person can apply it who can drive a nail."[105] It also stands up to snow and ice, if its widespread usage in Quebec (even on churches) is any indication. In rural areas, it may have replaced wooden shingles, rather than the costly slate of cities, and its widespread usage would seem inextricably linked with the mid-nineteenth-century rise of the iron and steel industry and of the railroad.

The existence of the *tin-roofed backland*, like that of our three other conjectural areas of shingle, clapboard, and brick, has yet to be established. Yet even as impressionistic approximations of reality, these four exercise a strong influence on the imagination, memorably simplifying the incomprehensibly vast and complex. Glib appraisals dog the landscape historian, but have their value. They are clear-cut, easily tested, and ironically can get at the truth more readily than truer but mealy-mouthed generalizations. It has been stated that, in Ontario, farmhouses to the east of London are most commonly red brick with yellow trim, while those to the west are yellow brick with red trim.[106] Whatever else, assertions such as this are boldly rousing; one champs at the bit to put them to the test. Even if fallacious, they possess no small merit in having engaged our critical faculties and made us scan the landscape more clearly and eagerly. (Not until I had written these words did I take note of a goodly cluster of tin-roofed houses in Bridgewater MA, a scant twenty-five miles from Boston.)

Stud Frame

The stud (alias braced or post-and-beam) frame derives from England and was long the established standard in New England, supplanted by the balloon frame for houses only after the 1840s, and for barns not until 1910. The milder English climate allowed these frames to remain exposed, the interstices being filled by brick or wattle-and-daub, and often plastered, creating the familiar half-timber house. Picturesque surfaces must not blind one to the most important fact: that the skeletal timber frame is the same as that of the seventeenth-and eighteenth-century houses of Massachusetts Bay. A total visual transformation is effected by the characteristic New England use of clapboards, which, while known in East Anglia (under the name of weatherboards), are much less common and confined mostly to outbuildings. While standard, the stud frame was not uniform. English house carpentry varied greatly from county to county, and some of this variety made its way across the Atlantic. Any potential for transplanted localisms, however, was complicated by the fact that the first settlers of New England towns rarely emigrated from one place, and the carpenters themselves were highly itinerant.

Abbott Lowell Cummings elucidates many of these local quirks of carpentry in his *Framed Houses of Massachusetts Bay*. Transverse ground-story summerbeams with carved posts concentrate heavily in the vicinity of Salem, Beverly, Danvers, and Marblehead and may stem from Dorset and Somerset. In Dorchester, the Blake house (ca. 1650) uses squint butt bridled scarfs in the top plates, and secures the chimney girts with four rather than two pins: signs that suggest the work of Buckingham or Somerset carpenters. The use of rising wind braces, believed to derive from Suffolk and Essex, has been identified in Dedham and its daughter settlements of Medfield and Deerfield. For the curious, Cummings diagrams these framing details with admirable clarity, but overall the idea of such transplanted English regionalisms is as tantalizing as the evidence is arcane. What Cummings's diagrams do bring home is that a house is a machine for living, a machine with dozens of elaborately named and intricately wrought and assembled wooden parts. People marvel that a Slater or a Lowell could have smuggled out the workings of English textile machinery in their heads: but here we have a "machine" of medieval origin the secrets of which were smuggled in the heads of hundreds of emigrant carpenters. It was an Old World machine that needed to be fine-tuned to the New. House frames were not static, but evolved with new materials and techniques. Some changes were subtle: the average interval between floor joists slowly widened from about eighteen to twenty-four inches from 1637 to 1725, as has been carefully plotted by Cummings. Others were more obvious.[107]

The ideal roof frame for a thatched house differed from that of a shingled one. Shingled roofs were lighter than thatch, permitting slenderer timbers; they also did not need to be so steeply pitched to shed the rain. Roof pitches fell from a high of 54°, in 1635, to 35°–42°, in 1715. The structure of roofs changed and varied more radically still. This is an admittedly difficult subject, made more difficult by a confusing and inconsistent terminology. First, a few definitions. The vertical members of the roof frame are called rafters, the horizontal ones are called purlins (horizontal and vertical being understood relatively, within the plane of the roof). In a purlin-framed roof, the roofing boards are nailed vertically to the (horizontal) purlins. In a rafter-framed roof, the roofing boards are nailed horizontally to the (vertical) rafters.

By a quirk of history, these two different types of roof frame were carried into interior New England by separate settlement streams, the purlin-framed roof from the Massachusetts Bay hearth, and the rafter-framed roof from the Connecticut hearth, resulting in what Thomas Hubka called "a major cultural boundary between eastern and western New England." (Northern Maine, with its largely post-1830 and postvernacular settlement, is rafter-framed, the modern standard.) Thomas Visser, in his work on barns, mapped this boundary[108] along a line that, curiously enough, roughly coincides with other well-marked cultural boundaries, such as the isogloss between broad A and flat A in *calf*. Reduced, then, to a simple formula, on vernacular houses and barns, the roofing boards lie vertically in eastern New England and horizontally in western New England. This rule of thumb will suffice for our purposes, but the subject is complex, and those interested would do well to consult the cited sources: the diagrams are critical, because the terminology is unwieldy, and varies greatly between authors.[109]

Plank Frame

The plank frame persisted as a less known alternative to the standard stud frame well into the late nineteenth century. It used heavy vertical planks set between sills and plates both as structural members and wall-sheathing. Initially, as at Plymouth, planks had been applied to a freestanding post-

and-timber frame; later, as in Vermont, all other framing members were eliminated, and the wall planks stood as the sole support between the sill and the plate. Planks might be of any width (the wider the better), but needed to be uniform in thickness: requirements well met by the typical output of sawpits or early up-and-down sawmills. The method was extravagant of wood—in Vermont, planks ran as thick as 4" and as wide as 26"—but technically simple.[110] It held strong advantages on settlement frontiers where old-growth timber was plentiful and house carpenters were few.

In the seventeenth century, plank frames, as opposed to stud frames, were what chiefly distinguished house carpentry in Plymouth Colony and the Providence Plantations from that of their dominant neighbors, Connecticut and Massachusetts Bay. (South of Warwick RI, stud-framing was employed under the influence of Connecticut.) Indeed, Richard Candee regarded it as one aspect of a "recognizable subculture," of this Plymouth-Providence sphere that a "more detailed map of the diffusion of building methods, furniture styles, and other possible material indices" might corroborate. Plank-framing was in use in Plymouth throughout the eighteenth century and is found in as many as 90% of all documented buildings from before 1725. While stud-framing agrees nicely with the East Anglian origins of the Bay Colonists, plank-framing is more of mystery, as it appears to have been unknown in seventeenth-century England. Its use by Dutch settlers at New Amsterdam, while not directly connected, may stem from the same origin: the gable-ends of Old World Dutch houses were planked, and the Pilgrims' twelve-year exile in Holland may have familiarized them with the practice. (Candee is said to have later deemphasized the Dutch influence.) The entire Wesleyan Grove campground (ca. 1860–90) of some three hundred tiny Carpenter's Gothic cottages at Oak Bluffs MA employed this technique, presumably introduced from Warren RI, where the first such cottage was built "knocked down" for shipment to the island. Martha's Vineyard as a whole is stud-framed.[111]

Outside the Plymouth-Providence cultural area (including Cape Cod), a late seventeenth- early eighteenth-century cluster of about two dozen plank frames has been mapped in Essex County MA, an area where stud frames are otherwise universal. This anomalous choice in framing may have been motivated by a local school of carpenters or by the availability of mill-sawn plank, perhaps from New Hampshire or Maine. Another sixty or so identified in northeastern Vermont date from about 1780 to 1850, most often on Cape Cod houses and related types. At this time, plank frames readily satisfied the new taste for finished rooms without exposed timbers: something stud frames could be made to do only with difficulty. The heart of this cluster lies on the western edge of Caledonia County, as the name suggests, a strong Scotch-Irish enclave: the dissemination of plank-framed (as well as square-log) construction, has been associated with this group. In time, plank-framing spread further afield to central New York, Ohio, and probably seeped into Quebec and Ontario in the early nineteenth century. Concentrations of plank-framed dwellings only lately come to scholarly attention have long been known to local tradesmen who opened walls for repairs. Sheathed by clapboards, frame types are much of a muchness and only by means of radiography is the full history and extent of the plank-framed house becoming known.[112]

Infiltrates and Imports

New England is not watertight, and its western fringe lies in a borderland infused with features from New York State. Our uncertainty as to whether the ultimate origins of the One-and-a-Half cottage lie in New England or New York testifies as much to the cultural ambiguity of this borderland as to the artificiality of the border. The peculiar mix of house types (Saltboxes, One-and-a-Halfs, Cape Cods, and some Dutch types) recorded with primitive fidelity in the works of Grandma Moses, who lived in Eagle Bridge NY (hard by the Vermont border), shows that influences ran both ways. It is largely in its artificial cross-border character that infiltration differs from the ordinary mechanisms of migration and diffusion. New York influences in Vermont may be reflected in gambrel-roofed houses that postdate the general New England fashion and in the use of the Greek Revival pavilion with ells house plan. In Connecticut, early houses in the vicinity of New Haven with flared or "bellcast" eaves, such as the Beech house in Montowese (1759), strongly resemble the Kings County cottage identified by Allen Noble as characteristic of Dutch-settled Long Island. So-called Dutch doors, a half-dozen of which are known in Wethersfield CT and presumably more widely in the Connecticut Valley bespeak a similar origin.[113]

The northern borderlands are not without cultural ambiguity, all the more because the Canadian border was not legally fixed until 1842 and the Quebec side was often settled late and not always by French Canadians. This uncertainty is most true of Maine, where in the upper Saint John Valley, the Acadian barn, Quebec barn, and Madawaska twin barn originated with French Canadian farmers from north of the border, and not necessarily all that long ago. The twin barn, mostly built in the 1920s, has been traced to the Drummondville-Sorel area of Quebec, where early examples are found and whence at least one builder is known to have emigrated. Its unusual H-plan consisted of two barns "joined at the hip," and like the Connected Farmstead, might redeploy existing buildings.[114]

Locally unique features might be imported from elsewhere, and their origins can often be specifically documented. The Old Manse in Deerfield was enlarged in 1768 by Jonas Locke of Woburn MA, and given a double-hip roof, rare in the Connecticut Valley, but common as a status mark on parsonages in eastern Massachusetts, where he came from. The house-wright Caleb Stearns built four houses ca. 1820 in Northfield MA with basement kitchens—highly unusual in an inland country town, but common in coastal cities such as Boston where Stearns had worked for a year under the architect Peter Banner. Thus, individual houses may reflect the craft knowledge and life histories of their builders, as does the stone Tracey-Fellows house (ca. 1780) in Groton VT, built by emigrant Scottish stoneworkers. By what turn of events did the brick Noah Merritt farm-house (ca. 1790) in Sudbury VT come to have the diamond-diapered gable-ends typical of Quaker houses in Salem County NJ?[115] And what of a cobblestone house in Brattleboro VT whose construction belongs more properly to western New York? Vermont, with its diverse influx of early settlers, is fertile ground for such architectural anomalies. Such imports can succinctly illuminate the first phase of the origin-and-diffusion process: they are origins without diffusion, and as such, are more clearly focused, with the history and motives behind them more readily pinpointed. While the basement kitchen introduced by Caleb Stearns did not play well in Northfield, his early usage there of the side-hall plan won converts among the townspeople, and lost all trace of novelty.[116] Thus were many once-notable local wellsprings of innovation drowned in their own success.

Urban Fabric

We have seen how the initial settlement geography was inscribed on the (not altogether) blank page of the landscape; now our concern is with the more complex patterns of the nineteenth and twentieth century, in which this page has become a palimpsest overwritten many times. Pattern is not something inherent or discoverable in a single townscape; it emerges and clarifies itself only as the shared legacy of many townscapes, in which similar arrangements and similar responses to historical forces can be discerned. Pattern, like beauty, is in the eye of the beholder, something that the observer, depending on his level of interest, knowledge, and acuity, extracts to a greater or lesser extent from the townscapes that come his way. Obviously without pattern, every passage from town to town, street to street, even house to house, would be attended by the shock and wonder of novelty. Instead we lapse into incuriosity bred by familiarity. So how to reopen our eyes to these townscape patterns, too often seen, and too rarely observed? Below we shall explore several approaches.

Time Styles

As we elsewhere observed, time styles, not place styles, are the great built-scape pattern-maker of New England. Because time styles are meant to flaunt status (no vernacular conformity here), they are highly volatile. To elucidate the style periods of American architecture is beyond the scope of this volume. Fortunately, the subject is admirably treated by many field guides, complete with timelines, thumbnail histories, and detailed keys to identification. As such books are largely national, and often elite in perspective, their assertions should be tempered by local observation: Richardsonian Romanesque flourished here more, and Art Deco less, than elsewhere in the country. Styles of genteel and common, or urban and rural dwellings might be decades out of step: Georgian lingered in inland Royalston MA and Norwich VT as late as 1830; a Greco-Italianate hybrid remained popular in small towns a half century after its prototypes fell from grace.[117] Our emphasis here is on how a knowledge of styles, once acquired, can enhance appreciation of the New England landscape.

Time styles, by definition temporal, are also necessarily spatial, if for no other reason than that buildings that exist in time must also exist in space, and New England space was not evenly and uniformly developed over time. The economics of any given era favored certain places over others, and then invested those places with the architectural fashions of the day, translating time styles into quasi–place styles. Like the flaws and figures in polished wood, the stylistic grain of the builtscape is diagnostic of its growth history. The present pattern results from two to three centuries of phased growth, centered on scattered and often short-lived points, and marked by a dozen chronologically sequential styles. Because it is economically determined, the pattern is more punctual, repetitive, and jarringly juxtaposed than that of the true place style, which in theory diffused slowly and somewhat concentrically outward from a style center or cultural hearth.

The presence or absence of a particular style in a town or village is a reasonable gauge of building booms and busts, themselves contingent on local economic trends and rises and falls in population. For this reason the population profile of the town under study is of inestimable value. An upsurge in the population of Bedford MA in the 1830s fueled by growth in the local shoe industry is visibly reflected in the housing stock. The cluster of mean, back-lane dwellings west of the common is otherwise a mystery, having mushroomed before the arrival of the railroad and in a place remote from the waterpower requisite for other branches of manufacturing. Conversely, Bolton MA, which lost almost half its population between 1860 and 1900, has few Victorian houses.[118]

Styles are more than arbitrary marks of fashion or status. They can simultaneously reflect and react against advances in technology: the picturesque styles of the latter nineteenth century were a Machine Age product of balloon frames, dimension lumber, and millwork, as well as a fanciful flight from the Machine Age itself. Styles have a philosophical or political symbolism that becomes more deeply invested over time. The Greek Revival evokes an era of high democracy; the Neoclassical, an age of imperialism; the Colonial Revival began nostalgically with the centennial and, as immigration reached a high-water mark in the early twentieth century, became the didactic expression of Americanism.

Particular styles imprinted themselves on certain types of buildings for highly circumstantial reasons. The disestablishment of the Congregational Church in Massachusetts in 1833, followed shortly by a rare disbursement of the federal surplus in 1836, meant that some towns, such as Salem and Southbridge MA, suddenly in need of a secular "townhouse" also had a windfall to build it with. For these and less arcane reasons, the Greek Revival then in vogue became the natural choice for town halls, as witnessed in Framingham, Gloucester, Lincoln, Sandwich, Sterling, and Stow MA (that in Sudbury is a twentieth-century imitation). It also marked "Whale Oil Row" in New London CT and "Captains' Row" in Charlestown MA, and other shipbuilding and seafaring centers from Nantucket MA to Richmond ME.[119] Many rural towns like Petersham MA also peaked both in population and prosperity about this time, and still bear this as their crowning style. Banks began to multiply, and the Greek Revival temple became stereotypical of such institutions until the Great Depression. Railroads boomed, and the Stick Style, alias Railroad Gothic, emerged as the right style (cheap and gaudy) at the right time (1870s) to satisfy the massive small-town and suburban depot-building programs after the Civil War. The color scheme of the depots was often the "livery" of the railroads they served. The short-lived mansard style, which fell from favor with the collapse of the French Second Empire (1870) and the Panic of 1873 had a functional appeal as the attic rooms furnished maids' quarters for the post-bellum nouveau riche. (Their combustibility was also implicated in the Great Boston Fire of 1872.) Clusters of mansard cottages for early rail commuters are often telltale signs of long-vanished depots, as that of the defunct Middlesex Central in Concord MA.[120]

The great age of public library building, much of it philanthropically funded, occurred in a stylistically volatile era (1875–1920), and consecrated no single style, though the Richardsonian Romanesque and Neoclassical (often associated with Carnegie libraries) stand out. Any style that might justify a monumental tower (as a Renaissance Revival campanile) was favored for firehouses, which needed them to dry hoses. The steep hip-roofed Norman farmhouse became the model for Catholic rectories ca. 1900–1920. Craftsman bungalows and Colonial Revival foursquares have been said to demarcate the streetcar routes of the early twentieth century.[121] Without a knowledge of architectural chronology the landscape too often appears to be an ugly and chaotic hodgepodge, and its appreciation solely an exercise in personal aesthetics. Yet when read aright, its patterns become a chronicle of settlement tides, internal development, and social aspirations. As with fashions in clothing, music, and cars, the chronology of time styles comes quickly once one was has observed sufficient examples in historical context.

Slow or staggered growth and later infill can create complex patterns in the builtscape of a neighborhood. Corner lots were highly esteemed and corner houses were often the most elaborate of their day, such as a turreted Queen Anne villa. Houses may alternate in style or plan along the street, either because a builder mechanically sought to create variety, or because the lots were developed in strategic phases. Periodic panics and depressions, such as those of 1873, 1893, 1907, and 1929, stopped house-building dead, and can account more trenchantly than fashionable whim for abrupt changes in the fabric of a neighborhood or street. One side of the street may be of eighteenth- and nineteenth-century construction, and the opposite purely twentieth, which suggests land long held in agriculture. Often the explanations for such patterns are not readily apparent, but for the sightseeker, the first step is to notice them.

The builtscape is a stylistically datable mosaic that has grown piecemeal over time, but more than mere dates are registered there. Rather, like a postmarked stamp, which records not only date, but also value and place, each house carries its own indicia of when built, wealth of owner, and the kind of community that gave it rise. Were it possible to evaluate every house as such a demographic data point, for example, as "circa 1850, mill-owner, factory village," and to plot these triple values on a map, it would read like a majestic Domesday Book written large on the land. Many local historical societies come close when they mark old houses with original owner's name, occupation, and year built. It is neither trivial nor irrelevant that an early nineteenth-century cordwainer on Cape Ann lived in a three-quarter Cape and a deacon in a hipped, four-chimneyed Center Hall. Such information sheds valuable light on that era, and how best to interpret the majority of the builtscape that is not so neatly ticketed. An eye alert to certain readily recognized index types, such as templefronts, mansards, bungalows, and ranches, can pick out the age grain of the landscape, even at thirty miles per hour, when close scrutiny of each and every house is quite impossible.

Growth Rings

Although not neat or round or single-centered as the growth ring of a tree, cities and towns do grow concentrically outward over time, with the age of each layer registered in its architecture. Any inbound trip can be read as a backwards march in time, with the newest houses on the outskirts growing older in style-marked ripples as one moves inward to the center. Five blocks of Main Street in Concord MA run the gamut from a Colonial–Queen Anne villa, to Greek templefronts, to Federal brick stores on the milldam, to Georgian houses east of the square, but of course much else conflicts with this pattern. Leapfrog and infill development can obscure growth rings, and continual real estate pressures in business districts mean that entry into a city along a main artery, or into a town by its "miracle mile" rarely exhibits the clear historical grain of an unspoiled village or residential avenue.

Most probably, a wooden downtown, unless spared by stagnation, will have yielded to bricks and mortar, although with the rise of out-of-town malls this occurs less and less. In the latter nineteenth century, old buildings were often moved rather than razed when town cores were rebuilt in fireproof brick, or ground was cleared for more pretentious residences.

Resettled on clearly newer foundations often only a few hundred yards away, such "back-streeted" buildings fringe downtowns in Arlington, Lexington MA, and elsewhere. Foster Street in Cambridge inherited a collection of old Colonial houses removed from Brattle Street to make way for new, and at times ironically, Colonial Revival mansions. In a further twist, this neighborhood is said to have served as a back-alley servants' row for the Brattle Street well-to-do.

Town growth is well charted on nineteenth-century maps. Often a series will be reproduced in local histories, such as a state-mandated survey of the early 1830s; a Walling or Beers map of the 1850s; or perhaps one of the highly popular "balloon" or bird's-eye views of the 1880s. Matching maps with old photos, whether taken at ground level or from a stack or steeple, can help the imagination raise these flat plans into the third dimension, and clarify how cartography falsifies the familiar world of three dimensions into two. A photo can show how rough-and-tumble a small town actually was in 1900 (the diagrammatic precision of the Sanborn insurance map notwithstanding). Maps and photographs together can teach us to recognize various common stages in the growth of townscape, a valuable skill as almost anywhere places crop up that seem curiously arrested in stages more typical of a half century or more ago.

The Green Between

The outermost growth ring of each town was traditionally agricultural. New England was honeycombed with thirty-square-mile townships, each centered on one or more villages and travel between these villages entailed passing through a thinly settled countryside of farms and woodlots—a countryside that with the decline of agriculture became increasingly wooded and uninhabited. There was, to mangle Saint Paul, no continuing city. The Green Between was the natural vessel that imparted form to village settlement that was otherwise largely haphazard, form at once beautiful and fragile, without author or guardian, and always but an ephemeral reflection of the value of an acre of land. Of the seven main-traveled roads into Concord village, only three (including one preserved as a national park) currently retain the traditional character of the Green Between. By contrast, controlled growth has meant that all five roads into Lincoln do.

On a metropolitan scale, Boston developed in the late nineteenth and early twentieth centuries its own elaborately specialized outermost ring of often conflicting uses. The outlands were blighted by icehouses, brickyards, quarries, and sandpits on the one hand, and blessed with estate districts, market gardens, greenhouses, rural hospitals, reservoirs, and forest parks on the other. Much of the agricultural component is now under houses, although relics such as Wilson farm (since 1884) in East Lexington and Allendale farm in Jamaica Plain still survive. Among the most poignant of such relics are the vestiges of turn-of-the-century greenhouse architecture: vast, flower-filled, crystal cantonments huddled around square-chimneyed boilerhouses of finicky brick. Such sooty chimneys thrust up amidst flat fertile fields are one of the great visual clichés of this outermost ring. In general, growth rings are best studied at the local level: metropolitan patterns more resemble the complex ripples of a handful of pebbles cast into a pond, and are as difficult to analyze.

Slash and Burn

Growth rings bring sharply into focus anything that breaks the grain: anomalous or anachronistic enclaves that stand out like knotholes in the historic pattern of town growth. Such gaps in the architectural record suggest areas long undeveloped, or quite as likely, ones that were razed flat. In geological terms, they are unconformities in the strata, marking either periods of nondeposition or of erosion. Intown shopping plazas have often been jarringly inserted into much older surroundings. Porter Square shopping center in Cambridge MA occupies the old Rand estate; Lord Pond plaza in Athol MA stands on a filled millpond; Cape Ann market (now Shaw's) in Gloucester MA was built on a freightyard painted by Edward Hopper. All such large lots, characterized by historically anachronous development cry out to be explained. Every parking lot yawns like a bomb crater, and should arouse as much wonder, precisely for what is *not* there. The interest of the sightseeker lies as much in the absent elements of a landscape as in those which are present.

The gravest anomalies were the "fire districts" of older cities, which produced great, sudden rents in the urban fabric. Cities were repeatedly ravaged; and Boston alone had counted eight "Great Fires" by 1711.[122] By the nineteenth century, special building ordinances often governed reconstruction in the aftermath, and these fire districts, when rebuilt, often set themselves apart from their surroundings not simply by their newness, but also by masonry construction and a degree of engineered urbanism. Overnight, fires homogenized central business districts that would otherwise have only slowly squeezed out older uses. This is often still somewhat perceptible in a homogeneity of architectural style and in the lack of pre-fire landmarks, although escalating land values and endless rebuilding may obscure the evidence. The Great Boston Fire of 1872 consumed 65 acres and 776 buildings in a swath from Filene's almost to Post Office Square. The district is now covered with lofty buildings and demarcates itself best as a vacuum of pre-1872 structures: historic landmarks such as the Old South Meetinghouse and the Old Corner Bookstore negatively define its limits by signaling places the fire did not reach. The burning out of older, wealthy, central city congregations was taken as a godsend to relocate to more affluent, residential ends of town. The towering parade of fashionable Protestant churches built in the 1870s in the Boston Back Bay was swollen by refugees from the Great Fire.

The Great Lynn Fire of 1889 devastated 31 acres and destroyed 384 buildings, burning out 80 shoe firms, 75 leather dealers, 1 church, 4 banks, 3 daily newspapers and leaving 150 families without homes. "An entire segment of the architectural history of Lynn's growth was wiped away in the space of seven hours." The fire district was rebuilt fairly quickly and uniformly as a three- to eight-story brick manufacturing and retail district, one that subsequently endured the collapse of the shoe industry and another devastating fire in the 1980s. While these events confuse the picture, travelers by train can see from the vantage of the Central Square viaduct a blighted yet overmonumental downtown that seems still to rise phoenixlike from the ashes of an empty plain. The Great Salem Fire of 1914 killed 3 people, burnt out 15,000 others, destroyed 1,800 buildings, and did $15,000,000 in damage, yet, by one account, "happily claimed no residences of exceptional historic interest or architectural merit." Indeed, to believe these same contemporary experts, the fire corrected the aberrations of intervening years by eliminating in particular the decadent and "illogical" Greek Revival; the rebuilding was laudably Colonial Revival in style. These opinions, from the otherwise respectable *Colonial Architecture of Salem* (1919) would be laughable if not so callous.[123]

Portsmouth NH was repeatedly scourged by fire in 1802, 1806, and 1813, and the Federal brick rebuildings that followed, often rowhouses and shops with their rooflines broken by parapet firewalls, are characteristic of the historic fabric of the city. Newburyport MA burned in 1811 and was rebuilt all at once in charming Federal brick. Gloucester MA still possesses two Federal brick blocks built in the aftermath of its 1830 Great Fire. On Martha's Vineyard, the Great 1883 Fire in Vineyard Haven destroyed 26 shops, 32 houses, and 12 barns, and was quickly replaced by a nondescript downtown in stark contrast to the architectural glories of Oak Bluffs and Edgartown.[124] Time heals all wounds, so that it is often all but impossible to trace the lines of the great nineteenth-century fires on the ground a hundred years after the fact. Even on "timeless" Nantucket, where the bounds of the Great 1846 Fire are clearly marked by stones in the pavement, it is difficult to discern any telltale rents in the fabric of the town today. However, the counterexample of Old Town Marblehead, which amazingly eluded the ravages of fire, opens one's eyes to what these old quirky wood-built ports must have looked like.

Fires need not be conflagrations. The scourge of fire, the devourer of all things wooden, has also randomly subtracted key and often beautiful elements of the builtscape. Modern fire protection, illumination, and heating have all appreciably reduced this threat such that for these and other reasons, a building type (such as a church) that might have once lasted only forty years now survives a century and a half or more. (Ironically, the 1900 blaze that destroyed the First Parish Church in Concord MA also inaugurated, inauspiciously, the town's new fire alarm system.)[125] The Slum Clearances of the 1950s and 1960s were in effect "controlled burns" meant to foster the same conditions for (hoped-for) phoenixlike renewal that the Great Fires gratuitously gave.

Déjà Vu

Above all, the sightseeker must cultivate a sense of déjà vu. This eerie feeling of familiarity and recognition when experienced in a strange locality needs to be minutely analyzed, because the commonality of two places is often underlain by formal and historical causes. My own strongest experience of déjà vu came on two visits to Hopedale MA, which revealed itself as the dead ringer, the virtual doppelgänger of my native North Easton, where I had grown up in the uncontradicted conviction of its own uniqueness. As a child, I had never so much as heard of Hopedale, yet as the crow flies the two towns are only twenty-three miles apart. Both are essentially model company towns, created by two ostentatious, philanthropic, and politically powerful nineteenth-century industrial dynasties— the Ameses of North Easton and the Drapers of Hopedale—who liberally endowed their manufacturing fiefdoms with monuments to their own greatness. In Hopedale, the millponds and defunct manufacturing core, the Draper estates on Adin Street, the family vaults in the rural cemetery, the General Draper High School, the manicured grass and shiny black enameled iron railings, that breath of mausoleum stillness in the memorial library, that hint of depreciating trusts and endowments ticking on and on even while the shops are long shut and boarded up, like a century-clock winding down—all this for me held a profound and familiar resonance.

Déjà vu need not always come with so startling a shock of recognition, but the facility of comparing old and new places, of superimposing them in one's mind, aids immensely one's understanding of the catalysts and processes of urban growth. It is always a fruitful exercise to take the locality currently under one's eyes, or one richly stored in memory, and to think of another as much like it as possible. To do this one must define its essential characteristics and then develop points of similarity with some place else. Seaports are ideal for this, as they often evince a set of sibling character traits, which are exciting to compare and contrast. The whole coast sparkles with them: Portsmouth, Newburyport, Gloucester, Salem, Marblehead. . . . John Coolidge was equally amazed by a uniformity in the public buildings of great mill cities like Lowell, Lawrence, and Manchester, a sameness bred by municipal rivalry.[126] A more fanciful game is to pretend one is a movie location scout: what contemporary place would you "cast" in the role of Revolutionary Boston? Marblehead, perhaps? Would Weybosset Street in Providence do duty for the late nineteenth-century "Commercial Palace" district of Washington Street in Boston? Thus will the sightseeker become alert to both synchronic counterparts (Hopedale today resembles North Easton today) as well as diachronic ones (World's End in Hingham today resembles the Shawmut Peninsula of four hundred years ago).

Systems

The New England landscape is so complex because the tables have been turned on it so many times. Over the centuries the landscape has been rebuilt, reordered, and refocused on meetinghouses, mill seats, rail points, interchanges, and so forth. Regarded as an organism, the remarkable thing is that this landscape has periodically, perhaps every fifty years, regenerated a new circulatory system and set of vital organs completely independent of its earlier and moribund one. Which system one inhabits often depends on one's social class. It is quite possible to pass one's entire life connected to the postwar suburban system and never venture into the older, prewar one. An excursion on a railroad train reveals a parallel universe invisible to those whose lives are conditioned by the automobile.

An abundance of open land, now rapidly vanishing, has permitted this endlessly wasteful recolonization or resettlement of the older core of the region. This anatomical redundance in the landscape makes for the considerable variety of growth points to be encountered in one's travels, not simply towns and villages and neighborhoods, but also increasingly specialized, nonresidential developments such as malls and office parks that so defy traditional ideas of settlement as to be coldly derided as "pods." These pods may not resemble towns and villages, but they rob them of much of their economic and social rationale. One's capacity to identify, name, and understand such settlement types is limited only by one's ability to discern cogent recurrent patterns in the landscape.

Artifacts are rarely unique; in one way or another they generally participate in some system or collective phenomenon of their age. Many of the most mammoth, costly, and obvious systems are linear engineering works: railroads, canals, sewers, highways, electrical grids, large sections of which were opened after impressive and short bursts of construction. The knowledge that the Fitchburg Railroad opened in 1845 after less than five years' labor guarantees a most precious fact: a functional rail line of some fifty miles was in place in 1845. While the line was undoubtedly rebuilt, upgraded, double-tracked, and possibly realigned, nonetheless along this long swath of Massachusetts from Boston to Fitchburg there still lies fertile ground for the sightseeker to locate early railroad artifacts, remote by many miles, but systematically linked. If firm identification of a borrow pit in Concord is hampered by the dug-up look of the kame-and-kettle terrain, there are plenty of other borrow pits up the line on which to practice one's eye. Similarly, while the Middlesex Canal (1803) does not survive as a working whole like an outdoor museumpiece, cumulatively, over its course of some twenty-five miles, diverse artifacts and remnant sections with which to mentally reconstruct it abound.

A system is an encyclopedia of itself; it is its own best reference book. On Massachusetts highways you can find bridges as old as 1912 and as new as last year, and guardrails that run the historical gamut from iron Y-struts with white wooden rails to corrugated steel posts with cables. The squat Renaissance Revival Brattle Court pumping station in Arlington MA (1909) may seem a lone survivor simply because for miles around no ancillary structures stand higher than a manhole lid. Yet it is but one part of a vast turn-of-the-century metropolitan waterworks. Out of sight for the sightseeker must not be out of mind. Like sporadic mushrooms that are but the visible fruiting bodies of unseen acres of mycorrhiza, solitary artifacts and structures are often indissolubly linked with objects remote and unhinted. The study of the mechanically interrelated whole of a railroad, from switchhouse, freighthouse, and crossing tender's shanty to bridges and whistleposts, or the small, self-contained world of a textile mill, with pickhouse, dyehouse, and weaving sheds, becomes a paradigm for looking at the less clearly etched social functions that integrate the landscape at large.

Lifewebs

Villages and neighborhoods are not merely random collections of buildings, but culturally interconnected assemblages; as much as a Paleo-Indian rock shelter, shell-midden, and footpath, they testify to vanished lives that once flourished there. The simplest to unravel I call "the Home-Work-Worship" web, which can be profitably applied to almost any nineteenth-century neighborhood, because it is grounded in buildings that remain standing for many years, and because it is a lifeweb in which almost everyone was typically caught. A three-decker tenement, a fortresslike factory, and a redbrick Gothic Catholic church (a "Chartres of the Slums" in Kerouac's phrase) might be the web of a French Canadian loomfixer, while a suburban Queen Anne villa, a Boston office, and a puddingstone Episcopal church was that of a Yankee wool broker.

In a once-remote corner of Middleton MA now grown to woods and tract houses, stands the Lieutenant John Flint farmhouse (1782) at the top of Flint Street, a stone's throw from an old burying ground of some thirty graves, all but one demonstrably, by last, middle, or maiden name, Flints, or Flint kin: from the aforesaid John in 1802 down to 1971. This is about as much of a mark on the landscape as any yeoman farmer was apt to make. Another assemblage in East Lexington MA consists of Pierces Bridge, a railroad dry bridge and once an important local milk shipping point, the Edward L. Tyler dairy farmhouse (1831), and a walled cow-lane with overgrown pasture nearby. Such lifewebs are nothing more than the symbolic nexus of random but permanent landscape elements. Yet together these elements can tell us far more than single artifacts, however pristine. Historically, the dimensions of such webs might be measured in yards, or increasingly, in miles, with an attendant loss or redefinition of "community."

{ *Six* }

GRAVESTONES

VERNACULAR gravestones are uniquely suited to spatiotemporal study as artifacts: they are numerous, well preserved, rich in cultural history, widely distributed, and easily accessible. They are also plainly dated, making chronologies easy to establish, and distinguished by ornamental motifs that facilitate classification and the identification of stonecutters. Their artistic variety is rarely appreciated, because they lie outside our daily orbit, near but far, sequestered in hundreds of neglected graveyards scattered over thirty thousand square miles of the pre-1820 settlement area. While graveyards may be proverbially dull and lifeless places, their spatiotemporal study reveals the gravestones themselves to be remarkably dynamic as artifacts. Over the eighteenth century, headstones grew in size; death's heads were transmuted into cherubs; epitaphs shifted emphasis from the corruptibility of the body to the incorruptibility of the soul. More than any other landscape artifact, gravestones manifest a pronounced subregionalism across distances as small as thirty miles.[1]

The earliest dated gravestone in New England is said to be the "E.L." stone (1642) in Ipswich MA; the Ephraim Huit stone in Windsor CT (1644) and the Ann Erinton stone in Cambridge MA (1653) are almost as old, but such early dates must be accepted cautiously, as "earliest dated" may not mean oldest. A two- to four-year delay between burial and commemoration was entirely usual in colonial times, and very early stone markers must also be suspected of having retrospectively replaced even earlier wooden ones in the form of posts-and-rails (bedheads or leaping boards as they were known in England), which are mentioned in seventeenth-century probate records. Slabs or "wolf stones" may have also been laid down to protect graves from the depredations of animals (Examples exist in Dorchester and Salisbury MA.) Large numbers of early graves were never permanently marked (as few as one in fifteen in eighteenth-century Falmouth MA),[2] or at best marked only by rudely initialed or dated fieldstones, such as the ca. 1730 graves of the first settlers at Westford MA. Many old burying grounds (for example, Concord

South, Wayland North, MA) contain puzzling expanses barren of stones, which suggest unmarked graves.

While stones may be absent, bones definitely are not, as digging new graves often reveals—something that has led one local historian to surmise that "the practice of marking graves in old times was the exception and not the rule." If in the rear, these may be potter's fields, as for example, at the Concord Old Hill Burying Ground, where a long posthumous marker to John Jack, 1773, self-emancipated slave, tends to corroborate the theory. The unmarked graves of some three hundred African Americans in the Ancient Burying Ground at Hartford CT have only recently been memorialized by a tablet. In the Old North on Nantucket, barren ground may testify to a Quaker aversion to gravemarking; the Quaker burying ground in Bolton MA is similarly all but devoid of stones. In the nineteenth century, poverty among immigrants also gave rise to impermanent types of markers now lost to us (likely replaced with stone). William Dean Howells noted ca. 1869 in the "Irish grave-yard" in North Cambridge MA both the still standing marble shafts and slabs as well as long-vanished wooden markers: "black headboards . . . adorned with pictures of angels once gilt, but now weatherworn down to yellow paint."[3]

It is natural to look for English prototypes for New England gravestones. However, as with houses, the common headstone in England was a thing in becoming at the time of the Great Migration. Use of headstones had declined in the late Middle Ages, and so few survive from the sixteenth and early seventeenth century as to suggest that perishable wooden markers were the rule, and that iconoclasm or other factors may have further inhibited the practice. For the many, the churchyard was a common grave, marked only by the churchyard cross. For the few, burial within the church—the closer to the altar the better—or in a private tomb, were the more customary alternatives. Such early modern gravestones as do survive often are vaguely lollipop-shaped, with a round or squarish head and a shank set into the ground. One source gives 1640 as the oldest known English headstone and 1623 as the oldest Scottish, which the author considers to be "perhaps one of the earliest bids of a 'person of plebeian sort' for the right of remembrance in 'God's Acre.' " Even near the metropolis of London, in All Saints Churchyard, Edmunton, the oldest surviving headstone, low, thick, and barely ornamented, is dated 1667, and none appeared thereafter until around 1720, when they became truly commonplace.[4]

English scholars, more concerned with the lavish memorials of the wealthy, have shown little interest in establishing the origins of the common headstone. Thus the earliest English "prototype" of the Connecticut plain-style stone that I find illustrated is dated 1707; while its Cambridgeshire location is a plausible source region, the New England examples with which it has close affinities date back to 1644. Likewise, the Boston "bedboard" that so dominated eastern New England from about 1670 to 1790, crops up in photographs of English churchyards from London to the Midlands and Wales, where it leaps out to New England eyes. Yet it is only incidentally illustrated, one minor subtype among many, without dates or other data. Signal artifacts of New England culture, as diverse as bedboard headstones and clapboards on houses, frequently can only be traced back to disappointingly minor English wellsprings, largely neglected by British scholarship.[5]

Indeed, the preponderance of the research is often so heavily on this side of the Atlantic that to rely on published evidence one must all but conclude that the common custom of rearing gravestones in New England and Old arose virtually simultaneously, and established itself more quickly here. Thus, the "mysterious" lack of significant numbers of reliably dated stones prior to the late seventeenth century should not be blindly ascribed to the crudeness of the New England frontier. Surprisingly, even in England, the custom of commemorating virtually everyone but paupers with individual gravestones is of no great antiquity. The penchant to seek out the "postmedieval" in early colonial artifacts must not be allowed to obscure the predominantly protomodern character of the civilization that was imported. However far from modern their death's-head imagery may appear to be, the mere fact of the gravestones themselves is remarkably modern. And as we have seen with house types, while the imported concepts were innovative, under the natural, economic, and social rigors of New England they were often quickly reduced in variety to virtual monotypes like the five-room houseplan or the "bedboard" headstone.

Thus in a number of critical respects the New England burying ground lacks a strong English prototype: while variety and ostentation were discountenanced, the impulse to individualization among the common people—one man, one marker—took root sooner and stronger. Furthermore, the seventeenth- and eighteenth-century English churchyard, with its richer variety and greater social hierarchy of monuments, was also quite literally a *church* yard, a concept alien to early Puritan New England, where funerals were civil, and burying grounds were secular and administered by the town.[6] The adjacency of church and burying ground, as in Concord and Lexington, came about long after the fact; even the King's Chapel Burying Ground in Boston predates the building of King's Chapel (originally Anglican) by a half century, and despite every appearance is not truly a churchyard. The English churchyard was more successfully transplanted to Anglican Virginia, where the Blandsford Churchyard (Petersburg) and Bruton Parish Yard (Williamsburg) are convincing examples enwalled with brick and furnished with nobly ponderous, often imported English tombs.

Attributes

As with any artifact, a gravestone consists of a set of variable stylistic and technical elements or attributes chosen by the maker. These include the kind, shape, size, and finish of the stone itself, the type of motifs employed (as death's-head or spirit figure), the ornamental borders, and all of the typographic qualities of the text: style of letters, numbers, symbols (as ampersands), the erratic or archaic punctuation and spelling, and so forth. Every stonecutter or family of stonecutters had a recognizable style that was no more than the sum of such choices expressed in stone. Every generation evinced its characteristic trends, trends ironically often overlooked because the stones are so clearly dated that the analysis of the other, subtler datable details is neglected. The theological implications of the epitaph, and even a "stylistic" analysis of the deceased's name itself may yield further datable attributes. For example, the pronounced secularization of first names correlates with contemporaneous changes in motifs (from death's-head to cherub and later willow-urn) and formulas (from "Here Lies" to "In Memory of") on gravestones. (In theory, names should stylistically jibe with the birth year, the stone itself with the death year.)

There are two general shape sequences, one for each main settlement hearth. That of Massachusetts was long dominated (1670–1800) by the round-shouldered "bedboard," believed to have derived its form from a stylized wooden post-and-rail gravemarker meant to mimic the round head of a bed, flanked by two "bedposts." The central arch of the stone (often carved with a death's-head) is called the *tympanum* or *lunette*; the flanking shoulders may be embellished with stars or rosettes. A "cradle-end" variant, which narrowed toward the base, has been specifically correlated with children's graves in Plymouth, but in Boston was used more generally. Rhode Island stones in the latter eighteenth century were marked by a distinctively higher tympanum.[7] About 1775, bedboards might be occasionally elaborated by shoulders on top of shoulders, or lunettes on lunettes. Such unimaginative embellishments may reflect stylistic conservatism or the limited capacity of the slate itself to take on more florid outlines. By 1800 the bedboard yielded to a similar but simpler square-shouldered stone.

Measurements (in the Boston area at least) show clearly that while the shape of the stone remained relatively constant, the average height steadily increased from under twenty inches in 1700 to over forty inches in 1820, with perhaps a dip about the time of the Revolution. Square-topped stones were fashionable from 1830 to 1851, and Gothic arched (marble) stones from 1846 to 1889 (dates from Arlington and Lexington MA). In eastern Massachusetts stones were typically cut in slate, which allowed for crisp lettering and intricately figured borders. Marble began to appear about 1810, so soft that weathering has often rendered inscriptions and dates illegible. (Granite, in the form of high-style monuments, not true gravestones, did not commonly appear until after 1840.)

The Connecticut shape sequence began in the mid-seventeenth century with simple round-topped stones (sometimes with low, square shoulders) in the *plain style*, with the bedboard appearing only belatedly around 1720 under the influence of imported Massachusetts stones. The *florid style*, with intricately scalloped tops, appeared ca. 1750 and recalls contemporary English baroque and rococo work. (It is tempting to correlate the appearance of these stones with the rise of the Connecticut Valley doorway.)[8] Both the plain style, simply lettered with little ornament, and the florid style, with its highly sculpted shapes, may have responded to the softer nature of the sandstone itself, which could not take crisp inscription but was easily carved. The florid style readily translated to marble as the Connecticut school of carvers ventured into northwestern Massachusetts and Vermont.

As an oversimplification, slate characterized eastern Massachusetts and Rhode Island; sandstone, the Connecticut Valley below Deerfield; and marble, northwest Massachusetts and southwest Vermont. Reference to the distribution map appendixed to Allan Ludwig's *Graven Images*, however, will show that the actual pattern was far more complex, particularly in central Massachusetts and eastern Connecticut, where a mix of cruder materials such as granite, schist, and quartzite was used. (Ludwig omitted the odd marble pocket at Lime Rock RI.) In practice, the working and aesthetic qualities of these stones varied from quarry to quarry. Windsor CT was noted for bright red sandstone, Portland CT for chocolate brown; Harvard MA for dark black slate, Braintree MA for purple-banded slate, and Wrentham–North Attleboro MA for shaley slates layered with gray, olive, rose, orange, or cobalt blue, which could be contrastively exposed to create a cameo effect. The brighter sandstones were sometimes stained to darken them, perhaps an acknowledgement of the prestige of dark slate.[9]

Without doubt, the most flagrant hallmark of a cutter or school of cutters is the dominant motif found at the top of the stone, which can range from the malevolent low-browed winged skull of the Lamsons of Charlestown to the cartoonlike spirit images of the hook-and-eye tradition of Gershom Bartlett and William Buckland in the Connecticut Valley. Prior to 1760, gravestone art was, overall, highly conservative. Schools of cutters arose with shared motifs, fostered by geographic proximity, joint rights in quarries, bonds of apprenticeship, and the local socioreligious climate that conditioned the buyer's taste in stones. Even the more individual and artistically versatile cutters maintained underlying traits of workmanship that serve to identify them. The localization of dominant motifs is acute; more than the shape or material of the stone, it is the egregious character of this imagery that can radically distinguish burying grounds less than thirty miles apart. Such localizations are not of course absolute; the winged skull of the Lamsons, either by widespread exportation or local imitation, can be found as far north as Nova Scotia and as far south as Virginia. Nonetheless, more than house types, placenames, townplans, or any other landscape artifact, the spatiotemporal variation in these motifs furnishes the most tangible expression of the vernacular *genius loci* of whatever odd corner of New England the sightseeker finds himself in. He need only stop his car (or bicycle) and look.

By Duval and Rigby's 1978 reckoning, there were 211 vernacular stone-cutters known by name and dates working in New England: 109 (MA); 74 (CT); 10 (VT); 8 (RI); 7 (NH); 3 (ME).[10] The state by state distribution is due partly to the uneven progress of research, but it also reflects the fact that vernacular stones are restricted to the pre-1820 settlement area. Most of the major cutters throughout southern New England have now been identified, and many have been amply profiled in the literature, such that virtually anyone anywhere can adopt a local stonecutter and trace his work.

Initially a stone might be ascribed to an anonymous master known only by his characteristic motif, and many of the now well authenticated gravestone-cutters, such as William Young, the erstwhile Thistle-Carver of Tatnuck, went by such sobriquets in an earlier stage of research. Other artisans, such as the Cat-Face Carver or the Lakenham Spider Man have yet to be identified.[II] Attribution is typically achieved from probate evidence. The cutter's likely body of work is identified in the field on stylistic grounds, and from these stones a list of "customers" compiled. Court documents can then be combed for estate records making mention of payment rendered for a particular stone to a particular cutter. Equally conclusive can be entries in a cutter's account books, or the discovery of the rare signed stone (stones sent to outlying areas might be signed to drum up business in new quarters; shop apprentices might be required to sign their work; pride in craftsmanship might likewise demand a *sculpsit*). Quarry marks meant only to identify blank stones may reveal a cutter's initials.

Photography is indispensable to anyone seriously committed to this study, since it enables direct, minute comparison of stones located many miles apart, and it indiscriminately records details to which the eye is conceptually blind. In the case of Jonathan and Moses Worcester, for example, the key distinction between the output of father and son was the latter's adoption of a tiny triangular point ("beard") on the chin of his skulls. Such contrastive details are more readily brought to light by the systematic scrutiny of photos than even by comparison of adjacent stones in the field. An exhaustive subregional study such as Peter Benes's *The Masks of Orthodoxy* required a photographic inventory of some four thousand Plymouth County stones to compile.

Thus the sightseeker has two options: read up on the chief figures in his area in the standard sources (especially Forbes, Benes, Caulfield) and seek out their work, or alternatively, single out the distinctive output of some local cutter and study it anonymously on his own. For three years I took delight in a stonecutter whose work is widely found in Middlesex County, and whom I provisionally called (on the basis of his singular motif) the Skull-and-Stars Man, before I finally identified "him" in the standard sources as Jonathan and Moses Worcester, father and son, of Harvard MA. In the chart (fig. 20) I offer a very small sample of New England stonecutters' motifs and an indication of their home base. The information is quite inadequate for the positive identification of individual cutters; it is meant simply to awaken the sightseeker's awareness and curiosity as he moves from town to town.

Joseph Lamson
(1656 - 1722)
Charlestown, Mass.

Captain John Homer
(1727 - c. 1803)
Boston, Mass.

Jonathan Worcester
(1707 - 1754)
Harvard, Mass.

(his son Moses worked
in the same vein)

William Young
(1711 - 1795)
Worcester, Mass.

Rockwell Manning
(1760 - 1806)
Norwich, Conn.

William Buckland
(1727 - 1795)
East Hartford, Conn.

FIGURE 20
Gravestone Motifs

Initial Impressions

Before attempting closer scrutiny, it is helpful to size up a graveyard by studying the general effect of the stones without regard to the inscriptions. Impressions so formed will guide one's line of inquiry. Early burying grounds were oriented according to Christian burial custom, which dictated that the dead lie with feet to the east, so as to rise facing the Resurrection at the Last Trump.[12] Unlike today, the grave lies behind the inscribed face of the stone, not in front. (Sometimes a smaller, initialed footstone was paired with the headstone.) Shape, size, and material of the stones give a composite character to a graveyard that varies from place to place and era to era and that impresses the sightseeker at a glance.

An old "First Parish" burying ground in an eastern Massachusetts town that was filled by the early nineteenth century will be all but universally round-shouldered bedboard or square-shouldered slate, the shorter and more predominant the bedboards, the earlier the graveyard; the taller the stones, the greater the percentage of square-shouldered slates and the admixture of marble stones or even obelisks, the later. The oldest corner of the ground and its outward expansion is in this way readily distinguished; within family plots the heights of adjacent stones may step higher with each succeeding generation. The time when the graveyard fell into disuse can often be correlated with the opening of a rural or garden cemetery elsewhere in town. An outlying parish, or later-settled upland town, is more likely to be marked by some square-shouldered slates and a variety of later nineteenth-century marble and granite monuments.

By way of illustration, the composition of the Sudbury burying ground (Old Revolutionary Cemetery) reveals that it was expanded southward ca. 1800, with eighteenth- and nineteenth-century halves divided by an invisible east-west axis. This axis significantly skirts the grave of one Simeon, servant (slave) deceased 1750, buried at what was then the rear boundary of the yard. A hasty tally shows that only 5% of the stones in the eighteenth-century half are of marble, compared to 30% in the nineteenth-century half. (People continued to be buried in the old half, so newer, that is marble stones, do crop up there.) Across the road, yet another expansion around 1850 took the form of a rural cemetery characterized by marble and granite monuments, one in which, perhaps unusually, the old custom of east-west–oriented graves was maintained. This geographic distribution of styles and types of monuments reveals chronological patterns not unlike those to be found on a larger scale in the housing stock of a neighborhood, village, or town, and affords many suggestive parallels to the study of the builtscape as a whole. The house and the long home of the grave have traditionally been the two lasting marks left by the average man on the land, and the study of their forms and patterns is a constant illumination. Indeed, the graveyard is the mirror of parish or town, and from the style, death symbolism, and status marks of their stones, the social importance of whole colonial communities can be read.

Except for smallpox burial grounds, isolated for fear of contagion, scattered family plots—such as the 150 in Parsonfield ME alone—were relatively rare in the old settled Bay core, a rarity that may be a cultural index of the nuclear village and hegemonic Congregationalism. Rhode Island, a haven for Baptists and other dissenters, may have more family or neighborhood cemeteries. In early Providence—churchless until 1700—family burial plots were located amid the orchards and gardens at the far end of the two thousand–foot houselots that stretched back from "the Towne Street." Foster RI has approximately 140 cemeteries, a phenomenal number for a town that, despite its fifty-two square miles, is even today not heavily populated (4,316 people). The town historian of Milford NH noted that while the custom of "family yards" prevailed in eastern New Hampshire, particularly in Rockingham and Strafford Counties, it was scantly observed in southern Hillsborough County, an area otherwise known to have been settled from Massachusetts and hence more strongly Congregational.[13]

Exotic Stones

Certain key questions serve to guide our study of graveyards and gravestones. Foremost, as we have seen, is the identification of individual stonecutters by motif and, it is hoped, by name. Yet almost equal in importance to the question of who made the stones is that of who bought them and why. Often it is the anomalies among the markers that arouse our curiosity and that are the most intriguing to analyze. Exotic materials can immediately draw the eye to the unusual imported stone, as a locally unique ca. 1750 red sandstone bedboard to Captain Ephraim Brown in Concord South Burying Ground. This stone is badly spalled, but the crowned soul effigy indicates it to be the work of an unidentified carver from the Hartford CT area,[14] while the floral borders recall folk decoration on the Hadley and Wethersfield chests for which the Connecticut Valley is noted. (Decorative points in common between carved furniture and gravestones have also been noted in Essex County,[15] while the fig borders of the Lamsons of Charlestown resemble what are possibly Jacobean furniture motifs.) How and why such exotics came to be erected where they were affords lively grounds for speculation. In this case it is significant that the deceased was of secular high status (a captain of militia), one who merited (or could afford) an imported stone, and who was sufficiently well placed to flout orthodoxy and presume election with a gravestone adorned with a crowned soul and not a death's-head, yet one that still held to the standard bedboard shape: vain, but not too vain. (The higher orders in the Boston area were the first to defy convention in this way, typically with a winged angel-face.)[16]

Occasionally, it becomes possible to divine why a particular gravestone choice was made. For example, two 1771 slates in the Nantucket Old North Burying Ground were erected to Amos Otis and Thomas Delap of Barnstable, cast ashore "in ye snowstorm." The workmanship of these stones is unique, as far as I know, on the island, and is that of the shop of Asaph and Ebenezer Soule Jr. of Plympton MA, stonecutters whose trade area included the upper Cape.[17] Clearly, the stones were procured by relatives in Barnstable and shipped over. In general, Nantucket's lack of any dominant eighteenth-century stone type results from the Quaker aversion to grave-marking, a lack of local material, and its geography, which isolated it from the eastern Massachusetts hearth and opened it to a wider range of over sea, rather than overland, cultural connections. Choices were made within a very different framework than in a mainland town. In the Nantucket Old North burying ground there are some twenty brownstones that testify to sea trade across the Sound with towns of the lower Connecticut Valley.

A more conjectural explanation for an exotic stone, but one that typifies the diverting possibilities of this line of speculation, can be found in the Old Burying Ground, Bedford MA. The 1759 Hugh Maxuell stone, with its heart-and-coffin motif, is unique in the yard and appears to be the work of John Wight of Londonderry NH, twenty-five miles distant. Londonderry was a Scotch-Irish town, and Wight, like William Young of Worcester MA, employed Scotch-Irish imagery and depended on the clannish patronage of his fellow countrymen and Presbyterian corelegionists.[18] Hugh Maxuell (Maxwell) is arguably in both its elements a non-Yankee, Scotch-Irish name, and the surmise that he was Scotch-Irish with surviving relatives in the vicinity of Londonderry (no other Maxuells lie in the yard) forms an attractive, albeit unproven, answer to how this exotic stone came to Bedford.

Patronage Areas

We first examined the overall "urban form" of the graveyard, and then focused on the rare stones that immediately draw the eye. We have attempted to understand why a nonlocal artisan was commissioned to commemorate a specific individual. A Boston cutter chosen for reasons of prestige? Or a cutter with intrinsic appeal to a certain cultural group, such as the Soules had for Plymouth Pilgrim descendants and John Wight for the Scotch-Irish? Now we turn our attention to the main body of the stones, which represent the work of the dominant cutters in the district. Once we have identified these artisans, we can census their work over a number of towns, and by tabulating the business done in each parish, locate the overlapping "patronage areas" they served. We can then debate the reasons for overall parish or particular family loyalties to specific sources, and inquire how a new cutter might penetrate another's traditional territory.

By way of illustration, the pie charts (fig. 21) show the differential patronage of the Lamsons of Charlestown and the Worcesters of Harvard in seven contiguous Middlesex towns in the mid-eighteenth century. (While the work of these two families was as different as night and day, the summary attributions necessary to these tallies may have inadvertently included other artisans who used Lamson-like or Worcester-like motifs.) The Lamsons were an eminent, multigenerational family of stonecutters active ca. 1700–1800, known for their winged-skull motif, fine slate, graceful lettering, and richly carved floral borders. Their work carried the prestige of the Boston Puritan oligarchy. The Worcesters, Jonathan (1707–1754) and Moses (1739–?), used a simpler skull-and-stars motif derived from the Merrimack (or Essex County) school of John Hartshorne and Robert Mulican. Father and son turned out a readily recognized and virtually changeless product from ca. 1730–1770, artistically cruder in every respect than the Lamsons, except perhaps in its iconic power to captivate modern eyes.[19] Situated twenty-seven miles apart, these two stonecutting families engaged in a silent trade rivalry for the gravestone market of central Middlesex County.

The evidence of the pie charts is that distance from workshop to burying ground was paramount in the choice of cutter. The closer to Harvard, the more and the earlier the Worcester stones that are to be found in a town. Worcester stones reached Sudbury Center (12 miles away) as early as 1732, and captured 60% of the market; they only belatedly penetrated

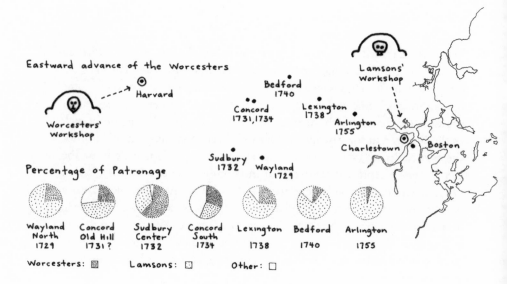

Eastward advance of the Worcesters

Worcesters' Workshop → Harvard

Bedford 1740

Concord 1731, 1734

Lexington 1738

Lamsons' Workshop

Arlington 1755

Charlestown Boston

Sudbury 1732

Wayland 1729

Percentage of Patronage

| Wayland North 1729 | Concord Old Hill 1731? | Sudbury Center 1732 | Concord South 1734 | Lexington 1738 | Bedford 1740 | Arlington 1755 |

Worcesters: ▨   Lamsons: ▢   Other: ☐

FIGURE 21

The Worcesters vs. the Lamsons:
A Silent Trade Rivalry Graven in Stone

Arlington (23 miles distant and deep in Lamson territory) in 1755 and represented a mere 3.3% of the stones. But factors other than mere distance clearly come into play. The high-quality slate and polished workmanship of the Lamsons must have come at a price; unfortunately I have no figures to set against the sum of 3 pounds 14 shillings paid to Jonathan Worcester's widow for gravestones (presumably a head and foot stone) in 1754 with which to pin down the cost factor.[20] The social prestige of Boston would weigh heavily in the choice of a Lamson stone; conversely, we must also wonder whether a Worcester stone might have been something more than a makeshift for the less well-to-do, whether it too made a social or religious statement, even somehow an anti-Boston statement, deliberately chosen. Why, for example, did the Meriams of Lexington suddenly break with family custom in the 1730s to do business with the Worcesters, who purveyed a less stylish and cruder product from nearly twice as far away? Having gone to the Lamsons since 1690 to commemorate eight of their dead, did they suddenly switch loyalties in 1738 simply to save money?

The differential patronage of the Lamsons in Concord to be observed between the Old Hill (47.1%) and South (22.9%) burying grounds—only three hundred yards apart—is symptomatic of a socioeconomic cleavage in the town, presumably geographic in origin, between the wealthier North and poorer South Quarter. The Old Hill is clearly the more prestigious burial place: it is older, contains the earliest marked graves (many crude, dateless, initialed fieldstones), occupies prominent high ground at the town center (once the site of the meetinghouse), and has on the whole the most elaborate stones, some with armorial bearings. The South Burying Ground was set aside specifically for the inhabitants of the later-settled South Quarter, whose poorer farms ultimately translated into poorer gravestones. The Wayland North and Sudbury Center burying grounds are only two miles apart, yet the acceptance of Worcester stones in each differed vastly (22% versus 60%). Division of the town of Sudbury in 1780 and name changes obscure the fact that this difference is one between the old East Precinct (now Wayland) with stronger loyalties to the traditional purveyor, the Lamsons, and a newer West Precinct (now Sudbury) that was formed ca. 1721, not long before the elder Worcester took up his trade.[21] The number of marked burials for the period in question (eighteen versus fifty) suggests that the West Precinct thrived while the East stagnated.

Taken together, the seven pie charts record the varying degrees of overlap between the Lamson and Worcester patronage areas in one section of central Middlesex County. More burying grounds would need to be censused and more charts compiled to tell the full story. In any case, after 1772, Moses Worcester ceased altogether to carve the traditional skull-and-stars stones and his work trails off in obscurity.[22] In the pre-Revolutionary decades the old stylistic hegemonies rapidly broke up, and even the monolithic run-of-the-shop output of the Lamsons becomes less easy to identify as the younger generations diversified to appeal to newer tastes. Comparable pie charts of stonecutters' market shares for the last third of the eighteenth century could be constructed only with difficulty and would be less easy to interpret.

Status

The most obvious indication of social status was the size of the gravestone. As was previously mentioned, over the hundred years from 1720 to 1820 stones generally doubled in size from 20 inches to about 40–45 inches. Stone heights for the Lexington Old and Concord Old Hill burying grounds were measured as systematically as possible (a tricky matter, as stones sink among other things) and the results graphed. Ministers, deacons, physicians, officers of the militia as recorded on their epitaphs figure in the high end of the distribution. Husbands may be taller than wives; children are virtually always small in size and the style of the stones often archaic.[23] It may be that prosperity or pretensions in some towns, such as Concord, locally drove up stone heights in the latter eighteenth century. (Tall round-shouldered stones are what is noteworthy here; tall square-shouldered stones postdate 1790 and are far more universal.)

In the course of focusing on the stone heights and status of individuals in graveyards in Concord, Lexington, and Arlington, light was also shed on the social hierarchy of each community and the aggregate standing of the town as a whole. Concord had at least two physicians, three colonels, and a "gentleman"; a major's, a deacon's, and an esquire's wife; as well as a proud gentry as testified by several armorial stones. Throughout the eighteenth century its stones were almost unexceptionally a cut above the others, and it seems to have made a quicker post-Revolutionary recovery, enjoying a greater measure of prosperity in the 1790s than the two other towns. Lexington's elite on the other hand were smaller fry: deacons, captains, a lieutenant, and an ensign. Arlington, until 1807 the outlying west parish of Cambridge, seems to have given the least scope to status-seeking as the largest stones there are more than a foot shorter than in Lexington or Concord.

High status (as of a minister) might also be indicated by a table tomb, often of imported Connecticut sandstone, or by a long inscription (charged by the letter), a Latin epitaph, or more rarely a coat of arms. Around 1750–1770 it might be indicated by a portrait stone; in orthodox Boston, by not adhering to the conventional winged death's-head. It might be indicated by a stone imported from a high-style area (as Boston), or by one elaborately fashioned or of exotic material (as marble). "Bakeoven" family tombs began to appear in country towns around 1810, sometimes dug in a hillside, but more often earth-covered banks of brick vaults lining the edge of the burying ground. This practice may have originated in overcrowded graveyards in Boston and other large towns and spread to the countryside.

Much as J. Ritchie Garrison observed of houses in Deerfield Street, social cohesion was emphasized by an overall similarity of form and style in gravestones, with status principally indicated by height. In eastern New England at any rate, the bedboard headstone gives a conformity that remarkably contrasts with the diversity in shape to be seen in contemporary English graveyards.[24] As with houses, there are significant exceptions, but in general this extraordinary conformity obtained among all levels of society until the early nineteenth century when industrialization brought both an increased disparity in wealth and a richness of material goods with which to manifest it. Thus the advent of the rural cemetery, to be discussed next, coincided with the rise of a wealthy class and an enhanced ability to display wealth in monumental stone.

At least in the case of the prototypical rural cemetery, Mount Auburn in Cambridge–Watertown MA, the social acceptability of this display was increased by the fact that here was no community of the dead policed by the social standards of the living, as was the town, village, or parish burying ground. Mount Auburn was not a cemetery specifically for any one, close-knit community, but for the proto-metropolis of Boston. These were private grounds. One need not concern oneself about the effect of one's costly monument on the sensibilities of one's pauper neighbor whose plot adjoined. It was an elysian fantasyland that seemed to justify, indeed demand, lavish display. However, when this concept was translated to closer-knit communities, as it was to the once-private Pine Grove Cemetery (1850) in Lynn MA, local opinion could prove caustic: "Nearly all the wealthy people in town have secured their lots in the new cemetery, and left the poor to find their own graves as they can."[25]

Rural Cemetery Monuments

So far our analysis of gravemarkers has confined itself to vernacular gravestones and burying grounds, such as they existed until about 1820. The willow-urn motif then prevalent marked the calm coda to more than a century of theological ferment as recorded in stone. It also, however, sounded the opening chord of the new sentiments on mortality embodied in the rural or garden cemetery. The willow-urn gravestone is a slate or marble reflection of embroidered or engraved mourning pictures, pictures whose idyllic scenes rural cemeteries sought to translate into real landscapes of somber groves, quiet glades, and romantic sheets of water. Rural cemeteries domesticated death, established a cult of heroes and ancestry, and created tranquil grounds in which these might be poignantly contemplated.[26] Rural cemeteries were not purely sentimental. There were also strong sanitary and practical reasons for removing the disposal of the dead from overburdened intown burying grounds.

Mount Auburn was established in 1831; by midcentury, many other rural cemeteries, patterned on it to a lesser or greater degree, had begun to appear throughout eastern Massachusetts towns, such as Wildwood (1852) in Winchester and Sleepy Hollow in Concord (1855). Remoter and less populous Harvard MA did not establish its own, Bellevue, until 1893, when the center burying ground had filled up; the scenic hillside won out on the strength of a petition raised by the townswomen, even after the first body had been laid to rest in the swampy flat previously agreed upon by the men.[27]

Five of the most common types of rural cemetery monuments will be examined here: chest tombs, pedestal tombs, obelisks, Celtic crosses, and Colonial Revival slate headstones. A brief word about some of the sociological, stylistic, and even technological forces that shaped these stones will aid in their appreciation. A great burgeoning of wealth occurred in nineteenth-century America, particularly after the Civil War. New money matured into old with each generational transfer, but even in staid Boston society, Gilded Age excesses were not wholly unknown. The practice, now abandoned, of giving birthplace and deathplace on epitaphs offers insight into the rise of cities and the creation of a wealthy urban class. Inscriptions at Mount Auburn on the order of "born at Uxbridge; died at Boston," yielded by century's end to "born at Boston; died at Manchester," testifying to founders of fortunes born in obscurity whose offspring died at summer estates. These were precisely the years of great immigration; as if in reaction, the ancestral pride of the old stock began to harden itself in monuments.

Much of the well-known sequence of nineteenth-century architectural styles is traceable in these monuments (and probably to some degree trends in furniture as well). Each style has its own resonance: Egyptian for eternity; Greek and Roman for civic virtue; Gothic and Romanesque for mystical religion; Colonial for American tradition; Neoclassical for dignified grandeur. Mausolea, because of their scale, are better able to fully reflect these styles. Smaller monuments, when they do not take the form of actual funerary artifacts, often resemble quaint bits of architectural salvage, or even furniture. In any event, the main architectural currents of the period, tempered by a seemly traditionalism, are not far to seek.

Technical trends are also discernible. Slate for stylish monuments went into almost total eclipse between 1845 and 1900. Common slate markers are occasionally found in country graveyards dated as late as the 1880s, but stylistically these had not evolved for half a century, and often belonged to matching sets bought much earlier. Marble was common ca. 1840–1890; its apogee was in the highly ornate, Gothic or baroque, three-foot Victorian "horrors," often inscribed Father or Mother. These melt like soap cakes and are hard to read and to date; incertitude also arises as to whether a family "set" was acquired upon the death of the first child or the first adult, often a difference of thirty years or more. (However, it is averred that, at least in the twentieth century "the tendency has long been to install monuments within a year of the first burial.")[28] Granite became truly competitive with marble after 1870, and was supreme by 1900.

Granite displayed evolving standards of workmanship based as much on the dictates of technology as of taste. Machine-tooled granite from 1840 to 1880 is often harshly geometric, with hard, planed surfaces, sharp angles, and stiff patternwork not unlike the wooden gingerbread of the period. In time, the gray Quincy granite that long set the standard slowly yielded to less somber hues. Evident from 1870 on is a rising use of pink and red granite. In the 1880s especially, highly polished red granite, either as urn-topped pedestals or "bedposts" (columns), was very fashionable. Technique might triumph over taste, as in a short-lived fad (ca. 1880–1900) for technological showpieces such as mirror-polished granite spheres as much as five feet across. By 1900, unpolished pale grays, verging on white, were valued as closer in tone to marble for Art Nouveau and Neoclassical work. This change parallels the new taste for golden oak over mahogany in furniture.

The term *chest tomb* conveniently designates a range of boxlike monuments whose forms allude to sarcophagi, coffins, ash-chests, sepulchral altars, and tomb bases for effigies. Some have precise classical antecedents. The sacrophagus of Cornelius Lucius Scipio Barbatus, unearthed in Rome in 1780, was a widely copied prototype. The first of its kind at Mount Auburn, an Italian import, was erected in 1832 for the phrenologist Gaspar Spurzheim, a celebrated traveling lecturer who died suddenly in Boston. It is frequently described as an altar tomb, although surmounted by what appears to be the carved likeness of the bolster-ended mattress of a Roman couch.[29] Another type, reminiscent of a Greek temple in shape, is a squat box with pilastered sides and ornate acroteria at the four corners of a pitched roof. The Henry Wadsworth Longfellow tomb (1882) at Mount Auburn is an austere adaptation.

Two chest tombs were medieval in origin. One, a paneled box embellished with quatrefoils and shields, resembles the base on which recumbent tomb effigies were supported. The other was the coped stone, a cross-ridged grave slab that took the shape of a Gothic transept roof.[30] The Mount Auburn examplar (1886) is that of "Nicholas Hoppin D.D.," an Episcopalian clergyman clearly styled as "Priest" in his epitaph; the monument is further inscribed with Latin prayers for the dead, making plain that the appeal of the coped stone was to very High Church Episcopalians.

Perhaps the most common type of chest tomb encountered is neither a facsimile nor a historical pastiche, but simply an oblong block of richly polished dark granite, with massive lid steadied on four squat corner columns, all tastefully dressed up in the Gothic or Romanesque manner. The basic design varied little and the type appeared widely in the 1870s and 1880s. The popular version of this stone, the "Royal Sarcophagus" of dark Barre granite, was sold by the Sears, Roebuck catalogue as late as 1908 in one- to four-ton sizes. By 1910, these Victorian monstrosities were scorned in the dismal trade as "Old Methusalems," and the chest tomb underwent a rapid evolution. Glass-polished black granite gave way to natural-finished light stone; coffinlike bulk melted to something more shapely and supple; the lid that once had hipped-roof lines softened to soap-cake curves; the lip of the lid narrowed to a decorative groove, then a mere band; tongues of Art Nouveau foliage began to lick at the corners. Freed of artifactual precedent, designs slowly lost the traditional solid volume of the chest tomb, became smaller, narrower, less ostentatious. In this trend is discernible the origins of the ubiquitous modern monument of thick granite, which seems to be as much a downsized chest tomb as the lineal descendant of the earlier flat slate vernacular gravestone.

The *pedestal tomb*, the vertical counterpart of the chest tomb, can be traced to the funerary plinths, pedestals, and columns of ancient Greece. These were surmounted by stone urns, lasting tokens of the earthen vessels in which the ashes of the deceased were buried.[31] The concept was revived in eighteenth-century England and both urn-capped pedestals and columns are common in Mount Auburn. The more modest variety rather resembles a chess piece (even more so in cross-topped Catholic versions). It flourished in the 1880s and can be found far and wide. Ultimately, the verticality of the pedestal tomb fell casualty to the twentieth century, which (despite skyscrapers) largely favored the horizontal in design.

The vertical extreme was the obelisk, which punctuates the skyline of most rural cemeteries, and must rank as the ne plus ultra of the nineteenth-century funerary monument. Raised in honor of Revolutionary heroes in Lexington (1799), Arlington (1818), at Bunker Hill (1825–), and finally in Washington, DC (1833–), its preeminence was soon established; around Boston it won favor as the fitting monument for patriots. Short on art and craftsmanship, and long on engineering and precise workmanship, it was an ideal vehicle to express the American genius of the time. Egyptian, it must not be forgotten, was the new style of the early nineteenth century: it proclaimed its modernity from the gates of progressive garden cemeteries, the facades of humane penitentiaries, and from the stations of that marvel of the age, the railroad, of which no fewer than five were built in the Egyptian Revival in Massachusetts alone. It also found favor with Freemasons who, under the spell of Rosicrucianism, had come to revere Egypt as the fount of ancient wisdom.[32] Freemasonry had a wide and influential adherence in the early republic, and its symbols appear in the eye and pyramid of the Great Seal of the United States.

The form and mass of the obelisk presented great technical challenges to the artificer; indeed, the earliest attempts were no more than flat slate gravestones cut to an obeliscoid outline, as in two 1798 examples in Concord and Plymouth MA. The first true obelisks were modest, mounted on pedestals and most often of marble, sometimes solid, sometimes hollow. Marble was the commoner and cheaper material before the Civil War. The sizeable granite obelisk as a private burial monument first appeared about 1834, flourished for the two decades after 1855, and was last seen about 1910. They ranged as high as twenty to twenty-five feet and must be accounted quarrying feats. Their severity might be relieved by red or pink granite, flared sides or canted edges; occasionally they were Christianized with a cross or dove.

The Celtic Cross, also known as the Irish High Cross or Wheelhead Cross, has pre-Christian origins in the Bronze Age. In medieval times they were raised to hallow and safeguard the precincts of monasteries, and were not gravemarkers at all. Their historic form was that of a limestone or granite shaft nine to twelve (at most twenty) feet high, richly carved on all four faces, with a stone ring encircling the juncture of the cross arm. Aesthetically, they were literal renderings in stone of jeweled crosses wrought in metal and wood. The stone bosses derived from rivetheads; the carved knotwork is the plaited interlace familiar in both Celtic and Saxon art.

While the crosses are most strongly identified with Ireland, they are found throughout the Celtic fringe of Britain; thus their revival as gravemarkers was not strictly Irish, or Catholic, but predominantly so. In the two Dublin garden cemeteries, Mount Jerome (Protestant) and Glasnevin (Catholic), both founded contemporaneously with Mount Auburn (1832), they were more commonly erected in the latter.[33] In turn-of-the century Boston, the crosses were deemed by the Irish as fitting memorials to a mayor, Patrick Collins, buried in Holyhood Cemetery in Brookline, and to a parish priest buried in Highland Cemetery in Norwood MA. Even among Irish Catholics they seem to have been regarded as lavish and their usage was rare: there is none in the Cambridge Catholic Cemetery, and only two, dated 1913, in Saint Bernard's in Concord. In Mount Auburn, where presumably they were not intended as specifically Catholic, they flourished mainly between 1895 and 1915. The Edward Ingersoll Browne monument does full justice to the form.

Among Yankee Protestants, the ethnic totem par excellence was, ironically, a revival of the Colonial slate gravestone, which only made its debut in Mount Auburn about 1887, having been banned there hitherto as the grim emblem of everything the cemetery founders abhorred. Some are ornate, with rose-carved borders, and others austere, chastely lettered on smooth slate. Two examples noted (1908 and 1916), while not literal copies, had been clearly inspired by the prestigious Lamson stones of the eighteenth century. The revived slates were not infrequently inscribed with crosses (a Puritan anathema)[34] and aesthetically were more Georgian than vernacular. (Marble versions may refer directly to English, not New England, prototypes.) Beyond their obvious reaction to Victorian excess, they were a badge of ancestry, one that testified on a more personal level to the spirit of Colonial Revivalism that marked the architecture of the period. They are still quite popular, examples as recent as 1997 being noted in Sleepy Hollow in Concord MA.

{ *Seven* }

EPILOGUE

O UR theme throughout has been spatiotemporal variation in the New England landscape, a concept succinct yet vast, which we have taken as our royal warrant to scour the countryside: searching houses, streets, graveyards, farms, and fields in quest of physical evidence. Our motto has been "artifacts for artifacts' sake," and indeed, had our inquiry been more dogmatic, we would have missed much. Ironically, to understand the limits within which things differ, we have often first had to study how much they are the same.

Now we must parse *spatio-temporal* and assess the relative impact of what we have called placemarks and timemarks on the landscapes we see. Placemarks should flow inland along the trunks and branches of the Kurathian hearth-streams, while timemarks should delimit the successive tidelines of settlement frontiers (or, more locally, the growth rings of a town). These complementary frameworks are spelled out in black and white in figures 2 and 3. While main highways reinforce the settlement paths, the settlement frontiers are demographic constructs unlimned by roads. You can retrace the historic upcountry travel-ways from Middlesex County MA to Hillsborough County NH, but you cannot follow the 1760 settlement frontier: it is as elusive as the lilac line in Spring. (The Old Post Road— U.S. 1A—may have been the first and last highway to skirt a settlement frontier, that of 1660, to hazard a guess.) The interconnectedness of Concord MA and New Ipswich NH is palpable, but what links Williamstown MA and Charlestown NH? Modern experience teaches that when, not where, a thing was made is the more probable explanation for any distinguishing attributes it may possess. Consciously or unconsciously, we soon learn to date, or at least chronologically sequence, objects on the basis of stylistic and technical criteria, a habit that helps us to make historic sense of the rapidly changing material world. So driven is this pace that time is now sold in ten-year installments and decadal nostalgia passes for culture. But life was not always this way.

While obvious in principle, it has profound implications that the oldest gravestones, houses, or boundaries in a town are no older than the settlement isochrone within which it lies. If we could construct type sequences for our six landscape artifacts between 1650 and 1850 and plot their distributions against the mapped settlement contours, then to some unknown degree they should chronologically sort themselves out between the earliest and latest settled bands. The most deeply etched would be the initial elements of settlement geography. Gravestones are more readily dated, houses more widely spread, boundaries may linger when placenames are forgotten. Timemarks register inconsistently, but taken together, vague complexes of "upcountry" and "downcountry" artifacts, temporally determined, surely exist, could we but tease them out. Toponymic evidence is primary, not because placenames speak to us in words, or are highly datable, or even immutable, but because they are the one readily interpreted landscape layer that cartographers have made it their daily business to map. If the Great-Big Line and the Burg-Boro Line follow settlement isochrones, then why not other unmapped cultural features such as central chimney houses, right-angled farms, or townplans?

The chronological gradient that has been artifactually ingrained in the landscape should be statistically discernible even where visually obliterated (or never in fact visible). This assertion could be infinitely qualified, but, on a basic level, it seems irrefutable that the vernacular landscape must grow younger as one goes northward, so much so that the perception that Vermont is quintessential New England (and not proto-Ohio) raises grave questions. (The hinterland has become the heartland.) Qualification of this assertion would only lead to an ever more sensitive appreciation of New England landscape history. It is an idea fertile in thought experiments to be tested and tinkered with.

Conventional sightseeing blithely lumps the whole historic landscape under "Old New England," an undifferentiated Yankee dreamtime that embraces two to three centuries and sixty thousand square miles. Sightseeking critically dissects this idyll along more calculated lines. The mind is quick to organize information according to the framework available, and a sightseeker is no more than a well-briefed sightseer with a focused agenda. The thrill of borders can be relied upon to spur the quest.

While the interpretation of temporal distributions (like that of geological strata) is fascinating, it is often as complex as it is conceptually obvious. (*Before* and *after* are intuitive notions.) The premise that artifacts are timemarks, while inexhaustible in its implications, cannot be seriously questioned, although the indelibility and legibility of their imprint on the much-erased palimpsest of the landscape has yet to be systematically measured. Chronological presentation has emphasized artifacts as timemarks as a matter of course; artifacts as placemarks more urgently occupy us here.

The distributions of thirteen artifacts discussed in the text have been schematically reduced to balloons and bubbles on the map (fig. 22). This is both to make plain that they are necessarily conjectural, and to simplify discussion. How best to delineate them is debatable: as indicated, some encircle the entire known distribution; others only the main body; still others the densest concentration. Artifacts such as the Cape Cod house and Saltbox, presumed to have been general within the settlement limits of their time, are not shown. Only one "Great Divide," the rafter-purlin roof line (no. 4), clearly demarcates the Bay and River hearths, but other artifacts cumulatively, if more vaguely, reinforce it. Such a trenchant line could only come about because for almost two centuries the framer of every New England roof was bound by custom in his choice of one of these two schemes. While the facts are perhaps not so black and white, it will be appreciated that the conditions conducive to such clear-cut divides are rare: few artifacts were so universal as roof frames, so long frozen in essentially two alternate forms, or so utilitarian as to be immune to fashionable change.

The map reveals few of the expected isoscenic bundles; rather they overlap or interlock or lie embedded. While balloons do not concentrically circumscribe neat territories, their outlines, taken together, reinforce certain fracture zones that might prove to be subregional boundaries with fuller and surer data. The -dale loop and stone mills give promise of one isoscenic bundle; the Tophet name-string, jut-bys, and yellow meeting-houses (not shown here) another; the Reverse Center Arc and the Connected Farmstead core, perhaps a third. The Connecticut Valley is a long balloon chain: that the Upper Valley is more properly joined with Worcester County (as was Kurath's view) than with the Lower is not immediately obvious. Some elongate bubbles, such as the Tophet name-string, while too small to delineate subregions, like straws in the wind do signal the direction in which they spread. Other bubbles are circular and attest only to local influences; in some cases (not mapped here), scattered bubbles may betoken population (or artisan) movements.

One obsession adopted uncritically from dialectology that merits re-examination is boundaries. It is paradoxical that while artifacts are held to emanate from the core so much concern is lavished on the exact demar-

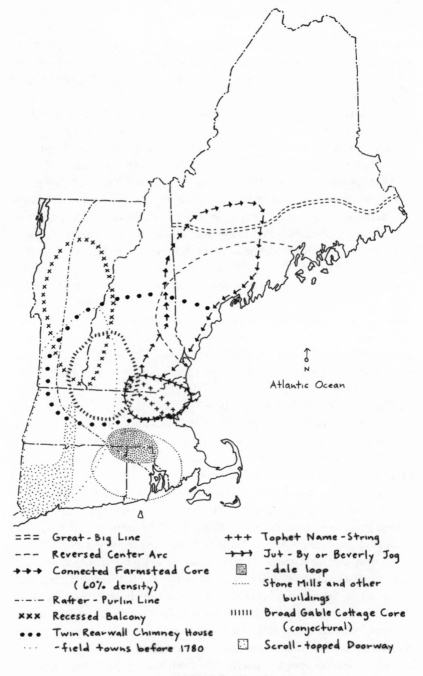

Atlantic Ocean

=== Great-Big Line
--- Reversed Center Arc
→+→ Connected Farmstead Core
      (60% density)
-··- Rafter-Purlin Line
xxx Recessed Balcony
••• Twin Rearwall Chimney House
··· -field towns before 1780

+++ Tophet Name-String
→→→ Jut-By or Beverly Jog
▨ -dale loop
····· Stone Mills and other
        buildings
||||| Broad Gable Cottage Core
        (conjectural)
▢ Scroll-topped Doorway

FIGURE 22
Distribution Balloons of 13 Landscape Artifacts

cation of their diffuse and intermingled peripheries. All eyes are on no-man's-land: it is the old thrill of borders. But political boundaries are legally (and militarily) enforced; dialect and artifactual boundaries are at best elaborate statistical constructs.[1] Perhaps we have not found the "right" isoscenes, or perhaps the balloons as drawn are gravely in error. But it would seem that we have mapped some of the most conspicuous isoscenes with rough accuracy. While the true distributions, when mapped, might be different, we shall assume that the general effect, and the attendant problem of interpretation, to be the same. The problem is that the isoscenes suggest a dynamic and not a static system. The balloons could be seen as a waxing and waning chain of lakes along the inland migration streams. The subregional model is too static to account for this; the boundaries of 1750, 1800, and 1850 could not be expected to have held their ground. Indeed, what might be called the statics—the fixed boundaries—of the Kurathian model have always been its most debatable aspect. The dynamics—the historic and geographic forces that shaped the boundaries—prove easier to corroborate and of greater value in interpreting artifact distributions.

Artifactual Dynamics

We earlier likened artifact distributions to spill-stains, and suggested that different cultural artifacts, under similar historical conditions, might diffuse along similar lines. Even where overall outlines differ, the forces at work may yet be similar. The two main external conditions that clearly gave shape and direction to the distributions were well known to Kurath: settlement paths and frontiers. However, as artifactual "spillways," these paths were not all-controlling; as much or more depended on the inherent diffusibility or exportability of the artifact itself. Some of these inherent qualities are so obvious as to be easily ignored: a gravestone cutter's work, as so much deadweight of stone, would predominate within a radius of cost-effective cartage; as an idea—an imitable or transportable motif—it could travel much further. The geometric face motif of the Merrimack River school spread from Haverhill MA by stonecutter migration and stylistic diffusion (1690–1810) both in the expected Kurathian directions into south central New Hampshire and Worcester County MA, but also unpredictably to southeastern Connecticut when John Hartshorn relocated there in 1719.[2]

While some artifacts were of a nature to counter the ingrained directionality of settlement expansion, others were not. The China Syndrome namefield expanded into virgin territory in Maine, not backward into already named country. The geometric revolution of the mid-eighteenth century could not unmake the cadastre of earlier-settled towns, although it might overlay it. The Connected Farmstead was a formula for thriftily reordering older buildings and could ripple out regardless of the grain of settlement; had it ordained new construction, its spread would have concentrated on the frontier or in areas prosperous enough to modernize. An innovation that built on earlier groundwork, whether literally, as the Saltbox on the Hall-and-Parlor, or more figuratively, as Unitarianism on Congregationalism, had the edge, and spread all the more readily into a historic range. Where such groundwork was lacking, its advance was slowed; countervailing custom or accomplished fact might stop it altogether.

An artifact had to find or make its niche, and its exportability was furthered by both its prestige and utility. It had to be the right thing in the right place at the right time, as is well illustrated by the selective, extraregional spread of New England house types after the Revolution. The Saltbox, obsolete, stayed home; the Center Chimney Large, well established, spread into upstate New York; the Upright-and-Wing, innovative, evolved en route and transformed the Old Northwest. The Cape Cod house and Connected Farmstead, while they lacked extraregional prestige, had regional utility and overspread vast areas of New England itself. Subregionalisms, by contrast, are often notably "not for export": the jut-by, the coffin door (although this reached Ohio), and the Broad Gable cottage stand out as aesthetically or functionally flawed, and not calculated to set the world on fire.

Artifactual dynamics poses many classic problems. The locus of diffusion of an artifact—the center of its outward spread—is one such. Thomas Hubka has charted in ovoid ripples of concentrically decreasing density (from 60% to 1%) the Connected Farmstead;[3] it is tempting to read here a record of its emanation from the southwestern corner of Maine, but this is only one interpretation of such a diagnostic pattern. Kniffen saw the westward march of the Upright-and-Wing in successively denser bands from peripheral New England (9%) to Michigan (73%).[4] In each case, the directionality implied by the gradations is interpreted in opposite ways. On a smaller scale, Benes found the "Templeton Run" to have spread linearly and chronologically in the "expected" historical outward direction of settlement. Yet the Friendship-Harmony name-string randomly crystallized in a way that defies notions of a neatly predictable locus of diffusion. The known dates and distributions of highly successful artifacts, both very old (Cape Cod house) and very new (Three-Decker) may fail to point conclusively to any one locus or pattern of dispersal. Even with widespread and securely dated artifacts such as gravestones, interpretations can radically differ. Deetz and Dethlefsen viewed the cherub as an innovative motif spreading slowly outward from Boston ca. 1700–1760 "at about a mile per year." Peter Benes reexamined the question in light of when cherubs became normative, not simply innovative, and saw a contrary flow: they diffused from Newport to Boston ca. 1720–1790.[5]

The outward flow of settlement entailed as its reciprocal an inward civic, cultural, and commercial pull toward hearth or early core. With changed conditions and improved communications, the influence of one core might grow outward and encroach on or "capture" the hinterlands of other cores. With the local decline of agriculture and the rise of coastwise trade ca. 1750, Truro and Eastham MA, as reflected in their gravestones, reoriented themselves from the Plymouth to the Boston sphere of influence. To judge by the spread of the Twin Rearwall Chimney house, Boston appears also to have captured to some degree the Upper Connecticut Valley with the advent of turnpike roads. The economic allegiance of Burlington VT reversed poles from Montreal to New York City when the Champlain Canal opened in 1822.[6] More gradually, New York weakened and in places overruled the influence of the rather focusless Connecticut hearth. The cultural dynamics of the coast has its own logic, which provides an interesting contrast to inland diffusion. While maritime connectedness might bring sophistication to seaboard towns, it also had, as an ironic corollary, the selective cultural insularity of Nantucket and Martha's Vineyard. A further question is whether vernacularisms arise at a particularly favorable stage of development, when time and place are "ripe," and whether innovative centers moved with the frontier line. Is it happenstance that the recessed balcony arose in the Upper Connecticut Valley north of and a half century after the scroll-topped doorway in the Lower? Does the locus of the Twin Rearwall in Middlesex County (ca. 1740) and of the Connected Farmstead in southwest Maine (ca. 1830) represent a shifted center of innovation for the Bay hearth? As the old cores matured, and became geographically and culturally remoter from the expanding frontier, they would become promulgators of polite and not pragmatic artifacts. The innovative center for New England extended was not in New England at all, but in the Mohawk Valley of New York.

To recall an earlier analogy, artifacts are like chess pieces. Having learned them by shape and by name, we come to recognize each as radically different in manner and range of movement and in strategic possibilities of play. Once we have mastered these dynamics, their dispersal across the board in midgame is no longer a random scatter, but a compelling pattern filled with signs of what came before. Words, houses, and gravestones are all different chess pieces on this board. Dialectology affords a ready and compelling model, but its dynamics are those of but one class of cultural artifact—language—each word of which has its own origins, its own diffusibility, its own survivability. Linguistic patterns may register the same cultural and historic forces in ways different from a house or gravestone. Only once the landscape has been fully inventoried, and its repertoire of artifacts firmly mapped, can a serious critique of the Kurathian subregions be attempted. Perhaps the recent creation of the Essex, Freedom's Way, and Blackstone River national heritage areas will prove a catalyst for the special blend of scholarly research and popular interpretation needed to elucidate what made subregionalism tick. *Sightseeking* offers a good beginner's toolkit.

Other Spill-Stains

In the meantime, certain other nineteenth-century demographic data, cultural and political, may shed light on the subregions that Kurath posited on historico-linguistic grounds. The distribution of New England colleges founded by 1825 may reflect a final focalization of subregional identity, although the location of some, as Williams and Dartmouth (*Vox Clamantis in Deserto*), was initially quixotic. Churches are a more numerous and sensitive index. Gaustad[7] maps U.S. churches by denomination in 1750, 1850, and 1950. The data for 1850 is most pertinent as it records the state of things at the end of the vernacular era. While it might seem sacrilegious to liken the spread of churches to that of houses, religion was clearly diffused by many of the same mechanisms as material things: family tradition, kinship networks, settlement paths and cultural-commercial axes, and the influence of subregional style centers. Theological ideas might also sweep through more rapidly to "burn over" an area, perhaps influence gravestones, yet leave other cultural indices intact. Plymouth County MA embraced Unitarianism but not the Twin Rearwall Chimney house.

Certain facts stand out clearly from the 1850 maps. Congregationalism is predictably coextensive with New England and New England extended, its long cultural shadow that fades westward from New York to Wisconsin.

Worcester County MA, and Cumberland County ME (that is, Greater Portland) register the greatest religious diversity, perhaps an index of the cultural ferment of innovative areas. Worcester stood at the crossroads of Massachusetts, Rhode Island, and Connecticut influences; Portland was a jumping-off place for the northern frontier. The Broad Gable cottage and the Connected Farmstead have, respectively, been identified with the two areas. Presbyterianism while minor in numbers has strong correspondences with the Scotch-Irish frontier.

The Unitarian-Episcopalian dichotomy strikes one as a last gasp of the seventeenth-century Bay-River rivalry. Unitarianism by 1805 had its American theological focus at Harvard and resembles in its northerly and westerly limits the distribution of another Federal-era Bay hearth (prestige) artifact: the Twin Rearwall Chimney house. The quip that Unitarians believe in "the fatherhood of God, the brotherhood of Man, and the neighborhood of Boston," is amply brought out, although *neighborhood* here comprehends the entire cultural hinterland. Episcopalianism, which is but Anglicanism reformed along American lines, has its colonial origins in a coastal necklace of forty-four churches that stretched from Connecticut to New Hampshire. In Connecticut, it won notably early converts at Yale (1722), and by 1750, churches had penetrated inland as far north as Windsor. This early groundwork underlies the 1850 distribution, the largest bloc of which encompasses all of central and western Connecticut, together with culturally dependent Berkshire County MA. This state of affairs was summed up by a contemporary observer: "that class of the population, somewhat elevated by taste and education, which in Massachusetts became Unitarians, have in our commonwealth [Connecticut] chosen to be Episcopalians."[8] The map (fig. 23), despite glaring anomalies, reveals the continued influence of the old hearths. (The distribution of tonic = soda, everyone's favorite Bostonism, has been overlaid to show that sacred and profane artifacts from different eras may nonetheless display remarkably similar patterns.)

Even in 1850, with but eighty-two in all New England, the downcountry distribution of Roman Catholic churches already prefigured the floodtide of Three-Deckers built between 1880 and 1930. By 1950, when their numbers had risen almost twentyfold, it was their density, not their geography that had changed. To what degree this vast urban archipelago, this "Catholicopolis," might qualify as a popular subregion can only be asked rhetorically here, but the answer has profound implications for how we are to understand subregions in the postvernacular world.

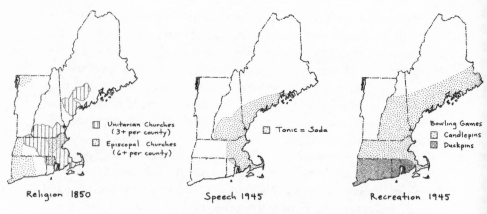

FIGURE 23

The Sacred and the Profane

The old Bay-River rivalry was perpetuated in culture high and low.
Church data derived from Gaustad 1976, 129, fig. 105; 69, fig. 56;
tonic data from Kurath 1973, 28, chart 5; and bowling data from
Harmon 1985, 117, fig. 2.

While religion was a bulwark of the old subregional order, politics, as mapped in the pages of Charles Paullin's historical atlas, better charts its decline. Only in the general elections of 1804 and 1808 can anything like the old east-west Bay-River cleavage be found, between Federalists in the west and Democratic-Republicans in the east (and western fringe). A few legislative landmarks continued to evince something of this polarity, but not for long: the Embargo Act (1807), the Declaration of War (1812), and lastly, the General Tariff Act (1828). By 1856, New England as a whole voted solidly Republican, and would do so through the elections of 1928. One of the strongest forces behind the slow merger of the two hearths over the first half of the nineteenth century can be measured in rate of travel maps. While in 1804 it took seven days to travel overland from New York City to Belfast ME, in 1830 it took four days, and in 1857 (after the advent of rail) only one. Expressed another way, in 1830 (before rail), it cost as much to ship one ton of goods thirty miles inland by road as it did to ship it three thousand miles from Europe by sea.[9] The isolation that fosters subregions had evaporated, a principal factor, along with industrialization and immigration, in the decline of vernacular culture by the mid-nineteenth century.

After 1850

Throughout, our primary focus has been on vernacular, as opposed to popular artifacts, with roughly 1850 as the watershed between the two. While the sheer volume and variety of the popular material makes it impossible to treat here fully, the six artifact classes continued to assert themselves in the landscape, yet by the early twentieth century it is difficult to identify new types as rich in regional significance as the old, or even as amenable to the same modes of analysis. The country was well settled, and the basic and universal needs represented by these artifacts had become socially organized on a vast scale. Nationalization of the economy and of culture meant that the scope of our subject had suddenly exploded well beyond the borders of New England. As late as 1880, the origin and diffusion of the Three-Decker could pose a problem as fascinating as any vernacular one; by 1910, however, innovative house types such as the bungalow and foursquare had clearly extraregional origins.

Cultural artifacts became increasingly standardized, periodized, and certainly not subregionalized. Even when modern placemarks are tentatively identified, their cultural significance seems diminished. Of three common bits of twentieth-century street furniture, sidewalk-makers' trademarks, public mailboxes, and fire hydrants, careful study shows that only the last holds much interpretive potential as a placemark. The first are merely brass business cards laid in concrete, the second are largely of interest because they underwent subtle design changes and are dated (the oldest discovered in the Boston postal district was 1927). Fire hydrants are the most intriguing because they show a stylistic and technical evolution; there are reasonable numbers of presumed antiques (stylistically, at least, as old as the 1890s) still on the street; they have very distinctive profiles, as the Chapman, the Ludlow Diamond, and the stereotypical A. P. Smith, all quite identifiable at a distance and as background detail in news photos, TV cop shows, and movies. Their places of manufacture trace out historical displacements in the iron industry, as New England foundries yield to ones in the Middle Atlantic states and finally the South.[10] In eastern Massachusetts at least, their colors vary from town to town, and amount to a municipal livery; indeed, in some towns they are the high school football colors. (Such color changes are often a cyclist's first indication that he has crossed town-lines.) Yet, when compared with gravestones (which they seem strangely to parody), hydrants are an industrial product devoid of sociocultural significance: literally and figuratively hollow.

Nonetheless, the inventiveness of the latter nineteenth century could still display some of the old vernacular dynamics, both on sea and land. Small sailing-craft enjoyed a golden age along the New England coast ca. 1850–1900, with the development of local work boats, each suited to its home waters, weather, and special task: boats that bore such evocative names as the Eastport Pinky, Maine Peapod, Piscataqua Skiff, Bank Dory, Boston Hooker (a Galway immigrant), Kingston Lobster Boat, Cape Cod (originally Martha's Vineyard) Catboat, Block Island Cowhorn, Noank Sloop, and New Haven Sharpie, most of which either developed or flourished about this time. Some were even built of odd local timber (hackmatack, red oak, cedar, yellow birch). What lessons in origins and diffusion, artifactual dynamics, and the conditions conducive to a ferment of local invention might be drawn here! One gropes for a landscape parallel: is this vernacularism run amok, or essentially modern in spirit, like the Yankee ingenuity of the small-town machine shop? Whatever its impetus, we know its death knell was the gasoline engine.[11] On land, the first night-owl workers' lunch carts spread northward from Providence RI up the mill towns of the Blackstone Valley to Worcester MA, where railroad car builders transformed them into the classic diner of the early to mid-twentieth century: a clear instance of axial diffusion.[12] Nor does it surprise that the quintessential blue-collar eatery should have arisen in a valley better remembered as the cradle of the American Industrial Revolution.

The ferment of this industrial culture saw the emergence of new urban forms of recreation and refreshment. Basketball was invented in Springfield MA (1891), and volleyball in nearby Holyoke (1895); their diffusion, ultimately worldwide, was aided by their Y.M.C.A. roots, as well as by American missionaries and the U.S. military.[13] Ironically, even as these games went global, the subregionalization of New England bowling games was under way (see fig. 23). Candlepins was invented in Worcester MA in 1880; aggressively promoted, it spread north and east, dislodging the earlier tenpins from Boston after 1903, and ultimately spreading coastwise as far as Nova Scotia. Meanwhile, duckpins, of unknown, but probably New England origin, gained a foothold in New Haven CT in 1896, and within ten years moved east through Rhode Island into southeastern Massachusetts. Statewide organization complicated the pattern, but here the old Kurathian subregions are still at work. Even the fall of Vermont to tenpins is predictable, given the game's nineteenth-century New York power base.[14] Turn-of-the century urban thirsts could be temperately and sociably quenched at the soda fountain or "spa," which evolved its own subregional terms and specialties: frappe, tonic (Boston); cabinet, coffee milk (Providence); egg cream (New York). Fun was not yet nationally organized.

In Fourth of July "Horribles" parades we can glimpse an ephemeral folk custom that successfully remolded itself in the twentieth century. "Antiques and Horribles," burlesquing the finery of the Massachusetts governor's honor guard, the Ancient and Honorable Artillery Company (established 1638), provided comic relief at mid-nineteenth-century military parades, largely within the Boston sphere of influence. With the decline of training days and militias, the custom evolved into one of costumed children's Fourth of July parades, which today enjoy an expanding popularity, particularly in coastal resort towns. (Many inland towns have ceased to hold Fourth of July parades altogether.) The scant recorded evidence (much based on Internet searches) for the present and former extent of the custom focuses heavily on the Massachusetts coast, particularly the North Shore (see Appendix). Some of the more peripheral are of recent or variant origin: that at Mystic Seaport CT is a museum recreation in the style of the 1870s; Marshfield MA celebrated its around Labor Day as a farewell event for summer people; New Bedford MA and increasingly many schools hold Halloween Horribles. The Norway-Paris Firemen's Carnival in Maine has hosted a midsummer "Stephen King Horribles Parade": the tradition has certainly kept up with the times. ("Parade of horribles" among lawyers has come to describe a rhetorically inflated litany of ills or abuses.) It is perhaps safe to say that while (sub)regional expression in the era of metropolitanization was far from dead, it was shaping things vastly different from the landscape. Nor is it surprising that in an era of mass production, factory workers would more often find cultural expression in the realms of play than in those of work, as in an era of artisans.

Conclusion

We have identified six main landscape elements and catalogued some hundred or more specific types, endeavoring whenever possible to explain their distribution. Spatiotemporal diversity has been our constant theme, hence a strong emphasis on exact places and dates and specific examples, meant to ground discussion in the actual landscape. Our approach has been to combine map study and field observation with an intense literature search, and inevitably the depth and scope of our coverage has been shaped by the body of published research available—at times inversely, for we have not sought to reduplicate what has been ably and accessibly treated elsewhere. Gravestones possess the most abundant body of literature, one to which we have done but summary justice. Houses come second, but New England house types have never been fully inventoried, nor have their distributions been firmly mapped. Architectural styles alone can claim exhaustive treatment in print, and even here localisms often appear only as footnotes in specialized sources. Houses are but the singlemost representative element of the building stock; barns, churches, and mills are among the many rich building types that space has not allowed us to touch upon. Placenames have been the subject of many gazetteers and compilations; our treatment of them while novel, indeed often heterodox, has proved surprisingly fruitful in new insights. Townplans and road-nets are known far more in ideality than reality, while boundaries (and the cadastre generally) remain the most obscure and least studied of all. The literature, while sufficient for interregional comparisons, often proves inadequate to highlight and explain intraregional differences within the thousand town mosaic of the New England landscape. Thus the effort to compile and explicate even a select catalog of landscape artifacts has been considerable, particularly as much under our purview has never been artifactually analyzed before, and many key questions have gone not only unanswered, but unasked.

Having blithely set out to "unriddle" the New England landscape, it is fair to ask after some three hundred pages how far we have succeeded in that endeavor. At times it would seem we have avoided the question altogether; many touristic icons of the cultural landscape (covered bridges, lighthouses, cranberry bogs) and the entire natural landscape have been neglected. Instead we have opted for a strict framework of analysis, and a quasi-archeological approach. Why? It is unlikely that anyone would spontaneously identify these six classes of artifact as the basis for his own subjective experience of "New Englandness." Yet these are the half dozen most pervasive and permanent building blocks of the settlement landscape. Implicit in the study of these building blocks is a concept of landscape that radically differs from the pictorial one that defines it as the amount of scenery framed within a single glance from a single point of view. The sightseer's small visual harmonies yield to the sightseeker's vast unseen patterns, whether these be of placenames, house types, or townplans.

This change in perspective is thrust upon us in part by two trends. First, widespread landscape degradation that often renders any one swatch of scenery not merely painful to contemplate, but deficient in the elements necessary for its historic interpretation based on internal evidence alone.

Second, a modern decentralization, which has meant that working wholes, still visible in nineteenth-century mill towns (as mill, mansion, and mill-hands' housing) no longer exist in one landscape, but instead are dispersed far and wide, over hundreds of square miles. Pattern-finding in such a vast and chaotic field is best accomplished by looking at a limited number of specific things: our *videnda*. We have concentrated on vernacular artifacts as these compose vestiges of a simpler, more spatially organized, and less populous world. Time itself has further narrowed the field; the vernacular houses, farmsteads, and placenames we examine are only the fraction that have survived to our day. Thus it becomes all the more necessary to study them collectively over a wide area, to construct a composite portrait of a whole that survives only as scattered and imperfect parts.

While the study of these building blocks may strike one initially as dull and inadequate to the overall task, once one's curiosity is aroused, there is no landscape in New England that will ever be dull again. Ordinary things —select but ubiquitous ordinary things—assume an unexpected importance, not only in the field, but in old photographs, sketches, and maps. Heretofore objects of nostalgic reverie, these become vital documents to be minutely scrutinized for key detail. Historic landmarks, praised and preserved as rare or unique, are taken down from their pedestals and reconnected with the traditions and landscapes that created them. Because there is nothing overtly boosterist or filiopietistic in this approach, two traditional weaknesses of local history are countered. And because there is nothing aesthetic either, a sprawl-engulfed farmhouse or burying ground can bear as vital a witness as one in unspoiled country. Where a photographer would have to crop his shot, or a travel writer semantically evade, the aboveground archeologist is free to seek and tell the truth. Since archeology is the study of ruins, the retrieval of the irretrievably lost, he can ply his trade while the landscape is ravaged all around him. The sight-seeker's role as rescue archeologist hardens him to the inevitable pain of landscape destruction: a pain doubly inevitable because only a sensitivity to landscape would dispose him to his task. It is as though a squeamish person became a doctor to armor himself against the sight of pain.

The universality of these six classes of artifact means that wherever one travels is a field for fresh, if highly focused, observation. Because vernacular artifacts derive from arguably simpler, paleotechnic times, when classic geographic controls played a far stronger role in man's activities than today, simplistic models of artifactual distribution seem easier to credit (though this presumption is perhaps naive). Yet unlike the simplistic models of the classical archeologist, because our subject is not prehistoric, but truly historic, we often have at our disposal a variety of records to cross-check our conclusions. (It is precisely this ability to validate independently archeological concepts such as frequency seriation that made the study of New England gravestones so attractive to Dethlefsen and Deetz.)[15]

Our six artifact classes, while chosen for New England, travel well, offering a portable framework for interregional (even international) landscape comparisons. Indeed, the essential complementarity of this approach with basic archeological surveys of Roman Italy or Bronze Age Gaul is no accident. A rich genre of such scholarly syntheses, disguised as popularizations, yet written by distinguished experts, has served as an ever-present model for this volume. In effect, this book strives to be a pseudo-popularization of a subject that has not yet fully taken coherent form in the scholarly literature. My excuse for this temerity is simple but cogent. I have written it because I am convinced no expert ever would. Fools rush in where angels fear to tread. While the subject of this book, in academic terms, may be interdisciplinary, for the curious New Englander in search of his own landscape heritage, it is literally a seamless whole. And ironically, while the number of academics who delve in this field may be small, the number of curious New Englanders who cherish a lively interest in their local history and surroundings is vast. Sightseeking merely raises the geographic horizon one critical notch, from the parochial to the supra-parochial, and transmutes local sentiment into dispassionate analysis.

Six artifacts, of course, do not the New England landscape make. However feasible or unfeasible the cultural geographer's dream to inventory all landscape artifacts of all periods, to identify their patterns, and to devise a tidy terminology with which to explain, order, and humanize an increasingly disorderly and alien landscape, within the compass of a single book by a single author, this is clearly not possible. (The *LANE*, after all, was the collective scholarly effort of a decade.) Yet such an objective clearly benefits from the experience gained in our more modest survey, for we have plumbed both the riches and the poverty of the artifactual approach to landscape study, and have trained ourselves in habits of thought and observation that will serve us well in interpreting landscapes new and old, in New England and beyond.

APPENDIX

Handlist of Reversed Centers

ABBREVIATIONS: (f) former, (d) dubious, (al) central.

The main Arc, east to west, counterclockwise: Lincolnville (f), Belmont, Montville, Vassalboro, Sidney, Minot, Lovell ME; Conway, Sandwich, Tuftonboro, Ossippee, Effingham, Barnstead, Strafford NH; Lebanon ME.

Others outside Arc:

Maine: Hancock (al), Guilford (f).
New Hampshire: Haverhill, Alstead (f).
Vermont: Rutland.
Massachusetts: Marshfield (f), Abington (f), Grafton (d).
Connecticut: Groton.
New York (grouped geographically): Moriches, Islip (al), Nyack (al); Berlin, Brunswick (f), White Creek, Cambridge; Centerlisle, Canisteo, Almond, Sherman.
Pennsylvania: Moreland.
Ohio: Belpre.

China Syndrome (all Maine)

1787–1799: Limerick, Norway, Paris, Belgrade, Dresden, Poland.
1800–1809: Denmark, Vienna, Rome, Athens, Palmyra, Palermo, Lisbon.
1810–1840: Sweden, Peru, Canton, Mexico, Carthage, Madrid, Moscow, Corinth, Troy, China, Bremen, Naples.

Not mapped: Hanover, Lubec, Sorrento, Bangor, Frankfort, Belfast, Brunswick.

Handlist of Horribles Parades

Both current and former (north to south): Southeastern New Hampshire (unspecified), Rockport, Gloucester, Manchester (current), Hamilton (ca. 1877), Danvers, Peabody, Beverly Farms, Salem Willows, Marblehead (1899–), Concord (ca. 1900), Lynn (1876), Winthrop (Point Shirley), Boston (1940), Needham, Northampton (ca. 1860), Hopkinton, Hingham (1948), Marshfield (ca. 1950?), Wellfleet (recent), New Bedford (Halloween; recent), Hyannis (1929), Oak Bluffs (1874), Pawtucket RI (nineteenth century), Providence RI (1858), Mystic Seaport CT (recent).

NOTES

PREFACE (p. xiii)

1. Dwight 1969, 4:2–3.
2. Quoted in Glen L. Harmon, "Regional Cooperation," *Boston Globe*, February 10, 1999, A23.

1. PROLOGUE (pp. 2–12)

1. As L. P. Hartley has memorably expressed it: "The past is a foreign country: they do things differently there." L. P. Hartley 1990, 3.
2. *British Landscapes through Maps* 1960–79. This monograph series admirably illustrates this method.
3. Fagan 1978, 32–35; Deetz 1967, 26–33, 45–52. The curve is "battleship-shaped" on the type of bar graph customarily used by archeologists to record the frequency of finds over time.
4. Kirk 1972, 195 (quote), 195–200. Hosley 1978, 69 n. 6; Kniffen 1965, 567–68, 572 (map); Visser 1997, 12–13.
5. Carver 1987, 21–29. Figure 1 adapted from Carver 1987, 25 (fig. 2.3), incorporating data from Kurath 1973; figure 2 adapted from Kurath 1973, plate 1; figure 3 adapted from Wood 1977, 35 (fig. 1.5).
6. Ibid., 25.
7. Kurath 1973, 31, 36.
8. Visser 1997, 14; Kurath 1973, 30.
9. Dwight 1969, 1:5.
10. Evans 1996, 682–717, 394, 318.
11. Ibid., 354–55, 343, 528, 380, 274.
12. Ibid., 9.
13. Ibid., 4.
14. A. Forbes 1995, 22–23.
15. Kirker 1969, passim; Bunting 1985, 41–42, 290 n. 11; Whitehill 1968, 99.

2. PLACENAMES (pp. 13–86)

1. Reaney 1961, passim.
2. Ibid., 50.

3. *Webster's Third New International Dictionary* 1969 (etymologies of Massachusetts and Connecticut).

4. Huden 1962, 46.

5. Goddard 1978, 76.

6. Russell 1980, 29 (map).

7. Goddard 1978, 74–76.

8. Gelling 1978, 215–16; Gould 1978, 27–29.

9. American Coast Pilot 1854, 247.

10. Sullivan 1990, 102, 111, 114–16.

11. *Boston Globe*, August 15, 1992, Metro 1.

12. Green and Sachse 1983, 75, 100.

13. American Coast Pilot 1806; American Coast Pilot 1854.

14. United States Coast Survey 1875, 174 (Big Ben quote).

15. Hendrickson 2000, 177, 212.

16. Wilkie and Tager 1991. See the John Seller map (1675) on the front endpaper; Fogelberg 1976, front endpaper map of Burlington.

17. Swan 1980, 7.

18. Lillywhite 1972, passim.

19. W. T. Davis 1883, 151.

20. Federal Writers' Project 1937, 258.

21. Gilman and Gilman 1966, 36–37.

22. Benes 1977, 198.

23. Banks 1966, 2:22.

24. W. T. Davis 1883, 153; E. M. Hunt 1970, 185; Attwood 1946, 224; Swift 1977, 41, 236, 541.

25. Hughes and Allen 1976, 117, passim; Swift 1977, 134.

26. Gould and Kidder 1852, 287.

27. Morison 1921, 163, 229; Phillips 1947, 233.

28. Attwood 1946, passim.

29. Cushing 1990, 57; Candee 1992, 47; Copeland and Rogers 1960, 44; Robinson 1976, 89.

30. *Boston: The Official Bicentennial Guidebook* 1975, 162, 167; Sinclair 1998, 359; Whitehill 1968, 7–8.

31. Field 1989, xvii, 32–33.

32. Hughes and Allen 1976, 49 (Bridgewater origin).

33. Frizzel 1955, 278, 311; Swift 1977, 532.

34. Lee 1967, 59, 64.

35. Swift 1977, 408.

36. *Best Read Guide: Martha's Vineyard* June 11–27, 1999, 43; Wright 1977–81, 1:24–26;

M. C. Coolidge 1948, 269, 273; Jane 1978, 15; Rhode Island Historical Preservation Commission 1982a, 55.

37. Massachusetts Federal Writers' Project 1941; E. M. Hunt 1970; Hughes and Allen 1976; Swift 1977; Attwood 1946; Chadbourne 1955. These standard sources for MA, NH, CT, VT, and ME have furnished much basic data throughout this section for which it has not always been feasible to cite page references. Care has been taken to acknowledge higher interpretations beyond the basic data. Also used have been the *Omni Gazetteer of the United States of America* (in 11 vols.), the DeLorme state atlas series, online databases, and numerous other maps and gazetteers.

38. Hunt 1970, 78, 109–10; Hughes and Allen 1976, 464.

39. Emery 1893, 149 (Taunton quote); Hughes and Allen 1976, 713.

40. Swift 1977, 511.

41. Fischer 1989, 44–47, 31–33, 38.

42. Hughes and Allen 1976, 767–69.

43. Hunt 1970, 120; Swift 1977, 491, 516.

44. Chadbourne 1955, 182.

45. Fite and Freeman 1926, 124–27 (includes map).

46. Anderson 1976, 29; M. C. Coolidge 1948, 24.

47. E. M. Hunt 1970, 102, 115–16; Swift 1977, 78.

48. Hurd 1890, 1:570–71.

49. Federal Writers' Project 1983, 514.

50. Swift 1977, 476, 501.

51. Massachusetts Federal Writers' Project 1941, 27, 40, 42, 5; Barber 1839, 242; Kirkpatrick 1975, 1:65; Hughes and Allen 1976, 530; Browne 1921, 1:62; Tilden 1887, 36.

52. *Boston Globe*, November 21, 1994, 1.

53. Rhode Island Historical Preservation Commission 1982a, 15; E. M. Hunt 1970, 179.

54. Stansfield 1983, 17.

55. Massachusetts Federal Writers' Project 1941, 49.

56. *British Landscapes through Maps* 1960–79; Hughes and Allen 1976, 115, 445, 668.

57. E. M. Hunt 1970, 28; Room 1986, 7–8, 129, passim.

58. *Paul Revere's Boston* 1975, (poem) 210; E. Forbes 1942, 406.

59. Name-types graphed were *-ton, -boro(ugh), -burg(h), -ville*, and surnames 1630–1920. RI was excluded.

60. Etymologically kindred names, such as Norwood MA and Suffield CT, are included in the statistical tallies; Hughes and Allen 1976, 137.

61. Hughes and Allen 1976, xiii, 688.

62. E. M. Hunt 1970, 131.

63. Swift 1977, 201–202, 92–93.

64. Ibid., 540–41.

65. E. M. Hunt 1970, 227.

66. Attwood 1946; See Zelinsky 1955, fig. 25, for settlement line.

67. Ibid., fig. 7, 329.

68. Strong 1890. Counts were determined from this concordance.

69. W. S. Powell 1968 provides basic data for North Carolina placenames.

70. Gowans 1962, 7.

71. Leech 1859; Merrill 1889, passim. Furnishes evidence of inconsistent usage.

72. Hughes and Allen 1976, 162.

73. Merrill 1889, 542; Swift 1977, 417.

74. Zelinsky 1955, 340–41, 348.

75. *Canada Gazetteer Atlas* 1980.

76. Whitney 1888, 110; Zelinsky 1955, 333.

77. Federal Writers' Project 1983, 580.

78. Whitney 1888, 110.

79. Kurath 1973, 24–25, 31.

80. Whitney 1888, 114–15.

81. Zelinsky 1967, 467.

82. C. S. Parker 1907, 84; E. A. Wright 1977–81, 1:26; Tilden 1887, 246–47.

83. Boston, Massachusetts, City of 1879, 4. This report and the annual reports of the Boston Street Laying Out Department 1896–1950 provided basic data for street names, dates, name changes, and so forth, used throughout this section.

84. Stewart 1967, 244.

85. Ibid., 203–204.

86. Ibid., 105–106, 245.

87. Ibid., 105–106.

88. Garland 1978, 357.

89. C. S. Parker 1907, 85–87.

90. E. N. Hartley 1990, 15, 275ff.

91. Bebbington 1972, 129, 60.

92. Hubka 1984, 99, 208 n. 11; Whitcomb 1988, 424–25.

93. Bebbington 1972, 19.

94. See Hubka 1984, 20 (fig. 18) and Benes 1979a, 60 (fig. 6) for range maps of the connected farmstead and yellow meetinghouse.

95. Perrin 1967, 97–100; Perrin 1981, 53–54.

3. BOUNDARIES AND TOWNPLANS (pp. 87–158)

1. Price 1995, 4.

2. Estopinal 1989, 167; Estopinal 1993, 12, 36–37.

3. F. C. Morris 1949, 7; Candee 1980, 18.

4. Boston City Planning Board 1942, 7; F. C. Morris 1949, 10.

5. Thoreau 1962, 2:1225–26; Lawson 1990, 67; James 1988, 210.

6. Walcott 1884, 7; Candee 1980, 22.

7. Fairbanks 1982, 1:38.

8. Lancaster 1972, 17.

9. C. M. Andrews 1889, 38.

10. C. M. Brown 1969, 75, 118.

11. Morris 1949, 10.

12. Brown, Robillard, and Wilson 1981, 144.

13. C. M. Brown 1969, 71.

14. Paullin 1932, 72–73, plate 97a.

15. Candee 1980, 20.

16. See Mooney and Sigourney 1980, 108–10, for an account of William Hussey Macy, the blind nineteenth-century Nantucket Register of Deeds and versifier. J. H. Andrews 1985, 5 gives an Irish example.

17. Candee 1980, 11.

18. Ibid., 38, 40; declination form USGS Lynn Quadrangle 1998; Smart 1962–67, xviii.

19. Pattison 1957, 78, 148–49 (quote); J. H. Andrews 1985, 302–307.

20. Brown, Robillard, and Wilson 1981, 151; Uzes 1977, 4–5.

21. Ibid., 1; Pattison 1957, 74; Gifford 1993, 148.

22. Uzes 1977, 2, 290; C. M. Brown 1969, 148, 150; Estopinal 1993, 62; Lawson 1990, 42.

23. Glassie 1975, 22.

24. C. M. Brown 1969, 8.

25. Goodspeed 1904, 14–15, 11.

26. Estopinal 1993, 144.

27. Gould and Kidder 1852, 39.

28. Estopinal 1993, 60, 61, 69.

29. C. M. Brown 1969, 152.

30. Nason 1877, 10.

31. Uzes 1977, 7; Brown, Robillard, and Wilson 1981, 151.

32. Cameron 1963, 2:450; Breed and Hosmer 1925, 1:212–21 describes method.

33. Smart 1962–67, xxiv.

34. Pattison 1957, 221, 215.

35. Bauer 1902, 45.

36. Ibid., 100–115, 42.

37. Breed and Hosmer 1908, 1:118–20; Breed and Hosmer 1958, 1:137.

38. Lancaster 1972, 26.

39. Jennings 1933, 17; Dort 1935, 62. Both Jennings and Dort seriously misreport the orientation. My figure is derived from the USGS Westport Quadrangle.

40. *Oxford English Dictionary*: see "clock," under: 4. Phrases; Gould and Kidder 1852, 137.

41. Garvan 1951, 45–49.

42. Reps 1965, 125–29; Massachusetts Historical Commission 1982a, 43.

43. Reed 1879.

44. Chamberlain 1925, 1–4.

45. Gannett 1990, 6 (quote), 7, 20–21, passim; Garvan 1951, 74–76.

46. Lawson 1990, 21–24, 50, 4 (quote), 24 (quote).

47. Gould and Kidder 1852, 36.

48. Ibid., 31, 53. The term "after divisions" is used here, in this sense.

49. McManis 1975, 53.

50. Old Bridgewater Tercentenary Committee 1956, not paginated; Russell 1976, 72.

51. Russell 1976, 21, 30–31, 54–55, 128–29; Dorchester 1859, 21.

52. Tilden 1887, 25–26, 28 (quote); Dorchester 1859, 21.

53. Worth 1901–13, 201.

54. Reps 1965, 137 (map) Waters 1905, 1:317 (quote).

55. Russell 1976, 25–27.

56. Ibid., 81–82.

57. Reps 1965, 117–18; Massachusetts Historical Commission 1982a, 44; Wood 1997, 41; Lockridge 1970, 82 (quote).

58. Russell 1976, 74.

59. Crosby 1946, 97–98, 126; Macy 1972, 35–36; Garrison 1991, 19.

60. S. C. Powell 1963, opp. 76 (map), 88 (quote).

61. Ibid., 77 (map), 95.

62. Ibid., after 60 (map).

63. Cambridge Historical Commission 1965–77, 5:9.

64. Lawson 1990, 2–3, 3 (quote); Chadbourne 1955, 59.

65. Sanderson 1936, 10.

66. MacLean 1988, 39.

67. Sanderson 1936, after 82 (map); see also MacLean 1988, 10–11, 30–31, 40–41, for maps that show how the various cadastral schemes in this area were interrelated.

68. Sanderson 1936, 18–20.

69. W. B. Stevens 1891, 12 (quote), 35 (quote).

70. Ibid., 12 (quote).

71. Cambridge 1901, 265–66; Cambridge 1896, 160–65; Sileo 1995.

72. Sileo 1995, 26, 59.

73. MacLean 1988, 31 (map); Fischer 1994, 223 (map).

74. Huntoon 1893, 1:2–6, discusses history of Dorchester New Grant.

75. Historical display maps, Trailside Museum, Milton, Massachusetts.

76. Garvin 1980, 47–68.

77. Ibid., 1980, 59; Gould and Kidder 1852, 24, 42.

78. Scofield 1938, 662–63.

79. Torbert 1935.

80. Turner 1984, 38–46; 46 (quote).

81. Garvan 1951, 21, 18–49; Reed 1879, 1836, 15–16 (quote).

82. Rawson 1942, 73–74; Frizzel 1955, 14; According to Griffin 1904, 350, the strong house at Fort Dummer was also called the citadel.

83. R. Parker 1975, 14, 47–48 (quote),.

84. Mourt 1963, 41 (quote); Morison 1956, 72, 190; Jewett and Jewett 1946, 17 (quote); Old Bridgewater Tercentenary Committee 1956, not paginated; Walcott 1884, 5, 18; Tilden 1887, 43–45; Banks 1966, 2:13; K. Stevens 1988, 44 (map); Russell 1976, 72, notes similarity of Concord and Sudbury sites.

85. Price 1995, 49–50; Fairbanks 1982, 29; McManis 1975, 57.

86. Wright 1936, 7–28; Federal Writers' Project 1983, 359; Judd 1976, 23 (two Hadley quotes); C. M. Andrews 1889, 4; Garvan 1951, 42 (Windsor quote); Thompson 1904–31, 1055–1060.

87. Barber 1839, 314; Swift 1969, 150; Melvoin 1989, 158.

88. Fairbanks 1982, 30 (1709 map); *DeLorme Massachusetts Atlas and Gazetteer* so maps Pleasant Street, Athol; Swift 1977, 52.

89. Reps 1965, 117–19; Plimoth Plantation brochure [2000?]; Whitehill 1968, 8 (Boston quote), 8–15 (includes maps).

90. Federal Writers' Project 1938b, 225; Candee 1992, passim; Downing and Scully 1952, 99 (Newport quote).

91. Aalen 1997, 61–62; Garvin 1980, 53–54.

92. Tilden 1887, 42.

93. L. K. Brown 1968, 40; Wharton 1974, 50–55.

94. Raup and Carlson 1941, 17–19.

95. Garvan 1951, 71, 63.

96. See *Dictionary of American Regional English* for four corners usage.

97. Bell 1986, 55–59, 58 (quote).

98. Griffin 1904, 306–307 (map), 21–23, 79, 96 et passim.

99. Benes 1978, 5.

100. Fleming and Halderman 1982, 154–56; Kurath 1973, 13.

101. List of frog ponds gleaned from Cushing 1990, 28, 93; Tree 1981, 249; Longsworth 1990, 64; Federal Writers' Project 1938a, 307.

102. Barry 1983, 136; Wharton 1974, 50–55; Garvin 1980, 53.

103. Boston Landmarks Commission 1983, 7–8, 14; Bunting 1985, 41; Barber 1839, 301 (Westfield quote).

104. Davis and Davis n.d., 9; Lunt 1976, 20–21; Ramsdell 1901, 361; Kurath 1939–43, 3:pt. 1, map 546 (Amherst usage); Longsworth 1990, 34; Reps 1965, 132 (Woodstock quote).

105. Jager and Jager 1976, frontispiece (photo); Lunt 1976, 15 et passim; Bigelow n.d., not paginated (photo).

106. Clayton and Peet 1933, plates 1–15; Gannett 1990, 15 (map of Cornwall), 18–21; Dyer 1994, 4; Brown 1974, endpaper map of Lane farm; Archeological Investigations 1990, 1:137, 360.

107. Dyer 1994, 86.

108. Allport 1990, 17–18.

109. Ibid., 121, 129; Neudorfer 1980, 54–55; Congdon 1946, 96.

110. Archeological Investigations 1990, 2:106–107; Dyer 1994, i–ii; Committee on Old Homes 1976, 48.

111. Hoskins 1955, 113; Allport 1990, 143, quoting John R. Stilgoe (eight-acre quote).

112. Greeley 1871, 100, 214, 313 (three quotes); Warren 1914, 378–81.

113. McHenry 1978, 10–12; Dyer 1994, 31.

114. Anderson 1976, 82; Whitcomb 1988, 380; Allport 1990, 89; Noble 1984, 2:127–28; Hubka 1984, 85; Lunt 1976, 1975.

115. Dyer 1994, 34, 143–44 (Rehoboth quote); Raup and Carlson 1941, (Sanderson farm maps); Allport 1990, 17.

116. Allport 1990, 187, 180; O'Keefe and Foster 1998, 13 (Petersham maps); Wikander, Terry, and Kiley 1964, 201.

117. For Petersham wall-net, see O'Keefe and Foster 1998, 13 and display, Fisher Museum, Harvard Forest. The Doolittle engravings are reproduced in NPS brochure, the locations are identified on fact sheet provided by park volunteer 10/98.

118. Longsworth 1990, 32, 61–63, 106; Fogelberg 1976 (Marion estate map); Dyer 1994, 4.

119. Allport 1990, 92–93, 108–10; Hunt 1906, 33 (sheep quote); Federal Writers' Project 1983, 522; Zimilies and Zimilies 1973, 164; Wooley and Raitz 1992, 86.

120. McHenry 1978, passim.

121. Herrick 1870, 283–84 (quote); Russell 1976, 191.

122. Richardson 1960, 2 (Bedford ditch quote); Sanderson 1936, 13 (quote).

123. Price 1995, 77–78.

124. For a brief treatment of central place theory and contact number, see Goodall 1987, 60–64, 90–91.

125. F. H. Williams 1890, 61 (tax-dodgers quote).

126. Thoreau 1962, 2:1753; MacLean 1988, 84–86.

127. Ewell 1904, 2, 5, 2 (three brief Byfield quotes); Hubka 1984, 15.

128. Wharton 1974, 46–47; M. C. Coolidge 1948, 27, 273; Costello 1975; Anderson 1976, 28; Darby 1994, 58, 173; Vanderhill and Unger, 1977.

129. Hudson 1968, 419–23.

130. Massachusetts Topographical Survey 1899, 21; Massachusetts Topographical Survey 1896, 11 (quote); Thoreau 1962, 1:275.

131. Wilkie and Tager 1991, 61.

132. Stekl and Hill 1972, 82–83; Pillsbury 1927, 981–82; *Boston Globe*, March 7, 2000, B1, B4; ibid., May 5, 2001, A1, A4.

133. Hughes and Allen 1976, 667.

134. Annin 1964, 6 (quote).

4. ROADS (pp. 159–81)

1. Livermore and Putnam 1888, 11.

2. Oliver 1912, 5.

3. Garvan 1951, 59–60.

4. Hoskins 1955, 139–54, 139 (quote), 154 (quote).

5. Lubar and Kingery 1993, 202.

6. Garrett 1988, 94.

7. Fuller 1964, 177–98, 178 (quoting from Quick 1925, 191).

8. Brown 1968, 30.

9. Hodgman 1883, 48, 49, 45–46, 49 (four brief quotes).

10. E. A. Wright 1977–81, 96–97 (four brief quotes).

11. Lunt 1976, 12–14; Bell 1986, 56–57; Hubka 1984, 85.

12. Livermore and Putnam 1888, 109.

13. Crosby and Goodwin 1928, 4–6 (two quotes).

14. Hodgman 1883, 43, 40 (two brief quotes).

15. Thoreau 1962, 2:1753–58 (Gleason map), 1721 (quote); ibid., 1: 268 (wood-path quote).

16. W. O. Stevens 1936, 208.

17. Macy 1972, 37; Worth 1901–13, 211.

18. Lowenthal 1956, 402–403.

19. Reps 1965, 125–31; Johnson 1988, 389.

20. Whitehill 1968, 74 123; Cambridge Historical Commission 1965–77, 3:17; ibid., 1: 40.

21. Thompson 1988, 1–10.; Lancaster 1972, 10.

22. E. P. Hoskins 1908, 8 (quote).

23. Rhode Island Historical Preservation Commission 1990, 6–7.

24. Reps 1965, 160–63, 297.

25. Zimilies and Zimiles 1973, III.

26. *Beyond the Neck* 1990, 11.

27. Warner 1978, 132–41.

28. Cutter and Cutter 1880, 163–64 (three brief quotes).

29. K. Stevens 1988, 54–56; Cambridge Historical Commission 1965–77, 4:61, 68.

30. Cambridge Historical Commission 1965–77, 2:30.

31. Garner 1984, 152–56, 206–14.

5. HOUSES (pp. 182–271)

1. Hadfield 1980, 20–31; Fleming and Halderman 1982, 124 (unsourced Henry James quote).

2. Gowans 1962, 13, 18, 23; Candee 1992, 7 (quoting Washington); Glass 1986, 73, 144.

3. Cummings 1979, 1; Candee 1969, pt. 2, 105.

4. Lewis 1994, 95–96 (five brief quotes); Zelinsky 1967, 469, 486.

5. Glass 1986, 3–4, 192–93 (quote).

6. Sizemore 1994, 51, 227 n. 1; Cummings 1979, 22, 26; Miner 1977, 124; Foley 1980, 35; Bibber 1989, 71.

7. Garrison 1991, 200.

8. Hubka 1984, 140, 175, 179–204; Whitcomb 1988, 322 (Bolton quote); Vorse 1942, 89 (Provincetown quote).

9. Ennals 1982, 15; C. Baisly 1989, 185; Vorse 1942, 87–88; Roberts 1993, 32; Brooks 1953, 58; Rhode Island Historical Preservation Commission 1990, 34; Forman 1966, 133; Hubka 1984, 139 (quote); Whitcomb 1988, 330–31; Ragan 1991, 86; Cambridge Historical Commission 1965–77, 2:37; Photograph caption, Concord Antiquarian Museum exhibit, February 1993.

10. Cambridge Historical Commission 1965–77, 5:58; *Boston Globe*, June 17, 1994, 1, 33.

11. Forman 1966, 131 (quote), 255, 104.

12. Garrison 1991, 153 (quote).

13. Cummings 1979, 22–23, 37–38; Wood 1997, 78; Candee 1992, 5–9; Noble 1984, 1: 21.

14. Shurtleff 1939, 87; Williams and Williams 1957, 50; Strickland 1950, 162–69; Foley 1980, 14; *Old House Journal* July–August 1998, 100.

15. Rhode Island Historical Preservation Commission 1982b, 7, 9, 12–13; Heath 1986, 229–30; Downing and Scully 1967, 28–32; Downing 1937, 20–21; Isham and Brown 1895, 17–18; Kelly 1963, 6–7; Noble 1984, 1:21; Rhode Island Historical Preservation Commission 1980a, 35.

16. Garner 1984, 94–95; Rhode Island Historical Preservation Commission 1980a, 9–10; Isham 1967, 7 (fig. 3); Fowler 1936, 278 (quote).

17. Kiely 1993, back cover; Forman 1966, 31–33, 79, 109; Aalen 1997, 150–52; Little 1981, 1, 22, 26, 27 (two quotes).

18. Brooks 1953, xv, 2, 40–45; Worthen 1976, 14–15; Massachusetts Historical Commission 1987, 202; Hubka 1984, 48; M. C. Coolidge 1948, 28; Gannett 1990, 10; Annin 1964, 5 (Richmond quote).

19. Baker 1980, 35 (quote); Deetz 1977, 194.

20. Thoreau 1965, 37–38 (quote); Walker 1987, 200–205; Vorse 1942, 91.

21. Young 1983, 43, 218: Kirkpatrick 1975, 2:207–208; Robinson 1976, 106–107; *Appalachian Trail Guide* 1978, 125; Sternagle and Cummings 1985, 234–35; 531; J. Coolidge 1942, 39.

22. Cummings 1979, 6–7, 23–24; Hubka 1984, 165, 89–90.

23. Schuler 1988, 5; Kelly 1963, 16; Cummings 1979, 33, 17; Mercer 1975, 71.

24. Eberlein 1928, 48–53 (photos); Simpson 1990, 16 (brief quote); Cummings 1979, 32–33; Rhode Island Historical Preservation Commission 1982a, 13.

25. Cousins and Riley 1919, 26; Massachusetts Historical Commission 1985, 205; Lunt 1976, 448; Lancaster 1972, 41–42; Forman 1966, 202, 134.

26. Massachusetts Historical Commission 1982a, 123; Hubka 1984, 35; Lancaster 1972, 25–26, 38, 44; Bedford Historical Society, architectural history handout, n.d.; Cummings 1979, 32.

27. Schuler 1988, 17; Connally 1960, 50; Scott 1985, 67.

28. Eberlein 1928, 48; Schwartz 1983, 31–32; Gowans 1962, 10–13; Sanchis 1977, 23–25; Schuler 1988, 43.

29. *Dictionary of Americanisms*; Jennings 1933, 111–12; Dort 1935, 85; Nutting 1923, 57; Gowans 1986, 140 (quote), 155; Schuler 1988, 5.

30. McAlester and McAlester 1984, 78; Candee 1992, 9.

31. Garvan 1951, fig. 49, 53, 55a; Massachusetts Historical Commission 1984, 150; Delue 1925, 208, 158 (photos); Candee 1992, 11; McAlester and McAlester 1984, 144–45; Massachusetts Historical Commission 1987, 176–77.

32. Candee 1992, 9, 68; Hubka 1984, 36–37; McAlester and McAlester 1984, 80; Noble 1984, 1:25–26.

33. Schuler 1982, 5; Dwight 1969, 4:50–51.

34. Garvan 1951, 116, passim; Cummings 1958, 21; Johnson 1988, 480; Jager and Jager 1976, 476–78; Hubka 1984, 109, 206; Lancaster 1972, xvi; Schuler 1982, 44.

35. Connally 1960, 51, passim; Schuler 1982, 141, 15, and various plans; Rhode Island Historical Preservation Commission 1982a, 54, 12–13, 60.

36. Connally 1960, 54; *Dictionary of American Regional English*.

37. Connally 1960, 51, 53; Massachusetts Historical Commission 1984, 158; Baisly 1989, 21–22; Schuler 1982, 13–15, 21, 78–79; Massachusetts Historical Commission 1982b, 136.

38. Mullins 1987, 145 (quote), 143–60 (this book consists of republished selections from the White Pine Series of Architectural Monographs); Schuler 1982, 14; Thoreau 1962, 2:1358; Jewett and Jewett 1946, 198–99; Massachusetts Historical

Commission 1985, 205; M. C. Coolidge 1948, opp. 44; McGowan and Miller 1996, 92–93.

39. Nash 1975, 108 n. 1 (quote), 109; C. Coolidge 1929, 7 (quote); Baisly 1989, 133; Ennals 1982, 5–21.

40. Connally 1960, 56; Schuler 1982, 100–101; Hubka 1984, 111; A. M. Forbes 1995, 51; Johnson 1988, 480; Massachusetts Historical Commission 1982b, 151.

41. Schuler 1982, 16, 129–30.

42. Foley 1980, 141; Bibber 1989, 116–19; Noble 1984, 1:104–106; Schuler 1982, 63; Schuler 1988, 43.

43. Noble 1984, 1:104 (quote); Bracz 1983, 74–75; 157 nn. 37–39, 195 (fig. 26).

44. Kniffen 1965, 558–59; Cunningham and Warner, 1984, 190; Stanchiw and Small 1989, 140; Walker 1981, 78–79.

45. Williams and Williams 1957, 81–84.

46. Massachusetts Historical Commission 1987, 175; Bibber 1989, 117.

47. Stanchiw and Small 1989; Small 1988; Garrison 1991, 159.

48. Stanchiw and Small 1989, 139 n. 3, 142; Small 1988; Garrison 1991, 159–61; McGowan and Miller 1996, 197, 112–13, 100–101.

49. Williams and Williams 1957, 148–53 and frontispiece, 152 (quote); Garvin 2001, 96.

50. Wood 1997, 80–83.

51. Lewis 1994, 87; Massachusetts Historical Commission 1982a, 122–23; Massachusetts Historical Commission 1985, 158; Hubka 1984, 35.

52. Downing 1937, 85; Candee 1992, 9; A. M. Forbes 1995, 17–18, 50.

53. McAlester and McAlester 1984, 138–67.

54. A. M. Forbes 1995, 17; Massachusetts Historical Commission 1982a, 126; Candee 1992, 12.

55. Candee 1992, 72, 55–56, 10; *Lexington Minuteman*, June 15, 2000, 1, 13; D. Thompson 1976, 57–58.

56. Bishir 1990, 290–91; McAlester and McAlester 1984, 96; Kniffen 1965, 533 n. 10; Massachusetts Historical Commission 1982a, 123; Rhode Island Historical Preservation Commission 1990, 9–11, 93.

57. Massachusetts Historical Commission 1982a, 126 (two quotes); Massachusetts Historical Commission 1982b, 139.

58. Cambridge Historical Commission 1965–77, 1:66–68; Candee 1992, 11, 57.

59. Massachusetts Historical Commission 1982a, 123, 126; Massachusetts Historical Commission 1982b, 139; Massachusetts Historical Commission 1985, 207, 210–11; Massachusetts Historical Commission 1984, 153, 159.

60. Candee 1992, 10–11; 61–62; Hubka 1984, 34–38, 37 (quote); Johnson 1988, 480.

61. Bibber 1989, 59–60; Noble 1984, 1:109.

62. Lewis 1994, 107 (quote), 96–98, 98–99 (two brief quotes); *Preservation* September–October 1998, 100.

63. Kniffen 1965, 559.

64. Johnson 1988, 482–83, 182, 124; Foley 1980, 136 (photo); Bibber 1989, 37, 45 n. 2; McAlester and McAlester 1984, 92–93.

65. Massachusetts Historical Commission 1985, 166; Visser 1997, 74 (photo); Bibber 1989, 41.

66. Hubka 1984, 20–22; Massachusetts Historical Commission 1982b, 273; Noble 1984, 2:65–66; Zelinsky 1958, 543, 549.

67. Hubka 1984, 13–14, 23–24, 131–33; Garrison 1991, 138; Zelinsky 1958, 544.

68. Hubka 1984, 121, 55, 114.

69. Rhode Island Historical Preservation Commission 1980b, 19; Candee 1992, 13, 149–50; Massachusetts Historical Commission 1982b, 157; Elliot 1994; Cambridge Historical Commission 1965–77, 5:78; Rhode Island Historical Preservation Commission 1982b, 27–28.

70. Massachusetts Historical Commission 1982b, 157; Cambridge Historical Commission 1965–77, 5:76.

71. Krim 1970, 48 (map) reprinted in Rooney 1982, 75; Cambridge Historical Commission 1965–77, 5:76 (quote).

72. Cambridge Historical Commission 1965–77, 1: 69 (quote); Massachusetts Historical Commission 1982a, 141; Tucci 1978, 120.

73. Tucci 1978, 101–30.

74. Milot 1993, 12–13; Prof. John R. Stilgoe similarly alluded to a French Canadian origin in lecture (VES 107), fall 1991, so the theory is more generally known; Marsan 1981, 266–67; Cambridge Historical Commission 1965–77, 5:78; Elliot 1994; Tucci 1978, 120; Brault 1986, 154 (quote), 185–88.

75. Kirkpatrick 1975, 2:176, 182 (quote), 183.

76. Joseph H. Lenney, personal communication.

77. Brault 1986, 1; Official Montreal tourist brochure, c. 1999; Ennals 1982, 19 (quote); Cambridge Historical Commission 1965–77, 5:78.

78. Wood 1997, 137, 145 (illustrations); Gowans 1992, 68–69.

79. Massachusetts Historical Commission 1982a, 123; Massachusetts Historical Commission 1982b, 137; Tolles 1979, 97–98, 104, 109; Congdon 1963, 42.

80. O'Gorman 1977, 113–14; Massachusetts Historical Commission 1982a, 127; Von Hoffman 1994, 6, 8–10.

81. Candee 1992, 77, 8; Myers 1974, 19, 215, 220; D. Thompson 1988, 57–58; Downing 1937, 132, 235, 327–29, 430; Longsworth 1990, 8–9, 15; McCallum 1996, 141, 145.

82. Miller 1983, 11, 63–65.

83. Ibid., 7 (quote), 12–13; 128–29.

84. Ibid., 48–49, 124–27; Massachusetts Historical Commission 1984, 169, 177.

85. Benes 1982, 1:30; Williams and Williams 1957, 65, 136–37; see *Dictionary of American Regional English*, "funeral door."

86. Miller 1983, 18, 28–29; Cummings 1979, 13, 115; Massachusetts Historical Commission 1984, 156, 161; Brooks 1953, 49–51, 9; Kelly 1963, 64.

87. McGowan and Miller 1996, 199–200, 42.

88. Congdon 1946, 57–58, 217; Zimilies and Zimilies 1973, 384; Hodgson 1965, 233; Johnson 1988, 394.

89. Bibber 1989, 66, 107, 188; Howells 1941, 202–204; Rhode Island Historical Preservation Commission 1990, 72; Massachusetts Historical Commission 1984, 168.

90. Anderson 1976, 121.

91. Benes 1978, 6, 12–13.

92. Congdon 1946, 8–15; Williams and Williams 1957, 77.

93. Schwartz 1983, 121; Baisly 1989, 89.

94. Massachusetts Historical Commission 1982a, 129; Kennedy 1989, 204, 10.

95. Hamlin 1944, 177; Massachusetts Historical Commission 1982a, 129; Federal Writers' Project 1973, 66; Bibber 1989, 34, 41 (quote), 44 (quote), 59–61; Massachusetts Historical Commission 1982b, 142.

96. Massachusetts Historical Commission 1982a, 10, 178; Massachusetts Historical Commission 1984, 156, 248, 169; Hitchcock 1968, 141; Zimilies and Zimilies 1973, 112, 141, 176; Rhode Island Historical Preservation Commission 1981, passim Massachusetts Historical Commission 1982b, 145, 222–24. Hambourg 1988, plate 176.

97. Johnson 1988, 260, 127, 227, 79, 19; Swan 1980, 369.

98. Stevens 1891, 24; Swift 1977, 152, 133–34.

99. Benes 1979a, 59–60; L. K. Brown 1968, 30; Mansur 1974, 114; Wharton 1974, 64; Benes 1978, 9–11.

100. Kurath 1939–43, map 249; Kurath 1973, 36; Williams, Kellogg, and Lavigne 1987, 83; Lewis 1994, 108 n. 1.

101. Glass 1986, 40, 91–93 (quote).

102. Scott 1985, 140; Schuler 1982, 12; Jorgensen 1978, 310–11; E. A. Wright 1977–81, 2:234; ibid., 1:423; Dwight 1969, 3:50; Morrison 1952, 32; Kelly 1963, 84–85. Cummings 1979, 143; Downing 1967, 161; Isham and Brown 1895, 78–80; Visser 1997, 30; Hubka 1984, 143; Gowans 1992, 52.

103. Lewis 1994, 91 (quote).

104. Johnson 1988, 405; Congdon 1946, 63, 67; Noble 1984, 1:30. Noble discusses these economics with respect to the Hudson Valley; I have read a similar argument advanced for a brick house in northern New England where distance from the nearest sawmill was the main factor.; Whitcomb 1988, 203, 354–55; Brooks 1953, 57 (quote), 76.

105. Bibber 1989, 83–84; Sears, Roebuck and Co. 1969, 589.

106. Gowans 1992, 140.

107. Cummings 1979, 102–107, 115, 240.

108. Hubka 1984, 41 (quote), 54–55; Visser 1997, 14.

109. Cummings 1979, 115–16, 99; Kelly 1963, 50–51.

110. Lewandoski 1985, 104.

III. Candee, 1969, part 3, 39–50, 50 (quote); Massachusetts Historical Commission 1987, 165; Weiss 1987, 43–49.

112. Massachusetts Historical Commission 1987, 164. Cummings 1979, 90–91; Lewandoski 1985, 104–21; Massachusetts Historical Commission 1984, 154.

113. Congdon 1946, 19; Johnson 1988, 222; Kelly 1963, 131; Noble 1984, 1:36–37; Benes 1982, 1: xvi, 39.

114. Lewandoski 1985, 121; Williams, Kellogg, and Lavigne 1987, 83; Konrad 1982, 32–33; Visser 1997, 96–97.

115. Massachusetts Historical Commission 1984, 156; Garrison 1991, 172–73; Congdon 1946, 89–90; Johnson 1988, 405; Schwartz 1983, 32.

116. Garrison 1991, 168–74.

117. Wood 1997, 124; Hubka 1984, 134–36.

118. Whitcomb 1988, 307.

119. Bibber 1989, 51.

120. Sammarco 1997, 58; A. M. Forbes 1995, 29, 35.

121. Massachusetts Historical Commission 1982a, 152; Massachusetts Historical Commission 1982b, 149, 152.

122. Wilkie and Tager 1991, 23.

123. Cushing 1990, 156–58, 157 (quote); Cousins and Riley 1919, 236–37, vii.

124. Randall 1982, 8, 16; *Best Read Guide: Martha's Vineyard*, June 11–27, 1999, 24–25.

125. A. K. Teele 1887, 320.

126. J. Coolidge 1942, 209.

6. GRAVESTONES (pp. 272–92)

1. Deetz 1977, 66.

2. Deetz 1977, 88; Duval and Rigby 1978, vii; H. M. Forbes 1927, 21–22, 50; Benes 1977, 5, 38, 41, 225 n. 9; Speight n.d., 4–5; Federal Writers' Project 1983, 407.

3. Hudson 1968, 586 (quote), 572–73; Forman 1966, 57; Whitcomb 1988, 122–23, 147; W. D. Howells 1969, 64–65.

4. Burgess 1963, 20, 107–108, 116–17; Bailey 1987, 27; Willsher and Hunter 1978, 2; Speight n.d., 4–5.

5. Ludwig 1966, 264; Bailey 1987, 52, 120, opp. 104; Cummings 1979, 127–30.

6. Tashjian and Tashjian 1974, 7.

7. Benes 1977, 42–42.

8. Deetz 1977, 115.

9. Ludwig 1966, map 2; H. M. Forbes 1927, 99; Rhode Island Historical Preservation Commission 1982b, 47, 60; H. M. Forbes 1927, 9–10.

10. Duval and Rigby 1978, 127–29.

11. Benes 1977, 195.

12. Ibid., 42.

13. Whitcomb 1988, 148; Robinson 1976, 48–49; Woodward and Sanderson 1986, 41–42; Rhode Island Historical Preservation Commission 1982a, 53, 3; Ramsdell 1901, 361; Gaustad 1976, 60.

14. Compare Ruth Wells stone 1744 in H. M. Forbes 1927, opp. 98, 101.

15. Tashjian and Tashjian 1974, 193–94.

16. Benes 1977, 49–50.

17. Ibid., 141.

18. Watters 1977, 2, 5; H. M. Forbes 1927, 80–85.

19. Tashjian and Tashjian 1974, 201–203; Ludwig 1966, 373.

20. H. M. Forbes 1927, 78.

21. Hudson 1968, 289–90.

22. Ludwig 1966, 373.

23. Deetz 1977, 82; Benes 1977, 47–48.

24. Garrison 1991, 153; Benes 1975, 63.

25. Cushing 1990, 62 (quoting Alonzo Lewis 1852).

26. Douglas 1977, 208–13; Linden-Ward 1989, 110, 120, 127.

27. Anderson 1976, 134–36.

28. Jackson and Vergara 1989, ix.

29. Anonymous 1908, 308; Linden-Ward 1989, 234; Curl 1980, 43.

30. Burgess 1963, 109, 125.

31. Curl 1980, 24.

32. Curl 1982, 83.

33. Richardson and Scarry 1990, 9–23; Curl 1982, 226.

34. Linden-Ward 1989, 219; Jackson and Vergara 1989, 90.

7. EPILOGUE (pp. 293–308)

1. Carver 1987, 10–16.

2. Benes 1977, 216 n. 27.

3. Hubka 1984, 20.

4. Kniffen 1965, 559.

5. Deetz and Dethlefsen 1967, 32 (quote), 33; Benes 1977, 166–67.

6. Deetz 1978, 50–51; Klyza and Trombulak 1999, 70.

7. Gaustad 1976.

8. Ibid., 128.

9. Paullin 1932, plate 138; Leblanc 1969, 16.

10. For a similar observation on foundry migration, see Stilgoe 1998, 6.

11. Chapelle 1951, passim.
12. *Boston Globe*, September 2, 1992, 35.
13. Arlott 1975, 62–70; 1072–76.
14. Harmon 1985, 109–24.
15. Deetz 1967, 30–33.

BIBLIOGRAPHY

Aalen, F. H. A., Kevin Whalen, and Matthew Stout, eds. 1997. *Atlas of the Irish Rural Landscape*. Toronto: University of Toronto Press.

Allport, Susan. 1990. *Sermons in Stone: The Stone Walls of New England and New York*. New York, Norton.

American Coast Pilot. 1806. Compiled by Lawrence Furlong. *American Coast Pilot*, 5th ed. Newburyport, Mass.: Edmund M. Blunt.

————. 1854. *American Coast Pilot*, 17th ed. New York: E. and G. W. Blunt.

Anderson, Robert C. 1976. *Directions of a Town: A History of Harvard, Massachusetts*. Harvard, Mass.: Harvard Common Press.

Andrews, Charles M. 1889. *The River Towns of Connecticut: A Study of Wethersfield, Hartford, and Windsor*. Baltimore: Johns Hopkins University Press.

Andrews, J. H. 1985. *Plantation Acres: A Historical Study of the Irish Land Surveyor and His Maps*. Belfast, U.K.: Ulster Historical Foundation.

Annin, Katherine Huntington. 1964. *Richmond, Massachusetts: The Story of a Berkshire Town and its People, 1765–1965*. Richmond, Mass.: Richmond Civic Association.

Anonymous. 1908. "An Exact Copy of the Famous Scipio Tomb." *Park and Cemetery and Landscape Gardening* 18, no. 2:308.

Appalachian Trail Guide. Massachusetts and Connecticut. 1978. 5th ed. Harpers Ferry, Va.: Appalachian Trail Conference.

Archeological Investigations of Minute Man National Historic Park 1990. 2 vols. Vol. 1, *Farmers and Artisans of the Historical Period*, edited by Alan T. Syneki; vol. 2, *An Estimation Approach to Prehistoric Sites*, edited by Duncan Ritchie et al. Cultural Resources Management Studies. Nos. 22–23. Boston, Mass.: U.S. Department of the Interior.

Arlott, John. 1975. *The Oxford Companion to World Sports and Games*. London: Oxford University Press.

Attwood, Stanley B. 1946. *The Length and Breadth of Maine*. Augusta, Maine: Kennebec Journal.

Bailey, Brian. 1987. *Churchyards of England and Wales*. London: Robert Hale.

Baisly, Clair. 1989. *Cape Cod Architecture*. Orleans, Mass.: Parnassus Imprints.

Baker, Vernon. 1980. "Archaeological Visibility of Afro-American Culture: An Example from Black Lucy's Garden, Andover, Massachusetts." *Archeological Perspectives on Ethnicity in America: Afro-American And Asian American Culture History*, edited by Robert L. Schuyler, 29–37. Farmingdale, N.Y.: Baywood Publishing.

Bancroft, George. 1875. *History of the United States from the Discovery of the American Continent*. Boston: Little, Brown.

Banks, Charles Edward. 1966. *The History of Martha's Vineyard, Dukes County, Massachusetts* (in 3 volumes; not continuously paginated). Edgartown, Mass.: Dukes Country Historical Society.

Barber, John Warner. 1839. *Massachusetts Historical Collections . . . Relating to the History and Antiquities of Every Town in Massachusetts. . . .* Worcester, Mass.: Dorr, Howland.

Barry, William A. 1983. *History of Framingham, Massachusetts.* Bowie, Md.: Heritage Books. Original edition, 1847.

Bauer, L. A. 1902. *United States Magnetic Declination Tables and Isogonic Charts for 1902 and Principal Facts Relating to the Earth's Magnetism.* United States Coast and Geodetic Survey. Washington, D.C.: GPO.

Bebbington, Gillian. 1972. *London Street Names.* London: B. T. Batsford.

Bell, Michael. 1986. *The Face of Connecticut: People, Geology, and the Land.* Bulletin 110. Hartford, Conn.: State Geological and Natural History Survey of Connecticut.

Benes, Peter. 1975. "Additional Light on Wooden Gravemarkers." Essex Institute Historical Collections 3:53–64.

Benes, Peter. 1977. *The Masks of Orthodoxy: Folk Gravestone Carving in Plymouth County, Massachusetts, 1689–1805.* Amherst: University of Massachusetts Press.

———. 1978. "The Templeton 'Run' and the Pomfret 'Cluster': Patterns of Diffusion in Rural New England Meetinghouse Architecture, 1647–1822." *Old-Time New England* 68, nos. 3–4:1–21.

———. 1979a. "Sky Colors and Scattered Clouds: The Decorative and Architectural Painting of New England Meeting Houses, 1738–1834." In *New England Meeting House and Church: 1630–1850,* edited by Peter Benes, 51–69. Dublin Seminar for New England Folklife, *Annual Proceedings.* Boston: Boston University.

———. 1979b. "Twin-Porch Versus Single-Porch: Two Examples of Cluster Diffusion in Rural Meetinghouse Architecture." *Old-Time New England* 69, nos. 3–4: 44–68.

———. 1982. *Two Towns: Concord and Wethersfield: A Comparative Exhibition of Regional Culture, 1635–1850.* Concord, Mass.: Concord Antiquarian Museum.

Best Read Guide: Martha's Vineyard. June 11–27, 1999. Free bi-weekly visitor publication.

Beyond the Neck: The Architecture and Development of Somerville, Massachusetts. 1990. Updated ed. Somerville, Mass.: City of Somerville.

Bibber, Joyce K. 1989. *A Home for Everyman: The Greek Revival and Maine Domestic Architecture.* Greater Portland Landmarks. Lanham, Md.: AASLH Library.

[Bigelow, Edwin L]. N.d. *1761–1976–1776.* Manchester, Vt.: Manchester Historical Society.

Bishir, Catherine W. 1990. *North Carolina Architecture.* Chapel Hill: University of North Carolina Press.

Boston City Planning Board. 1942. *Report on a Geodetic Survey for Boston, 1935–1941.* Boston: City Planning Board.

Boston Landmarks Commission. 1983. *The South End: District Study Committee Report.* Boston: Boston Landmarks Commission.

Boston, Massachusetts, City of. 1879. *Report of the Joint Standing Committee on Ordinances on Nomenclature of Streets.* Document 119. Boston: City of Boston.

———. Various years. Street Laying-Out Department. *Annual Report.* Boston: City of Boston.

Boston Social Register. 1899/1915/1930. New York: Social Register Association.

Boston: The Official Bicentennial Guidebook. 1975. New York: E. P. Dutton.

Bracz, Nancy J. 1983. "The Evolution of the Greek Revival Style in the Domestic Architecture of Ohio's Western Reserve, 1820–1860." Ph.D. diss. University of Ohio.

Bradford, William. 1912. *History of Plymouth Plantation, 1620–1647.* 2 vols. Boston: Massachusetts Historical Society.

Brault, Gerard J. 1986. *The French-Canadian Heritage.* Hanover, N.H.: University Press of New England.

Breed, Charles Blaney, and George L. Hosmer. 1908. *The Principles and Practice of Surveying.* 2 vols. 3d ed. New York: John Wiley and Sons.

————. 1925. *The Principles and Practice of Surveying.* 2 vols. 5th ed. New York: John Wiley and Sons.

————. 1958. *The Principles and Practice of Surveying.* 2 vols. 9th ed. New York: John Wiley and Sons.

Bridge, Ruth, ed. 1977. *The Challenge of Change: Three Centuries of Enfield, Connecticut History.* Enfield Historical Society. Canaan, N.H.: Phoenix.

British Landscapes Through Maps. 1960–79. Sheffield, Eng.: The Geographical Association.

Brooks, Robert R. R., ed. 1953. *Williamstown: The First Two Hundred Years, 1753–1953.* (William H. Pierson advised on architectural matters.) Williamstown, Mass.: McClelland.

Brown, Curtis M. 1969. *Boundary Control and Legal Principles.* 2d ed. New York: John Wiley and Sons.

Brown, Curtis M., Walter G. Robillard, and Donald A. Wilson. 1981. *Evidence and Procedures for Boundary Location.* 2d ed. New York: John Wiley and Sons.

Brown, Louise K. 1968. *Wilderness Town: The Story of Bedford, Massachusetts.* Privately printed [Minuteman Publications].

————. 1974. *A Revolutionary Town.* Canaan, N.H.: Phoenix.

Browne, George Waldo. 1921. *The History of Hillsborough, New Hampshire, 1735–1921.* Manchester N.H.: John B. Clarke.

Bunting, Bainbridge. 1985. *Harvard: An Architectural History.* Cambridge: Harvard University Press.

Burgess, Frederick. 1963. *English Churchyard Memorials.* London: Lutterworth.

Cambridge, City of. 1896. *The Register Book of the Lands and Houses in the "New Towne" and the Town of Cambridge . . . Generally Called "The Proprietors Records."* Cambridge, Mass.

————. 1901. *The Records of the Town of Cambridge (Formerly Newtowne) Massachusetts, 1630–1703.* Cambridge, Mass.

Cambridge Historical Commission. 1965–77. *Survey of Architectural History in Cambridge.* 5 vols. Cambridge: M.I.T. Press.

Cameron, Kenneth Walter. 1963. *Transcendental Climate: New Resources for the Study of Emerson, Thoreau and their Contemporaries.* Vol. 2. Hartford [Conn.?]: Transcendental Books.

Canada Gazetteer Atlas. 1980. [Montreal]: Macmillan of Canada.

Candee, Richard M. 1969. "A Documentary History of Plymouth County Architecture, 1620–1700." *Old-Time New England* 59, no. 3:59–71; 59, no. 4:105–111; 60, no. 2:37–53.

———. 1980. "Land Surveys of William and John Godsoe of Kittery, Maine: 1689–1769." In *New England Prospect: Maps, Place Names, and the Historical Landscape*, edited by Peter Benes, 9–46. Dublin Seminar for New England Folklife, *Annual Proceedings*.

———. 1992. *Building Portsmouth: The Neighborhoods and Architecture of New Hampshire's Oldest City*. Portsmouth, N.H.: Portsmouth Advocates.

Carrott, Richard G. 1978. *The Egyptian Revival: Its Sources, Monuments, and Meaning, 1808–1858*. Berkeley and Los Angeles: University of California Press.

Carver, Craig. 1987. *American Regional Dialects: A Word Geography*. Ann Arbor: University of Michigan Press.

Caulfield, Ernest. 1951–78. "Connecticut Gravestones." Connecticut Historical Society Bulletin 16–43 (published in 15 installments).

Chadbourne, Ava Harriet. 1955. *Maine Place Names and the Peopling of its Towns*. Portland, Maine: Bond Wheelwright.

Chamberlain, Allen. 1925. *Beacon Hill: Its Ancient Pastures and Early Mansions*. Boston: Houghton Mifflin.

Chapelle, Howard I. 1951. *American Small Sailing Craft: Their Design, Development, and Construction*. New York: Norton.

Chase, Theodore, and Laurel K. Gabel. 1990. *Gravestone Chronicles: Some Eighteenth-Century New England Carvers and Their Work*. Boston: New England Historic Genealogical Society.

Clayton, C. F., and L. J. Peet. 1933. *Land Utilization as a Basis of Rural Economic Organization*. Bulletin 357. University of Vermont Agricultural Experiment Station. Burlington, Vt.

Committee on Old Homes of Know Your Town and the South Hadley Historical Society. 1976. *A History of South Hadley's Old Homes Built before 1850*. South Hadley, Mass.: Hadley Printing.

Congdon, Herbert Wheaton. 1946. *Old Vermont Houses*. New York: Knopf.

———. 1963. *Early American Homes for Today: A Treasury of Decorative Details and Restoration Procedures*. Rutland, Vt.: Charles E. Tuttle.

Connally, Ernest Allen. 1960. "The Cape Cod House: An Introductory Study." *Journal of the American Society of Architectural Historians* 29, no. 2: 47–56.

Coolidge, Calvin. 1929. *The Autobiography of Calvin Coolidge*. New York: Cosmopolitan Book.

Coolidge, John. 1942. *Mill and Mansion: A Study of Architecture and Society in Lowell, Massachusetts, 1820–1865*. New York: Columbia University Press.

Coolidge, Mabel Cook. 1948. *The History of Petersham, Massachusetts*. Petersham Historical Society.

Copeland, Melvin T., and Elliott C. Rogers. 1960. *Saga of Cape Ann*. Freeport, Maine: Bond Wheelwright.

Costello, David L. 1975. *The Mohawk Trail, Showing Old Roads and Points of Interest*. N.p.: n.p.

Cousins, Frank, and Phil M. Riley. 1919. *The Colonial Architecture of Salem*. Boston: Little, Brown.

Crosby, Everett U. 1946. *Nantucket in Print*. Nantucket, Mass.: Tetaukimmo.

Crosby, W. W., and George E. Goodwin. 1928. *Highway Location and Surveying*. Chicago: Gillette.

Cummings, Abbott Lowell. 1958. *Architecture in Early New England*. Sturbridge, Mass.: Old Sturbridge Village.

————. 1979. *The Framed Houses of Massachusetts Bay, 1625–1725*. Cambridge: Harvard University Press.

Cunningham, Janice P., and Elizabeth A. Warner. 1984. *Portrait of a River Town: The History and Architecture of Haddam, Connecticut*. Middletown, Conn.: Greater Middletown Preservation Trust.

Curl, James Stevens. 1980. *A Celebration of Death*. London: Constable.

————. 1982. *The Egyptian Revival*. London: Allen and Unwin.

Cushing, Elizabeth Hope. 1990. *The Lynn Album: A Pictorial History*. Norfolk, Va.: Donning.

Cutter, Benjamin, and William R. Cutter. 1880. *History of the Town of Arlington, Massachusetts . . . 1635–1879*. Boston: David Clapp and Son.

Darby, Marge, et al. 1994. *A Guide to Nashaway: North Central Massachusetts*. Harvard, Mass.: New Guide Group.

Davis, Earl N., Jr., and Mary O. Davis. N.d. *Rochester Remembers, 1781–1981/Rochester, Vermont/A Collection of Photographs: Some Old—Some New*. Randolph, Vt.: Herald Printery.

Davis, William T. 1883. *Ancient Landmarks of Plymouth*. Boston: A. Williams.

Deetz, James. 1978. "Late Man in North America: Archaeology of European Americans." In *Historical Archaeology: A Guide to Substantive and Theoretical Contributions*, edited by Robert L. Schuyler, 48–52. Farmingdale, N.Y.: Baywood Publishing Co.

————. 1977. *In Small Things Forgotten: The Archaeology of Early American Life*. Garden City, N.Y.: Anchor Press/Doubleday.

————. 1967. *Invitation to Archaeology*. Garden City, N.Y.: Natural History Press.

Deetz, James, and Edwin S. Dethlefsen. 1967. "Death's Head, Cherub, Urn and Willow." *Natural History* (March): 28–37.

Delue, Willard. 1925. *The Story of Walpole, 1724–1924*. Norwood, Mass.: Ambrose.

Dorchester Antiquarian and Historical Society. 1859. *History of the Town of Dorchester, Massachusetts*. Boston: Ebenezer Clapp, Jr.

Dort, Wakefield, ed. 1935. *Westport in Connecticut's History*. Bridgeport, Conn.: Warner Bros.

Douglas, Ann. 1977. *The Feminization of American Culture*. New York: Knopf.

Downing, Antoinette Forrester. 1937. *Early Homes of Rhode Island*. Richmond, VA.: Garrett and Massie.

Downing, Antoinette F., and Vincent Scully. 1952. *The Architectural Heritage of Newport, Rhode Island, 1640–1915*. New York: Clarkson N. Potter.

————. 1967. *The Architectural Heritage of Newport, Rhode Island, 1640–1915*. 2d rev. ed. New York: Clarkson N. Potter.

Duval, Francis Y., and Ivan B. Rigby. 1978. *Early American Gravestone Art in Photographs*. New York: Dover.

Dwight, Timothy. 1969. *Travels in New England and New York*. Edited by Barbara Miller Solomon. 4 vols. Cambridge: Harvard University Press.

Dyer, E. Otis. 1994. *Swamp Yankee*. Taunton, Mass.: William S. Sullwood.

Eberlein, Harold Donaldson. 1928. *Manor Houses and Historic Homes of Long Island and Staten Island*. Philadelphia: J. B. Lippincott.

Elliot, Lynn. 1994. "Massachusetts Triple-Deckers." *Old-House Journal*, back cover.

Emery, Samuel Hopkins. 1893. *History of Taunton, Massachusetts*. Syracuse, N.Y.: D. Mason.

Ennals, Peter. 1982. "The Yankee Origins of Bluenose Vernacular Architecture." *American Review of Canadian Studies* 12, no. 2: 5–21.

Ernst, Joseph W. 1979. *With Compass and Chain: Land Surveyors in the Old Northwest, 1785–1866*. New York: Arno.

Estopinal, Stephen V. 1989. *A Guide to Understanding Land Surveys*. Eau Claire, Wisc.: Professional Education Systems.

————. 1993. *A Guide to Understanding Land Surveys*. 2d ed. New York: John Wiley and Sons.

Evans, Nancy Goyne. 1996. *American Windsor Chairs*. New York: Hudson Hills.

Ewell, John Louis. 1904. *The Story of Byfield, a New England Parish*. Boston: George E. Littlefield.

Fagan, Brian M. 1978. *Archaeology: A Brief Introduction*. Boston: Little, Brown.

Fairbanks, Jonathan L., curator. 1982. *New England Begins: The Seventeenth Century*. 3 vols. Boston: Museum of Fine Arts.

Federal Writers' Project. 1937. *Maine: A Guide "Down East."* Boston: Houghton Mifflin.

————. 1938a. *Connecticut: A Guide to Its Roads, Lore, and People*. Boston: Houghton Mifflin.

————. 1938b. *New Hampshire: A Guide to the Granite State*. Boston: Houghton Mifflin.

————. 1973. *Vermont: A Guide to the Green Mountain State*. St. Claire Shores: Mich.: Somerset. Original edition, 1937.

————. 1983. *The WPA Guide to Massachusetts*. New York: Pantheon Books. Original edition, 1937.

Field, John. 1989. *English Field-names: A Dictionary*. Gloucester, U.K.: Alan Sutton. Original edition, 1972.

Fischer, David Hackett. 1989. *Albion's Seed: Four British Folkways in America*. New York: Oxford University Press.

————. 1994. *Paul Revere's Ride*. New York: Oxford University Press.

Fite, Emerson D., and Archibald Freeman. 1926. *A Book of Old Maps: Delineating American History from the Earliest Days Down to the Close of the Revolutionary War*. Cambridge: Harvard University Press.

Fleming, Ronald Lee, and Lauri A. Halderman. 1982. *On Common Ground: Caring for Shared Land From Town Common to Urban Park*. Harvard, Mass.: Harvard Common Press.

Fogelberg, John E. 1976. *Burlington : Part of a Greater Chronicle*. Burlington, Mass.: Burlington Historical Commission.

Foley, Mary Mix. 1980. *The American House*. New York: Harper and Row.

Forbes, Anne McCarthy. 1995. *Narrative Histories of Concord and West Concord.* [Survey of Historical and Architectural Resources, Concord, Massachusetts.] Concord Historical Commission.

Forbes, Esther. 1942. *Paul Revere and the World He Lived In.* Boston: Houghton Mifflin.

Forbes, Henriette Merrifield. 1927. *Gravestones of Early New England and The Men Who Made Them, 1653–1800.* Boston: Houghton Mifflin.

Forman, Henry Chandlee (misspelled as Foreman on title page). 1966. *Early Nantucket and its Whale Houses.* New York: Hastings House.

Fowler, A. N. 1936. "Rhode Island Mill Towns" (part of The Monograph Series: Records of Early American Architecture). *Pencil Points.*

Frizzel, Martin McD., et al. 1955. *Second History of Charlestown, N.H., the Old Number Four.* Town of Charlestown.

Fuller, Wayne E. 1964. *RFD: The Changing Face of Rural America.* Bloomington. Indiana University Press.

Gannett, Michael R. 1990. *The Distribution of the Common Land of Cornwall, Connecticut, 1738–1887.* Cornwall, Conn.: Cornwall Historical Society.

Garland, Joseph E. 1978. *Boston's North Shore.* Boston: Little, Brown.

Garner, John S. 1984. *The Model Company Town: Urban Design and Private Enterprise in Nineteenth-Century New England.* Amherst: University of Massachusetts Press.

Garrett, Wilbur E. 1988. *National Geographic Historical Atlas of the United States.* Washington, D.C.: National Geographic Society.

Garrison, J. Ritchie. 1991. *Landscape and Material Life in Franklin County, Massachusetts, 1770–1860.* Knoxville: University of Tennessee Press.

Garvan, Anthony N. B. 1951. *Architecture and Town Planning in Colonial Connecticut.* New Haven: Yale University Press.

Garvin, James L. 1980. "The Range Township in Eighteenth Century New Hampshire." In *New England Prospect. Maps: Place Names, and the Historical Landscape,* edited by Peter Benes, 47–68. Dublin Seminar for New England Folklife, *Annual Proceedings.*

Garvin, James L. 2001. *A Building History of Northern New England.* Hanover: University Press of New England.

Gaustad, Edwin Scott. 1976. *Historical Atlas of Religion in America.* Rev. ed. New York: Harper and Row.

Gelling, Margaret. 1978. *Signposts to the Past: Place-names and the History of England.* London: J.M. Dent and Sons.

Gifford, William H. 1993. *Colebrook: "A Place Up Back of New Hampshire."* Colebrook, N.H.: News and Sentinel.

Gilman, Stanwood, C., and Margaret Cook Gilman. 1966. *Land of the Kennebec "Ye Great and Beneficial River," 1604–1965.* Boston: Branden.

Glass, Joseph W. 1986. *The Pennsylvania Culture Region: A View from the Barn.* Ann Arbor, Mich.: UMI Research Press.

Glassie, Henry. 1975. *Folk Housing in Middle Virginia: A Structural Analysis of Historic Artifacts.* Knoxville: University of Tennessee Press.

Goddard, Ives. 1978. "Eastern Algonquian Languages." In *Handbook of North Amer-*

ican Indians: Northeast, edited by William C. Sturtevant, 15: 70–77. Washington, D.C.: Smithsonian Institution.

Goodall, Brian. 1987. *The Penguin Dictionary of Human Geography*. Harmondsworth, Eng.: Penguin Books.

Goodspeed, Alexander McL. 1904. "Benjamin Crane and Old Dartmouth Surveys." *Old Dartmouth Historical Sketches*, no. 8. New Bedford, Mass.: Old Dartmouth Historical Society.

[Gould, Augustus A., and Frederic Kidder]. 1852. *The History of New Ipswich, from its First Grant in MDCCXXXVI to the Present Time*. Boston: Gould and Lincoln.

Gould, Nicholas. 1978. *Looking at Place Names*. Havant, Eng.: Kenneth Mason.

Gowans, Alan. 1962. "New England Architecture in Nova Scotia." *Art Quarterly* 25, no. 1:6–33.

———. 1986. *The Comfortable House: North American Suburban Architecture, 1890–1930*. Cambridge: MIT Press.

———. 1992. *Styles and Types of North American Architecture: Social Function and Cultural Expression*. New York: IconEditions.

Great Britain. Ordnance Survey. 1992. *Gazetteer of Great Britain: All Names from the 1:50,000 Scale Landranger Map Series*. 3d ed. London: Macmillan.

Greeley, Horace. 1871. *What I Know of Farming*. New York: G. W. Carleton.

Green, Eugene, and William Sachse. 1983. *Names of the Land: Cape Cod, Nantucket, Martha's Vineyard, and the Elizabeth Islands*. Chester, Conn.: Globe Pequot.

Griffin, S. G. 1904. *A History of the Town of Keene*. Keene, N.H.: Sentinel.

Hadfield, John, ed. 1980. *Shell Book of English Villages*. London: Joseph.

Hambourg, Serge, et al. 1988. *Mills and Factories of New England*. New York: Abrams.

Hamlin, Talbot. 1944. *Greek Revival Architecture in America*. New York: Oxford University Press.

Harding, Walter. 1982. *The Days of Henry Thoreau: A Biography*. New York: Dover. Original edition, 1962.

Harmon, John E. 1985. "Bowling Regions of America." *Journal of Cultural Geography* 6, no. 1:109–24.

Hartley, E. N. 1990. *Ironworks on the Saugus*. Norman: University of Oklahoma Press. Original edition, c. 1957.

Hartley, L. P. 1990. *The Go-Between*. Chelsea, Mich.: Scarborough House. Original edition, 1953.

Heath, Kingston. 1986. "John Dewey Revisited: Teaching Vernacular Architecture by Doing." (abstract). In *Perspectives in Vernacular Architecture*, 2:229–30. Annapolis, Md.: Vernacular Architecture Forum.

Hendrickson, Robert. 2000. *The Facts on File Dictionary of American Regionalisms*. New York: Facts on File.

Herrick, William A. 1870. *The Powers, Duties, and Liabilities of Town and Parish Officers in Massachusetts*. Boston: Little, Brown.

Hitchcock, Henry-Russell. 1968. *Rhode Island Architecture*. Cambridge: MIT Press. Original edition, 1939.

Hodgman, Edwin R. 1883. *History of the Town of Westford, in the County of Middlesex, Massachusetts, 1659–1883*. Westford, Mass.: Westford Town History Association.

Hodgson, Alice Doan. 1965. *Thanks to the Past: The History of Orford, New Hampshire.* Orford, N.H.: Historical Fact Publications.

Hoskins, Elmore P. 1908. "Some of the Streets of the Town of New Bedford." *Old Dartmouth Historical Sketches*, no. 19. New Bedford, Mass.: Old Dartmouth Historical Society.

Hoskins, W. G. 1955. *The Making of the English Landscape.* London: Hodder and Stoughton.

Hosley, William N., Jr. 1978. "The Rockingham Stonecarvers: Patterns of Stylistic Concentration and Diffusion in the Upper Connecticut Valley, 1790–1817." In *Puritan Gravestone Art*, edited by Peter Benes, *II*:68–89. Dublin Seminar for New England Folklife, *Annual Proceedings*.

Howells, John Mead. 1941. *The Architectural Heritage of the Merrimack: Early Houses and Gardens.* New York: Architectural Book.

Howells, William Dean. 1969. *Suburban Sketches.* Freeport, N.Y.: Books for Libraries.

Hubka, Thomas C. 1984. *Big House, Little House, Back House, Barn: The Connected Farm Buildings of New England.* Hanover: University Press of New England.

Huden, John C. 1962. *Indian Place Names of New England.* New York: Museum of the American Indian.

Hudson, Alfred Sereno. 1968. *History of Sudbury, Massachusetts, 1638–1889* Sudbury, Mass.: Sudbury Press. Original edition, 1889.

Hughes, Arthur H., and Morse S. Allen. 1976. *Connecticut Place Names.* Hartford: Connecticut Historical Society.

Hunt, Elmer Munson. 1970. *New Hampshire Town Names and Whence They Came.* Peterborough, N.H.: Noone House.

Hunt, Thomas F. 1906. *How to Choose a Farm.* New York: MacMillan.

Huntington, James Lincoln. 1949. *Forty Acres: The Story of the Bishop Huntington House.* New York: Hastings House.

Huntoon, Daniel T.V. 1893. *History of the Town of Canton, Norfolk County, Massachusetts.* Cambridge, Mass.: John Wilson and Son.

Hurd, D. Hamilton. 1890. *History of Middlesex County, Massachusetts, with Biographical Sketches of its Pioneers and Prominent Men.* 3 vols. Philadephia; J. W. Lewis.

Isham, Norman Morrison. 1967. *Early American Houses and a Glossary of Colonial Architectural Terms.* New York: Da Capo. Original edition, 1928; glossary 1939.

Isham, Norman M., and Albert F. Brown. 1895. *Early Rhode Island Houses: An Historical and Architectural Study.* Providence, R.I.: Preston and Rounds.

Jackson, Kenneth T., and Camilo J. Vergara. 1989. *Silent Cities: The Evolution of the American Cemetery.* New York: Princeton Architectural Press.

Jager, Ronald, and Grace Jager. [1976?] *Portrait of a Hill Town: A History of Washington, New Hampshire, 1876–1976.* Washington, N.H.: Washington History Committee.

James, Edward. 1988. *The Franks.* Oxford: Blackwell.

Jane, Nancy. 1978. *Bicycle Touring in the Pioneer Valley.* Amherst: University of Massachusetts Press.

Jennings, George Penfield. 1933. *Greens Farms, Connecticut: The Old West Parish of Fairfield.* Greens Farms, Conn.: Modern Books and Crafts.

Jewett, Amos Everett, and Emily Mabel Adams Jewett. 1946. *Rowley, Massachusetts:*

"Mr Ezechi Rogers Plantation," 1639–1850. Rowley, Mass.: Jewett Family of America.

Johnson, Curtis B. 1988. *The Historic Architecture of Rutland County*. Division of Historic Preservation. Montpelier, Vt.: State of Vermont.

Jorgensen, Neil. 1978. *A Sierra Club Naturalist's Guide to Southern New England*. San Francisco: Sierra Club Books.

Judd, Sylvester. 1976. *The History of Hadley, Massachusetts*. Somersworth, N.H.: New Hampshire. Reprint of 1925 ed.

Kelly, J. Frederick. 1963. *The Early Domestic Architecture of Connecticut*. New York: Dover. Original edition, Yale University Press, 1924.

Kennedy, Roger G. 1989. *Greek Revival America*. New York: Stewart, Tabori and Chang.

Kiely, Gayle. 1993. "Whale Houses of Nantucket, Massachusetts." *Old-House Journal* (September–October): back cover.

Kirk, John T. 1972. *American Chairs: Queen Anne and Chippendale*. New York: Knopf.

Kirker, Harold. 1969. *The Architecture of Charles Bulfinch*. Cambridge: Harvard University.

Kirkpatrick, Doris. 1975. *The City and the River* (vol. 1) and *Around the World in Fitchburg* (vol. 2). Fitchburg Historical Society.

Klyza, Christopher McGrory, and Stephen C. Trombulak. 1999. *The Story of Vermont: A Natural and Cultural History*. Hanover: University Press of New England.

Kniffen, Fred. 1965. "Folk Housing: Key to Diffusion." *Annals of the Association of American Geographers* 55, no. 4: 549–77.

Konrad, Victor A. 1982. "Against the Tide: French Canadian Barn Building Traditions in the St. John Valley of Maine." *American Review of Canadian Studies* 12, no. 2: 22–36.

Krim, Arthur J. 1970. "The Three Decker as Urban Architecture in New England." *Monadnock* 44 (map, p. 48).

Kurath, Hans, ed. 1939–43. *Linguistic Atlas of New England*. 3 vols. in 6. Providence: Brown University.

——. 1949. *A Word Geography of the Eastern United States*. Ann Arbor: University of Michigan Press.

——. 1973. *Handbook of the Linguistic Geography of New England*. 2d ed. New York: AMS Press. Original edition, 1939.

Lancaster, Clay. 1972. *The Architecture of Historic Nantucket*. New York: McGraw-Hill.

Lawson, Charles E. 1990. *Surveying Your Land: A Common-Sense Guide to Surveys, Deeds, and Title Searches*. Woodstock, Vt.: Countryman.

Leblanc, Robert G. 1969. *Location of Manufacturing in New England in the 19th Century*. Geography Publications at Dartmouth No. 7. Hanover: Dartmouth College.

Lee, Deane, ed. 1967. *Conway, 1767–1967*. Town of Conway, Mass.

Leech, D. D. T., comp. 1859. *List of the Post Offices in the United States, with the Names of Postmasters . . .* Washington, D.C.: John C. Rives.

Lewandoski, Jan Leo. 1985. "Plank Framed House in Northwestern Vermont." *Vermont History* 53, no. 2: 104–21.

Lewis, Peirce. 1994. "Common Houses, Cultural Spoor." In *Re-Reading Cultural Geography*, edited by Kenneth E. Foote, chap. 7. Austin: University of Texas Press.

Lillywhite, Bryant. 1972. *London Signs: A Reference Book of London Signs from Earliest Time to About the Mid-Nineteenth Century*. London: Allen and Unwin.

Linden-Ward, Barbara. 1989. *Silent City on a Hill: Landscapes of Memory and Boston's Mount Auburn Cemetery*. Columbus: Ohio State University Press.

Little, Elizabeth A. 1981. *Historic Indian Houses of Nantucket*. Nantucket Algonquian Studies No. 4. Nantucket, Mass.: Nantucket Historical Association.

Livermore, Abiel Abbot, and Sewall Putnam. 1888. *History of the Town of Wilton, Hillsborough County, New Hampshire*. Lowell, Mass.: Marden and Rowell.

Lockridge, Kenneth A. 1970. *A New England Town: The First Hundred Years*. New York: Norton.

Longsworth, Polly. 1990. *The World of Emily Dickinson*. New York: Norton.

Lowenthal, David. 1956. "The Common and Undivided Lands of Nantucket." *Geographical Review* (July): 339–403.

Lubar, Steven, and W. David Kingery. 1993. *History from Things: Essays on Material Culture*. Washington, D.C.: Smithsonian Institution Press.

Ludwig, Allan I. 1966. *Graven Images: New England Stonecarving and its Symbols, 1650–1815*. Middletown: Wesleyan University Press.

Lunt, Anne D., ed. 1976. *A History of Temple, New Hampshire*. Dublin, N.H.: William L. Bauhan.

MacLean, John C. 1988. *A Rich Harvest: The History, Buildings, and People of Lincoln, Massachusetts*. Lincoln, Mass.: Lincoln Historical Society.

Macy, Obed. 1972. *The History of Nantucket*. 2d ed. Clifton, N.J.: Augustus M. Kelley. Reprint. Previous editions, 1835, 1880.

Mansur, Ina. 1974. *A New England Church, 1730–1834*. Freeport, Maine: Bond Wheelwright.

Marsan, Jean-Claude. 1981. *Montreal in Evolution*. Montreal: McGill-Queen's University Press.

[Massachusetts Federal Writers' Project]. 1941. *The Origin of Massachusetts Place Names of the State, Counties, Cities, and Towns*. New York: Harian.

Massachusetts Historical Commission. 1982a. *Historic and Archaeological Resources of the Boston Area: A Framework for Preservation Decisions*. Boston: Massachusetts Historical Commission (January).

———. 1982b. *Historic and Archaeological Resources of Southeast Massachusetts: A Framework for Preservation Decisions*. Boston: Massachusetts Historical Commission (June).

———. 1984. *Historic and Archaeological Resources of the Connecticut Valley: A Framework for Preservation Decisions*. Boston: Massachusetts Historical Commission (February).

———. 1985. *Historic and Archaeological Resources of Central Massachusetts: A Framework for Preservation Decisions*. Boston: Massachusetts Historical Commission (February).

———. 1987. *Historic and Archaeological Resources of Cape Cod and the Islands: A*

Framework for Preservation Decisions. Boston: Massachusetts Historical Commission.

Massachusetts Topographical Survey. 1884–1900. *Reports of the Topographical Survey Commission and Town Boundary Triangulation of Massachusetts, 1884–1900*. Boston: Wright and Potter.

McAlester, Virginia, and Lee McAlester. 1984. *A Field Guide to American Houses*. New York: Knopf.

McCallum, Kent. 1996. *Old Sturbridge Village*. New York: Abrams.

McGowan, Susan, and Amelia F. Miller. 1996. *Family & Landscape: Deerfield Home-lots from 1671*. Deerfield, Mass.: Pocumtuck Valley Memorial Association.

McHenry, Stewart G. 1978. "Eighteenth-Century Field Patterns as Vernacular Art." *Old-Time New England* (Summer–Fall): 1–21.

McManis, Douglas R. 1975. *Colonial New England: A Historical Geography*. New York: Oxford University Press.

Melvoin, Richard I. 1989. *New England Outpost: War and Society in Colonial Deerfield*. New York: Norton.

Mercer, Eric. 1975. *English Vernacular Houses: A Study of Traditional Farmhouses and Cottages*. London: Her Majesty's Stationery Office.

Merrill, Georgia Drew. 1889. *History of Carroll County, New Hampshire*. Boston: W. A. Fergusson.

Miller, Amelia F. 1983. *Connecticut River Valley Doorways: An Eighteenth Century Flowering*. Dublin Seminar for New England Folklife, Occasional Publication. Boston: Boston University for the Dublin Seminar for New England Folklife.

Milot, Richard. 1993. "La Maison Vernaculaire: Son Evolution dans les Cantons-de-l'Est durant la Première Ere Industrielle." *Continuité*, no. 56 (March/April/May): 9–13.

Miner, Robert G. 1977. *Early American Homes of New York and the Mid-Atlantic States*. New York: Arno.

Mooney, Robert F. and Andre R. Sigourney. 1980. *The Nantucket Way*. Garden City, N.Y.: Doubleday.

Morison, Samuel Eliot. 1921. *The Maritime History of Massachusetts, 1783–1860*. Boston: Houghton Mifflin.

———. 1956. *The Story of the "Old Colony" of New Plymouth (1620–1692)*. New York: Knopf.

Morris, Fred C. 1949. *Your Land: Surveys, Maps and Titles*. Virginia Polytechnic Institute Engineering Experiment Station Series Bulletin No. 71.

Morris, Richard B., ed. 1953. *Encyclopedia of American History*. New York: Harper and Brothers.

Morrison, Hugh. 1952. *Early American Architecture from the First Colonial Settlements to the National Period*. New York: Oxford University Press.

Mourt. 1963. *A Journal of the Pilgrims at Plymouth: Mourt's Relation*. New York: Corinth Books.

Mullins, Lisa C. 1987. *New England by the Sea*. [From material originally published as the White Pine Series of Architectural Monographs, edited by Russell F. Whitehead and Frank Chouteau Brown]. Pittstown, N.J.: Main Street.

Myers, Denys Peter, comp. 1974. *Maine Catalog: A List of Measured Drawings, Pho-*

tographs, and Written Documentation in the [HABS] Survey, 1974. Augusta, Maine: Maine State Museum.

Nash, Hope. 1975. *Royalton, Vermont.* The Town of Royalton, South Royalton Women's Club, Royalton Historical Society.

Nason, Elias. A. 1877. *History of the Town of Dunstable, Massachusetts.* Boston: Alfred Mudge and Son.

Neudorfer, Giovanna. 1980. *Vermont's Stone Chambers: An Inquiry into Their Past.* Montpelier, Vt.: Vermont Historical Society.

Noble, Allen G. 1984. *Wood, Brick, and Stone: The North American Settlement Landscape.* Vol. 1, *Houses.* Vol. 2, *Barns and Farm Structures.* Amherst: University of Massachusetts Press.

Nutting, Wallace. 1923. *Connecticut Beautiful.* Framingham, Mass.: Old America Company.

O'Gorman, James F. 1997. *Living Architecture: A Biography of H. H. Richardson.* New York: Simon and Schuster.

O'Keefe, John, and David R. Foster. 1998. "An Ecological History of Massachusetts Forests." *Arnoldia* (summer): 2–31.

Official Arrow Atlas. [1993?]. *Metro Boston-Eastern Massachusetts.* Expanded ed. Taunton, Mass.: Arrow Map.

Old Bridgewater Tercentenary Committee. *Old Bridgewater Tercentenary (1656–1956) June 13–17, 1956.* Brochure, no pagination. Bridgewater, Mass.: The Committee, 1956.

Oliver, Basil. [1912]. *Old Houses and Village Buildings in East Anglia, Norfolk, Suffolk, and Essex.* London: B. T. Batsford.

Omni Gazetteer of the United States of America. 1991. 11 vols. Edited by Frank R. Abate. Detroit, Mich.: Omnigraphics.

Parker, Charles S. 1907. *Town of Arlington Past and Present.* Town of Arlington Centennial Celebration.

Parker, Rowland. 1975. *The Common Stream.* New York: Holt, Rinehart and Winston.

Pattison, William D. 1957. *Beginnings of the American Rectangular Land Survey System, 1784–1800.* Chicago: University of Chicago Press.

Paul Revere's Boston, 1735–1818. 1975. Boston: Museum of Fine Arts.

Paullin, Charles O. 1932. *Atlas of the Historical Geography of the United States.* Washington, D.C.: Carnegie Institution of Washington.

Perrin, Richard W. E. 1967. *The Architecture of Wisconsin.* Madison, Wisc.: State Historical Society of Wisconsin.

———. 1981. *Historic Wisconsin Buildings: Surveys in Pioneer Architecture, 1835–1870.* 2d rev. ed. Milwaukee, Wisc.: Milwaukee Public Museum.

Phillips, James Duncan. 1947. *Salem and the Indies: The Story of the Great Commercial Era of the City.* Boston: Houghton Mifflin.

Pillsbury, Hobart. 1927. *New Hampshire: Resources, Attractions, and its People.* 5 vols. New York: Lewis Historical Publishing.

Powell, Summer Chilton. 1963. *Puritan Village: The Formation of a New England Town.* Middletown: Wesleyan University Press.

Powell, William S. 1968. *North Carolina Gazetteer.* Chapel Hill: University of North Carolina Press.

Price, Edward T. 1995. *Dividing the Land: Early American Beginnings of Our Private Property Mosaic*. Chicago: University of Chicago Press.

Quick, Herbert. 1925. *One Man's Life: An Autobiography*. Indianapolis, Ind.: Bobbs-Merrill.

Ragan, Ruth Moulton. 1991. *Voiceprints of Lincoln: Memories of an Old Massachusetts Town and Its Unique Response to Industrial America: An Oral History*. Lincoln, Mass.: Lincoln Historical Society. Boston: Northeastern University Press.

Ramsdell, George A. 1901. *History of Milford*. Concord, N.H.: Rumford.

Randall, Peter E. 1982. *Portsmouth and the Piscataqua*. Camden, Maine: Down East Books.

Raup, Hugh M., and Reynold E. Carlson. 1941. *The History of Land Use in the Harvard Forest*. Harvard Forest Bulletin No. 20. Petersham, Mass.; Harvard Forest.

Rawson, Marion Nicholl. 1942. *New Hampshire Borns a Town*. New York: E. P. Dutton.

Reaney, P. H. 1961. *The Origin of English Place-Names*. London: Routledge and Kegan Paul.

Reed, John. 1879. *A History of Rutland, Worcester County, Massachusetts*. Worcester, Mass.: Tyler and Seagrove. Original edition, 1836.

Reps, John W. 1965. *The Making of Urban America: A History of City Planning in the United States*. Princeton: Princeton University Press.

Rhode Island Historical Preservation Commission. 1980a. *Historic and Architectural Resources of Scituate, Rhode Island*. Preliminary Survey Report. Providence, R.I.: Rhode Island Historical Preservation Commission.

———. 1980b. *Smith Hill, Providence*. Statewide Historical Preservation Report, P-P-4. Providence, R.I.: Rhode Island Historical Preservation Commission.

———. 1981. *Providence Industrial Sites*. Statewide Historical Preservation Report, P-P-6. Providence, R.I.: Rhode Island Historical Preservation Commission.

———. 1982a. *Foster, Rhode Island*. Statewide Historical Preservation Report, P-F-1. Providence, R.I.: Rhode Island Historical Preservation Commission.

———. 1982b. *Lincoln, Rhode Island*. Statewide Historical Preservation Report, P-L-1. Providence, R.I.: Rhode Island Historical Preservation Commission.

———. 1990. *Historic and Architectural Resources of Bristol, Rhode Island*. Providence, R.I.: Rhode Island Historical Preservation Commission.

Richardson, Hilary, and John Scarry. 1990. *An Introduction to Irish High Crosses*. Dublin: Mercier.

Richardson, Laurence Eaton. 1960. *Concord at the Turn of the Century*. Concord, Mass.: Concord Antiquarian Society.

Roberts, David. 1993. "A Walk in Newfoundland: Looking for Ghosts on the Wild Shores." *Outside Magazine* (August).

Robinson, William F. 1976. *Abandoned New England: Its Hidden Ruins and Where to Find Them*. Boston: New York Graphic Society.

Room, Adrian. 1986. *A Dictionary of Irish Place-Names*. Belfast, U.K.: Appletree Press.

Rooney, John F. et al. 1982. *This Remarkable Continent: An Atlas of United States and*

Canadian Society and Cultures. Society for the North America Cultural Survey. College Station, Tex.: Texas A&M Press.

Russell, Howard S. 1976. *A Long, Deep Furrow: Three Centuries of Farming in New England*. Hanover: University Press of New England.

————. 1980. *Indian New England Before the Mayflower*. Hanover: University Press of New England.

Sammarco, Anthony Mitchell. 1997. *The Great Boston Fire of 1872*. Images of America Series. Dover, N.H.: Arcadia.

Sanchis, Frank E. 1977. *American Architecture: Westchester County, New York: Colonial to Contemporary*. Croton-on Hudson, N.Y.: North River.

Sanderson, Edmund L. 1936. *Waltham as a Precinct of Watertown and as a Town, 1630–1884*. Waltham, Historical Society, Publication No. 5.

Schuler, Stanley. 1982. *Cape Cod House: America's Most Popular Home*. Exton, Pa.: Schiffer.

————. 1988. *Saltbox and Cape Cod Houses*. West Chester, Pa.: Schiffer.

Schwartz, Helen. 1983. *The New Jersey House*. New Brunswick: Rutgers University Press.

Scofield, Edna. 1938. "The Origin of Settlement Patterns In Rural New England." *Geographical Review* 27, no. 4: 652–63.

Scott, Jonathan Fletcher. 1985. *"The Early Houses of Martha's Vineyard."* Ph.D. Thesis, University of Minnesota.

Sears, Roebuck and Co. 1969. *1908 Sears, Roebuck Catalog: A Treasured Replica from the Archives of History*. Edited by Joseph J. Schroeder Jr. Chicago: Gun Digest.

Shurtleff, Harold. 1939. *The Log Cabin Myth: A Study of the Early Dwellings of the English Colonists in North America*. Edited by Samuel Eliot Morison. Cambridge: Harvard University Press.

Sileo, Thomas P. 1995. *Historical Guide to Open Space in Lexington*. Acton, Mass.: Concepts Unlimited.

Simpson, Ruth M. Raisey. 1990. *Out of the Saltbox*. Atlanta, Ga.: Cherokee. Original edition, 1962.

Sinclair, Iain. 1998. *Lights out for the Territory*. London: Granta Books.

Sizemore, Jean. 1994. *Ozark Vernacular Houses: A Study of Rural Homeplaces in the Arkansas Ozarks, 1830–1930*. Fayetteville: University of Arkansas Press.

Small, Nora Pat. 1988. "Hemenway Plan." *Old-House Journal* (July–August): back cover.

Smart, Charles E. 1962–67. *The Makers of Surveying Instruments in America since 1700*. Troy, N.Y.: Regal Art.

Speight, Martin E. n.d. *The Monuments in Edmonton Churchyard*. Edmonton Hundred Historical Society Occasional Paper, n.s., No. 17. Edmonton, U.K.

Stanchiw, Myron, and Nora Pat Small. 1989. "Tradition and Transformation: Rural Society and Architectural Change in Nineteenth-Century Central Massachusetts." *Perspectives in Vernacular Architecture* 3: 135–48.

Stansfield, Charles A., Jr. 1983. *New Jersey, A Geography*. Boulder, Colo.: Westview.

Steinmetz, Michael. 1989. "Rethinking Geographical Approaches to the Common House: the Evidence from Eighteenth-Century Massachusetts." In *Perspectives in Vernacular Architecture*, edited by Thomas Carter and Bernard L. Herman, 3: 16–26. Columbia: University of Missouri Press.

Stekl, William F., and Evan Hill. 1972. *The Connecticut River*. Middletown: Wesleyan University Press.

Sternagel, Mary E., and Henry S. C. Cummings Jr. 1985. *Middlefield History*. Middlefield, Mass.: Middlefield History Fund.

[Stevens, Kevin]. 1988. *Winchester, Massachusetts: The Architectural Heritage of a Victorian Town*. Winchester, Mass.: Winchester Historical Society.

Stevens, William B. 1891. *History of Stoneham, Massachusetts*. Stoneham, Mass.: F. L. and W. E. Whittier.

Stevens, William Oliver. 1936. *Nantucket: The Far-Away Island*. New York: Dodd, Mead.

Stewart, George R. 1967. *Names on the Land: A Historical Account of Place-Naming in the United States*. Boston: Houghton Mifflin. Original edition, 1945.

Stilgoe, John R. 1982a. *Common Landscape of America, 1580–1845*. New Haven: Yale University Press.

———. 1982b. "Town Common and Village Green in New England: 1620–1981." In *On Common Ground: Caring for Shared Land from Town Common to Urban Park*, by Ronald Lee and Lauri A. Halderman, 7–36. Harvard Mass.: Harvard Common Press.

———. 1998. *Outside Lies Magic: Regaining History and Awareness in Everyday Places*. New York: Walker.

Strickland, Charles R. 1950. "The First Permanent Dwellings at Plimouth Plantation." *Old-Time New England* 60, no. 3: 162–69

Strong, James. 1890. *The Exhaustive Concordance of the Bible*. New York: The Methodist Concern.

Sullivan, Robert F. 1990. *Shipwrecks and Nautical Lore of Boston Harbor*. Chester, Conn.: Globe Pequot.

Swan, Marshall W. S. 1980. *Town on Sandy Bay: A History of Rockport, Massachusetts*. Cannan, N.H.: Phoenix.

Swift, Esther M. 1969. *West Springfield, Massachusetts: A Town History*. West Springfield Heritage Association. Town of West Springfield.

———. 1977. *Vermont Place-Names: Footprints of History*. Brattleboro, Vt.: Stephen Greene.

Tashjian, Dickran, and Ann Tashjian. 1974. *Memorials for Children of Change*. Middletown: Wesleyan University Press.

Teele, Albert Kendall. 1887. *The History of Milton, Mass.: 1640 to 1887*. Boston: Rockwell and Churchill.

Teele, John Whittemore. 1985. *The Meeting House on the Green: A History of the First Parish in Concord and its Church*. Concord, Mass.: The First Parish.

Thompson, Deborah, ed. 1976. *Maine Forms of Architecture*. Camden, Maine: Downeast Magazine.

———. 1988. *Bangor, Maine, 1769–1914: An Architectural History*. Orono: University of Maine Press.

Thompson, Francis M., et al. 1904–31. *History of Greenfield, Shire Town of Franklin County, Massachusetts, 1682–1900*. 4 vols. Greenfield, Mass.: T. Morey and Son.

Thoreau, Henry David. 1962. *Journal of Henry David Thoreau*. Edited by Bradford Torrey and Francis H. Allen. Reprinted in 2 vols. New York: Dover. Original edition, 1906.

———. 1965. *Walden and Civil Disobedience*. New York: Airmont.

Tilden, William S. 1887. *History of the Town of Medfield, Massachusetts, 1650–1886*. Boston: Geo. H. Ellis.

Tolles, Bryant F., Jr. 1979. *New Hampshire Architecture: An Illustrated Guide*. Hanover: University Press of New England.

Torbert, Edward L. 1935. "The Evolution of Land Utilization in Lebanon, New Hampshire." *Geographical Review* 25:209–30 and plate.

Tree, Christina. 1981. *Massachusetts: An Explorer's Guide*. New rev. ed. Woodstock, Vt.: Countryman. Original edition, 1979.

Tucci, Douglas Shand. 1978. *Built in Boston: City and Suburb, 1800–1950*. Boston: New York Graphic Society.

Turner, Paul Venable. 1984. *Campus: An American Planning Tradition*. Cambridge: MIT Press.

United States Coast Survey. 1875. *Coast Pilot for the Atlantic Sea-board: Gulf of Maine and its Coast from Eastport to Boston 1874*. Washington, D.C: GPO.

United States Hydrographic Office. 1917–26. *British Islands Pilot*. 2d ed. 7 vols. Washington, D.C.: GPO.

Uzes, Francois D. 1977. *Chaining the Land: A History of Surveying in California*. Sacramento, Calif.: Landmark Enterprises.

Vanderhill, Burke G., and Frank A. Unger. 1977. "Georgia's Crenelated County Boundaries." *West Georgia College Studies in the Social Sciences* 16: 59–72.

Visser, Thomas Durant. 1997. *Field Guide to New England Barns and Farm Buildings*. Hanover: University Press of New England.

Von Hoffman, Alexander. 1994. *Local Attachments: The Making of an American Urban Neighborhood, 1850–1920*. Baltimore: Johns Hopkins University Press.

Vorse, Mary Heaton. 1942. *Time and the Town: A Provincetown Chronicle*. New York: Dial.

Walcott, Charles H. 1884. *Concord in the Colonial Period*. Boston: Estes and Lauriat.

Walker, Les. 1981. *American Shelter: An Illustrated Encyclopedia of the American Home*. Woodstock, N.Y.: Overlook.

———. 1987. *[Tiny] Tiny Houses*. Woodstock, N.Y.: Overlook.

Warner, Elizabeth A. 1990. *A Pictorial History of Middletown* [Conn.] Greater Middletown Preservation Trust. Norfolk, Va.: Donning.

Warner, Sam Bass, Jr. 1978. *Streetcar Suburbs: The Process of Growth in Boston (1870–1900)*. 2d ed. Cambridge. Harvard University Press. Original edition, 1962.

Warren, G. F. 1914. *Farm Management*. New York: Macmillan.

Waters, Thomas Franklin. 1905. *Ipswich in the Massachusetts Bay Colony*. Vol. 1 of 2 vols. Ipswich, Mass.: Ipswich Historical Society.

Watters, David H. 1997. "Fencing ye Tables: Scotch-Irish Ethnicity and the Gravestones of John Wight." *Historical New Hampshire* 52, nos. 1–2: 2–17.

Weiss, Ellen. 1987. *City in the Woods: The Life and Design of an American Camp Meeting on Martha's Vineyard*. New York: Oxford University Press.

Weslager, C. A. 1969. *The Log Cabin in America.* New Brunswick. Rutgers University Press.

[Wharton, Virginia, ed.]. 1974. *The Bedford Sampler: Bicentennial Edition.* Bedford, Mass.: Friends of the Bedford Public Library.

Whitcomb, Esther Kimmens. 1988. *About Bolton.* Bowie, Md.: Heritage Books.

White, C. Albert. 1983. *A History of the Rectangular Survey System.* U.S. Department of the Interior, Bureau of Land Management. Washington, D.C.: GPO.

Whitehill, Walter Muir. 1968. *Boston: A Topographical History.* 2d enlarged ed. Cambridge: Harvard University Press.

Whitney, J. D. 1888. *Names and Places: Studies in Geographical and Topographical Nomenclature.* Cambridge, Mass.: University Press.

Wikander, Lawrence E., Helen Terry, and Mark Kiley, comps. 1964. *The Hampshire History: Celebrating 300 Years of Hampshire County Massachusetts.* Northampton, Mass.: Hampshire County Commissioners.

Wilkie, Richard W., and Jack Tager. 1991. *Historical Atlas of Massachusetts.* Amherst: University of Massachusetts Press.

Williams, Fred H. 1890. *Argument . . . Before the Legislative Committee on Towns, February 21, 1890, in Favor of the Incorporation of the Town of Beverly Farms.* Boston: Samuel Usher.

Williams, Henry Lionel, and Ottalie K. Williams. 1957. *Old American Houses: How to Restore, Remodel, and Reproduce Them.* New York, Bonanza.

Williams, Norman, Jr, Edmund H. Kellogg, and Peter M. Lavigne. 1987. *Vermont Townscape.* New Brunswick, N.J.: Center for Urban Policy Research.

Willsher, Betty, and Doreen Hunter. 1978. *Stones: A Guide to Some Remarkable Eighteenth Century Gravestones.* Edinburgh, UK: Canongate.

Wood, Joseph S. 1997. With a contribution by Michael P. Steinitz. *The New England Village.* Baltimore: Johns Hopkins University Press.

Woodward, William McKenzie, and Edward F. Sanderson. 1986. *Providence: A Citywide Survey of Historic Resources.* Providence, R.I.: Rhode Island Historical Preservation Commission.

Wooley, Carolyn Murray, and Karl Raitz. 1992. *Rock Fences of the Bluegrass.* Lexington: University Press of Kentucky.

Worth, Henry Barnard. 1901–13. *Nantucket Lands and Landowners.* Nantucket Historical Association. *Bulletin* 2, nos. 1–7.

Worthen, Mark. 1976. *Hometown Jamaica: A Pictorial History of a Vermont Village.* Brattleboro, Vt.: Griswold.

Wright, Eugene A. 1977–81. *Tales of Old Plympton.* 2 vols. Plympton, Mass.: Plympton Historical Society.

Wright, Harry Andrew. 1936. *The Genesis of Springfield: The Development of the Town.* Springfield, Mass.: Johnson's Bookstore.

Young, Allen. 1983. *North of Quabbin: A Guide to Nine Massachusetts Towns.* Athol, Mass.: Millers River.

Zea, Philip, and Nancy Norwalk, eds. 1991. *Choice White Pines and Good Land: A History of Plainfield and Meriden, New Hampshire.* Portsmouth, N.H.: Peter E. Randall.

Zelinsky, Wilbur. 1955. "Some Problems in the Distribution of Generic Terms in the

Place-Names of the Northeastern United States." *Annals of the American Association of Geographers* 45:319–49.

———. 1958. "The New England Connecting Barn." *Geographical Review* 48:540–53.

———. 1967. "Classical Town Names in the United States: The Historical Geography of an American Idea." *Geographical Review* 67, no. 4:461–95.

Zimilies, Martha, and Murray Zimilies. 1973. *Early American Mills*. New York: C. N. Potter.

INDEX

Aboveground archeology, xiv, 11, 307
Acton MA, 44
Acworth NH, 167
African Americans: gravestones, 273, 280; houses, 200; stone walls, 149
Age grain of settlement: in architecture, 264; in landscape artifacts, 293–294; in street names, 71–72; in town boundaries, 151–152; in town names, 52; lacking in hydrographic names, 23. *See also* Settlement frontiers
Agriculture: decline of, 123, 164–166, 168–169, 171–172, 265; enclosed field, 104–105; openfield, 104–107
Alexandria NH, 67
Alfred ME, 240
Algonquians: Abenaki suffixes, 21; dialects, 18; epidemics among, 103, 198; -*et* suffix, 20–21; fishing weirs, 103; geographic legacy of, 103; old fields, 103; paths 103, 171; placenames, 15, 17–21; Q-K names, 19–20; Rhode Island stone walls, 149; and 'Sconset Whale house, 198; village sites, 103
Alna ME, 29
American Indians. *See* Algonquians
Ames, family of shovel makers, 76, 78, 268–269
Amherst MA, 138, 148
Andover MA, 143, 200
Ansonia CT, 85
Antiques and horribles parades, 305, 310
Appalachian Trail, xiii, 201
Arlington MA, 131, 144, 179; placenames, 54–55, 69, 75
Artifacts, 1–2; chess piece analogy, 1, 2, 299–300; clay pot analogy, 2, 137, 143, 182; in cultural landscape, 2, 10, 306–308; dynamics of, 297–301
Ashburnham MA, 34, 54
Ashby MA, 32, 54

Athol MA, 169, 177, 244; mud huts, 201; placenames, 14, 22, 70, 129
Auburn MA, 16
Auburn ME, 16
Auburn NH, 16

Balcony, recessed, 242 *illus.*, 245–246
Ball, placenames with, 64–66
Balloons and bubbles, 4, 295–297, 296 *map*
Bangor ME, 175
Barn: French Canadian, 259–260; New England, 232
Barnstead NH, 53
Barre MA, 56, 217–219
Barrington NH, 118
Basketball, 304
Bath ME, 246
Battleship-shaped curve, 2, 67, 250
Becket MA, 136
Bedford MA, 131–132, 139, 155, 227, 262; ditches in, 150–151; Job Lane house, 205; yellow paint, 250
Belchertown MA, 128, 134–135
Bell legends, 44–45
Belmont NH, 45
Bennington VT, 35, 43, 118–119
Bernard, Governor Francis, 42
Beverly jog, 197 *illus.*, 203–205
Beverly MA, 153, 203
Blackstone Valley, 50, 300, 304
Bloomfield CT, 41, 53
Bloomfield VT, 41
Blue Hills Reservation, 93, 117–118
Bolton MA, 254
Borning room, 203
Boston MA: crooked streets, 109; crossroads, 129; fire, 267; frog pond, 138; gravestone prestige, 283; grid, 175; placenames, 69–70, 75; street name changes, 80–81; Three-Decker, 234; tonic, 302 *map*, 304–305

349